SURFACE PHYSICS

Theoretical Models
and
Experimental Methods

SURFACE PHYSICS

Theoretical Models
and
Experimental Methods

M. V. Mamonova, V. V. Prudnikov
and I. A. Prudnikova

CISP

CRC Press
Taylor & Francis Group
Boca Raton London New York

CRC Press is an imprint of the
Taylor & Francis Group, an **informa** business

CRC Press
Taylor & Francis Group
6000 Broken Sound Parkway NW, Suite 300
Boca Raton, FL 33487-2742

First issued in paperback 2019

© 2014 by Taylor & Francis Group, LLC
CRC Press is an imprint of Taylor & Francis Group, an Informa business

No claim to original U.S. Government works

ISBN-13: 978-0-367-37937-7

Visit the Taylor & Francis Web site at
http://www.taylorandfrancis.com

and the CRC Press Web site at
http://www.crcpress.com

Contents

Preface

The investigation of the physical properties of the surface of solids is an important direction of research [180, 191, 203, 498].

The demands of production (formation of thin film structures in microelectronics, improvement of the technological methods of controlling the properties of surfaces of components of friction sections in order to increase their efficiency, wear resistance and durability) requires detailed consideration of the factors influencing the intensity of interaction of dissimilar materials which make contact with their surfaces.

The phenomenon of the formation of a bond between the surface layers of dissimilar condensed solids, brought into contact, is referred to as adhesion. Adhesion depends on the nature of the contacting bodies, the properties of their surface and the contact area. From the physical viewpoint, adhesion is determined by the forces of intermolecular interaction, and the presence of ionic, covalent, metallic and other types of bond. It is therefore necessary to determine the characteristics of adhesion interaction of different materials from the viewpoint of both applied and fundamental science of surface phenomena [80, 82, 96].

However, the value of adhesion strength depends not only on the type of bond between the solids which make contact, but also on the methods of measuring adhesion strength and also the separation method [12, 217]. For example, by separating the film from the substrate at different rates, it is possible to obtain different values of adhesion strength. At the present time, there are no methods of non-destructive inspection of adhesion strength which would give reliable results. The results of measurement of adhesion strength may also be influenced by the stress state of the interface between the film and the substrate as a result of thermal or shrinkage phenomena in the materials. Spraying may result in the formation of the thinnest oxide layer on the surface of the substrate which greatly influences the adhesion of the coating. The unavoidable roughness of the surfaces of the media, which make contact with each other, determines the formation of a gap between the surfaces. All these factors indicate that it is quite difficult to obtain reliable experimental values of adhesion strength of coatings, determine directly the role of small gaps and special features of prediction of their effect on bonding of the solids.

Therefore, the role of the theoretical approach to determination of the adhesion characteristics of different coatings becomes more important. The criterion of accuracy of a specific model of adhesion interaction and of the values of the adhesion characteristics, predicted by the model, may be the comparison of the calculated and measured values of the surface energy of materials, because the methods of experimental determination of the surface energy provide more reliable results than the data obtained in determination of adhesion strength [230].

The problem of calculating the surface energy of the metal is itself complicated and the theory which would provide reliable results for a large number of metals has not as yet been proposed. Difficulties in description are associated with the considerable heterogeneity of the system in the subsurface region and also with the structural distortions of the surface layer of metals formed as a result.

Recently, special attention has been given to the development of the methods of analytical description of the surface properties and adhesion processes, based on the determination of the energy state of a system of solids using the methods of quantum mechanics [13, 183, 220]. This approach is quite general, makes it possible to link functionally the adhesion energy with the internal parameters of the contacting materials and, when sufficiently developed, can be used to predict the adhesion strength. The phenomenon of adhesion of metallic films is described quite efficiently by the method of electron density functional (EDF) [101, 181, 220, 299, 361]. At the present time, the development of the theory of the heterogeneous electron gas has already been used as a basis for the application of the EDF to different problems relating to the surface properties of metals. However, regardless of a large number of investigations in this area, the application of the EDF theory to the investigation of the surface is far from complete. For example, the variation of the properties of solids, associated with the existence and interaction of surfaces, has made it possible to describe more or less adequately only several simulation systems and simple metals. In the case of metals of other groups, reliable results are obtained using a number of trial empirical parameters, and it should be mentioned that there are almost no theories capable of describing adequately the surface and adhesion properties of semiconductors, complex compounds and alloys. At the same time, the further development of various applications of the density functional theory or of its modification is essential both for the development of the theory of surface phenomena and for practical application of the calculation results and predictions of the theory in a number of technological and production processes.

In this monograph, we present relatively simple and efficient approaches and methods developed by us within the framework of the electron density functional theory and dielectric formalism. These methods make it possible, in good agreement with experiments, to calculate the surface and adhesion characteristics of a wide range of materials: metals, alloys, semiconductors and complex compounds. The proposed methods have been used in the adhesion theory of dry friction of the metallic surfaces developed by us which can be used to calculate the contributions to the friction coefficient as a result of electrodynamic effects of dissipation of energy and the effects of adhesion attraction of the surfaces of friction pairs. The principles and methods for selecting the optimum pairs of materials for non-lubricating friction sections taking into account the nature of their adhesion interaction have been developed. Consequently, it has been possible, already in the stage of technological development of the friction sections, to select materials and coatings characterized by the lowest friction coefficient between them. The original results obtained by us in the investigation of the physics of surfaces in 20 years of research have been published in a number of publications [25–35, 61–68, 140–158, 192–196].

THERMODYNAMICS OF SURFACE PHEHOMENA

Introduction

The basic principles of classical thermodynamics were established primarily as a result of phenomenological observations made over two centuries. These experiments, performed almost exclusively for the bulk phases, showed that the macroscopic system in the absence of external disturbances tends to an equilibrium state, characterized by a small number of thermodynamic variables. Studies of surface phenomena and, in particular, features of the influence of solid surfaces started much later, respectively, in the field and less accumulated experimental facts.

In 1877, Josiah Willard Gibbs published in 'Proceedings of the Connecticut Academy,' the second part of his monumental work 'Equilibrium of Heterogeneous Substances.' This study, which is rightly regarded as an outstanding achievement of the XIX century science, laid the mathematical foundations of thermodynamics and statistical mechanics [51]. Part of this work was a chapter on the influence of fracture surfaces on the equilibrium of heterogeneous systems, in which Gibbs described the thermodynamics of interfaces for contacting gases and liquids, followed by generalization to the case of contact of a liquid with a solid. Subsequent publications by various authors in this area largely explained his rather complicated way of describing the features of thermodynamics of surface phenomena.

Excluding the special properties of the interface, for contacting phases of a heterogeneous system it is assumed that the energy E and free energy F of the system are equal to the sum of energies or, respectively, free energies of individual phases. In addition, each of the phases is regarded as homogeneous up to the interface and, therefore, E and F of each phase are considered proportional to its mass.

However, in the surface layer of the phase with the thickness of the order of the phase molecular interaction (10^{-9} m), which is the boundary phase, where the molecules interact not only with molecules of their phases but also with the surrounding layer of a foreign phase, the physical properties of matter different from the bulk properties of the phase.

Since the surface of a body increases as the square of the size of the body and the volume as the cube of these dimensions, then for macroscopic systems the surface effects compared to the bulk effects can, as a rule, be neglected. However, if it is interesting to study the properties of the system near the interface or the substance is in the dispersed (fine-crushed) state with a developed surface, in these cases the surface effects become important. In addition, consideration of the surface effects is important for understanding some features of the first-order phase transitions with the formation of nuclei in the temperature range of existence of metastable states.

Of great interest for modern studies are the surface phase transitions, as the strong fluctuations in two-dimensional systems can significantly change the nature of the phase transition compared to the bulk phase transitions. The most striking example is the possibility of melting of two-dimensional systems through the second-order transition, while in a three-dimensional crystal melting is always a first-order transition.

1.1. Surface tension and surface stress

1.1.1. The surfaces of liquids

In the description of the thermodynamics of surface phenomena J.W. Gibbs regarded a surface layer as a new surface phase, which differs from the bulk phases that its thickness is very small compared to the length of the two other dimensions and, therefore, he considered a surface layer as a geometric separating surface with the possibility of applying general thermodynamic equations to this layer.

The size of the surface Λ of the phase is in addition to volume V a new parameter characterizing the state of the system. The increase in the surface of the system at constant temperature, volume and number of particles is accompanied by the expenditure of work, as in the formation of a new surface some of the particles from the volume should go to the surface, which is associated with the work against the forces of molecular interaction.

We denote the generalized force, thermodynamically conjugate with surface Λ, by σ. Then the elementary work by increasing the surface by $d\Lambda$ (At T = const, V = const and N = const) is equal to

$$\delta A = -\sigma d\Lambda, \tag{1.1}$$

and the differential free energy of the system with T, V, N and Λ changing is

$$dF = -SdT - pdV + \mu dN + \sigma d\Lambda. \tag{1.2}$$

For the case of multicomponent systems, when the change of the state of the system by changing the number N_i of particles of i-th component is characterized by chemical potentials μ_i of the components, expression (1.2) can be written in a generalized form

$$dF = -SdT - pdV + \sum_i \mu_i dN_i + \sigma d\Lambda. \tag{1.3}$$

The value of σ, which characterizes the balance between the two contacting phases, is called the surface tension (surface tension coefficient) and is equal to the force acting on the unit length on the surface, or to a change in free energy increasing the surface area by unity: $\sigma = (\partial F / \partial \Lambda)_{T,V,N_i}$. The surface tension σ can be regarded as an excess of free energy per unit surface area.

Consider the equilibrium conditions in a system consisting of two phases, separated by the interface. We know that when we neglect surface phenomena, the equilibrium conditions of the two phases of the same substance have the form

$$T_1 = T_2, \quad p_1 = p_2, \quad \mu_1 = \mu_2. \tag{1.4}$$

The same arguments that lead to the equations (1.4) for the equilibrium of two phases with of surface phenomena taken into account, give:

$$T_1 = T_2, \quad \mu_1 = \mu_2. \tag{1.5}$$

As for the conditions of mechanical equilibrium of two phases, where at the interface we must now take into account the forces of surface tension, the equilibrium between them occurs, generally speaking, at different pressures in the phases. We find this condition of equilibrium for two phases: 1 – liquid and 2 – vapour, based on the minimum free energy at $T = $ const and $V = $ const.

The differential free energy of the system, taking into account the interface between the phases, when the temperature and chemical potential in the phases are the same, has the form

$$dF = -p_1 dV_1 - p_2 dV_2 + \sigma d\Lambda. \tag{1.6}$$

At equilibrium, $dF = 0$, so

$$\sigma d\Lambda - p_1 dV_1 - p_2 dV_2 + \sigma d\Lambda = 0, \tag{1.7}$$

as well as the $V_1 + V_2 = V = $ const, then

$$p_1 = p_2 + \sigma \frac{d\Lambda}{dV_1}. \tag{1.8}$$

The value of $d\Lambda/dV_1$ determines the curvature of the interface. If this surface is spherical, then

$$\frac{d\Lambda}{dV_1} = \frac{d(4\pi r^2)}{d\left(\frac{4}{3}\pi r^3\right)} = \frac{2}{r} \tag{1.9}$$

(r is considered positive if the curvature is directed into phase 1).

In the case of an arbitrary surface

$$\frac{d\Lambda}{dV_1} = \frac{1}{r_1} + \frac{1}{r_2},$$ (1.10)

where r_1 and r_2 are the principal radii of curvature of the surface.

Thus, in the equilibrium of spherical liquid droplets with the vapour the pressure in the droplet p_1 and vapour pressure p_2 are related by

$$p_1 = p_2 + \frac{2\sigma}{r}.$$ (1.11)

This shows that at the interface of two phases (liquid–vapour) there is a pressure jump, equal to $\sigma(1/r_1 + 1/r_2)$, or $2\sigma/r$ in the case a spherical surface. This value is called the surface pressure (Laplace pressure). In the case of a flat interface between the liquid and the vapour ($r \to \infty$) the Laplace pressure is zero and the condition of mechanical equilibrium at $r \to \infty$ coincides with the same condition without surface phenomena:

$$p_1 = p_2 = p_\infty.$$ (1.12)

Formula (1.6) shows that at constant volume of a liquid droplet ($V_1 = $ const) its equilibrium shape is determined by the minimum surface Λ. Consequently, the liquid under the action of only surface tension forces takes a spherical shape since at the given volume the sphere has the minimal surface.

The thermodynamic potentials of the system, as additive quantities, are characterized by the fact that they are homogeneous functions of the first degree of their own additive variables. In particular, the free energy is a homogeneous function of the first degree of variables V, N_i, and Λ. As a result, on the basis of the formula (1.3) we can write an expression for the free energy of the system

$$F = -pV + \sum_i \mu_i N_i + \sigma\Lambda.$$ (1.13)

If we subtract the bulk free energy of the contacting phases from the total free energy of the system

$$F_n = -pV_n + \sum_i \mu_i N_{in}, \quad n = 1, 2,$$ (1.14)

we obtain an expression for the surface free energy in the form

$$F_s = \sigma\Lambda + \sum_i \mu_i N_{si},$$ (1.15)

where N_{si} is the surface excess of i-th component.

If now we subtract from the equation (1.3) similar expressions for the differentials of the bulk free energy of the coexisting phases

$$dF_n = -S_n dT - p dV_n + \sum_i \mu_i dN_{in}, \quad n = 1, 2, \tag{1.16}$$

we arrive at the well-known Gibbs thermodynamic equation of the interfacial layer:

$$dF_s = -S_s dT + \sigma d\Lambda + \sum_i \mu_i dN_{si}, \tag{1.17}$$

where S_s is the surface part of the entropy. A comparison of the expressions (1.16) and (1.17) shows that the thermodynamic relation for the surface characteristics is similar to the thermodynamic relation for the bulk phases with the accuracy up to the change of pressure by surface tension, taken with the opposite sign, and the volume – by the surface area. This analogy allows us to consider the surface characteristics as belonging to a third (surface) phase. This interpretation of surface features is often used both in textbook and scientific literature.

1.1.2. The surfaces of solids

The thermodynamics of the surface properties of solids is more complicated to describe as the process of formation of the surface in the solid and its change is much more complicated than changes of the surface of the liquid medium. At formation of the surface in the liquid the structure of the bulk phase does not change: changes take place only in that portion of the liquid which changes to the surface phase. In solids processes can take place in which the formation of a new surface can also affect the bulk phase. As also noted by J.W. Gibbs [51], in solid bodies there are two basic ways of forming a new surface.

In the first method the produced surface is identical in its nature with the original. This method is implemented, for example, in splitting solids. Since some work is spent splitting the surface, the increase in the free energy of the system should be proportional to the square of the surface, which allows to write the following thermodynamic relation:

$$dF = -SdT - pdV + \sum_i \mu_i dN_i + \sigma d\Lambda, \tag{1.18}$$

introducing the surface tension σ for the solid. It is obvious that (1.15) remains valid for the surface free energy in the case of solids.

The second way to change the surface is associated with tensioning the solid. Unlike the first method, in this case the number of surface atoms remains unchanged but the distance between them changes. Let the work expended on deformation be fully described in the framework of the theory of elasticity. The change in free energy, associated with deformation, can then be written in the form

$$dF = -SdT - pdV + \sum_i \mu_i dN_i + \sum_{\alpha,\beta} \int \sigma^b_{\alpha\beta} du^b_{\alpha\beta} dV + \sum_{\alpha,\beta} \sigma^s_{\alpha\beta} du^s_{\alpha\beta}, \tag{1.19}$$

where $\sigma_{\alpha,\beta}^{b}$ and $u_{\alpha,\beta}^{b}$ are the tensors of bulk stress and strain, $\sigma_{\alpha\beta}^{s}$ $u_{\alpha\beta}^{s}$ are the tensors of surface stress and strain. The integration is over the entire volume, so that in (1.19) the term $-pdV$, referring to the contact with the solid gas or liquid phase, is retained The value of $\sigma_{\alpha\beta}^{s}$ is a force reduced to unit length which acts from the side of the atoms of the solid in the direction perpendicular to the intersection of the plane defined by the normal along the axis α, with the surface in the direction β. The strain tensor is associated with the relative change in the surface by a simple relation:

$$\frac{d\Lambda}{\Lambda} = \sum_{\alpha} du_{\alpha\alpha}^{s},$$
(1.20)

and, thus, the last term in (1.19) determines the change in free energy due to deformation changes in the surface area.

There is some connection between the surface tension σ and surface stress $\sigma_{\alpha\beta}^{s}$. Thus, according to [214], they are related by

$$\sigma_{\alpha\beta}^{s} = \sigma\delta_{\alpha\beta} + \frac{d\sigma}{du_{\alpha\beta}^{s}},$$
(1.21)

which shows that in general the surface tension and surface stress are not identical.

For isotropic surfaces with the diagonal tensor of the surface stress, characterized by a single scalar magnitude of the stress σ^{s}, the relation (1.21) takes the form

$$\sigma^{s} = \sigma + \Lambda\frac{d\sigma}{d\Lambda}.$$
(1.22)

A special case occurs when the value of surface tension γ does not depend on small deformation, i.e. $d\sigma/du_{\alpha\beta}^{s} = 0$. It is realized only if the deformation is purely plastic, as in liquids, and the system is free to rearrange in response to the action of the acting perturbation. The surface tension and stress are the same only in this case. However, it should be noted here that if deformation of the liquid occurs very rapidly, so that the surface structure does not have time to relax to an equilibrium state, then the surface stresses begin to depend on the strain, and in this case $d\sigma/du_{\alpha\beta}^{s} \neq 0$

A similar situation occurs in the case of crystalline solids. If the deformation is sufficiently slow, and atoms in solids have a high mobility (for example, at high temperatures), the deformation can be reduced to plastic, and then $d\sigma/du_{\alpha\beta}^{s} \to 0$. Therefore, using the relations (1.21) and (1.22) should be used talking into account the relaxation times of processes in solids. The second term in (1.21) and (1.22) is not always small and may play an important role in the formation of the equilibrium surface structure of crystals. A particularly interesting case is when the value of $d\sigma/du_{\alpha\beta}^{s}$ or $d\sigma/d\Lambda$ is negative. This means that an increase in the surface

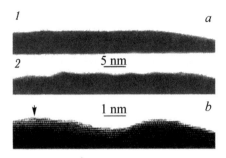

Fig. 1.1. The formation of wavy structures on the surface of the Au (111): *a* – free homogeneous surface in the initial state (*1*), after relaxation (*2*), *b* – the surface with a surface dislocation (indicated by arrow) [245].

area is accompanied by a decrease in surface stresses. In this case we can expect the formation of a wavy surface structure of crystals or the formation of surface dislocations [214]. A colorful illustration of this phenomenon has been obtained on the surface of Au(111) by high-resolution electron microscopy [245]. Figure 1.1*a* and shows changes in the surface after its formation and exposure for a sufficient time for its reorganization: see the formation of 'hills' and 'valleys' due to the dependence of the stresses on the surface area. The same type of ripple on the surface formed after the formation of surface dislocations (Fig. 1.1*b*).

There are various methods for determining the surface tension (or surface stresses) of solids. Each of them has both advantages and disadvantages, so we will not select any one of them. They are described in detail in [50, 165, 389]. The essence of these methods is as follows.

In the 'zero creep' method the sample (long string, foil) is heated to a sufficiently high temperature at which it begins to shrink in length under the influence of surface stresses. This effect was already observed by M. Faraday in 1857 in heating a gold foil to a temperature close to the melting point. It is known as creep strain and is due to rapid diffusion of atoms under the action of surface forces. By applying an external force to the sample without changing the sample shape, we can determine the surface tension. These experiments are quite difficult to perform with sufficient accuracy. However, this method yields some rough estimate of the surface tension in respect of the order of magnitude.

In the method of destruction (splitting) of the crystal the layer is separated by cleavage from the crystal. Surface tension is defined by the work expended in splitting, i.e. breaking bonds on the surface. In this regard, for the surface tension we write $\sigma = E_b \left(z_s / z \right) n_s$, where E_b is the bulk binding energy, z_s / z is the relative number of bonds per surface atom and broken as a result of splitting the crystal, n_s is the surface density of atoms. To estimate σ, we substitute typical values of $E_b \approx 3$ eV, $z_s / z \approx 0.25$, $n_s \approx 10^{15}$ atom/cm². As a result, we obtain $\sigma \approx 1.2$ J/m².

In the method of dissolving the powder we compare the solution heat of a thick crystal and of the powder of the same mass with surface area Λ_p determined independently. Then the surface tension is defined by the relation $\sigma = \Delta Q/\Lambda_p$, where ΔQ is the difference of the heats of solution.

In the method of the neutral droplet the surface tension is determined by the shape of the equilibrium droplet of another 'standard' material, located on the surface of the studied solid.

Note also the method of 'a healing scratch', in which the surface tension is determined on the basis of data on the time of 50% 'healing' of the scratch a given width made on the crystal surface. It is also necessary to know the surface diffusion coefficient, which also determines the rate of 'healing'. The measurements of surface tension for crystals are shown in Table 1.1.

1.1.3. The anisotropy of the surface tension of solids

The surface tension of a flat solid surface depends on its crystallographic orientation. This is because the binding forces in the crystal between the atoms differ for different crystallographic directions, so an important role in the formation of the surface, characterized by rupture of bonds in the direction perpendicular to the surface area, is played by the orientation of the surface relative to the existing bonds in the crystal. It is obvious that the minimum value of surface tension (surface energy) will be measured on those surfaces which have the largest number of strong bonds and which, at the same time, break (intersect) the smallest number of such bonds. For example, for a simple cubic lattice the faces of the cube (100) (see Fig. 1.2 [420]) are parallel to two systems of bonds with neighbours and intersect the third system, as well as all the systems following the immediate neighbours. The faces (110) parallel to a system of bonds with the nearest neighbours and a system of bonds with the next-nearest neighbors. The contribution

Table 1.1. Experimental data on surface tension of solids [203]

Material	Temperature, °C	σ, J/m^2
Ag	930	1.14 ± 0.09
Al	180	1.14 ± 0.2
Au	1040	1.37 ± 0.15
Cu	900	1.75 ± 0.09
Pt	1310	2.3 ± 0.8
W	1750	2.9 ± 0.3
Zn	380	0.83
KCl(100)	25	0.11
MgO(100)	25	1.2
NaCl(100)	25	0.27
Mica	25	0.31

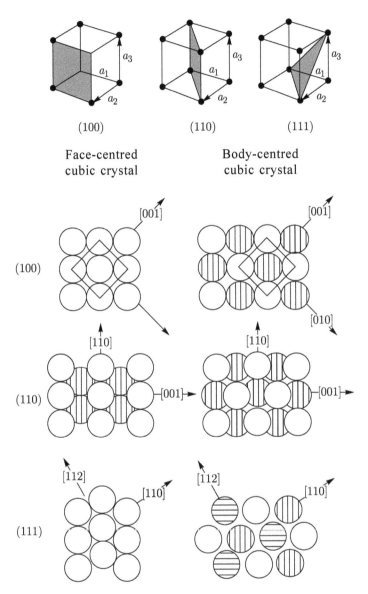

Fig. 1.2. The surface structure of a cubic crystal for the faces (100), (110) and (111). Vertical and horizontal lines shade the atoms from second and third layers, respectively [420].

to the surface tension for a given face comes from the two systems of bonds that intersect the face with the nearest neighbors and three systems of bonds with the next-nearest neighbors. In the plane (111) there is no system of bonds with the closest neighbors, and there is only a system of bonds with the next-nearest neighbors. This face of the system intersect

Fig. 1.3. The steps that form vicinal surface (*a*) and transverse section (*b*) [203].

all three systems of bonds with the nearest neighbours, so the (111) face is characterized by the highest surface tension. For the faces containing the two systems of bonds with immediate neighbors, the surface density of atoms is the largest, so these faces are called close-packed. For a cubic lattice the (100) face is close-packed. It also possesses the lowest surface tension (surface energy). Figure 1.2 shows for comparison, the structure of the faces (100) and (110) for a face-centered cubic lattice.

Consider how the surface energy changes with a weak deviation from the close-packed surface of the face (such surfaces are called *vicinal*). A small deviation of the intersecting face from close packed results in the formation of a system of steps or terraces on the surface (Fig. 1.3).

Ideally, the steps are straight areas of atomic height, separated by wide terraces na, where the a is the lattice constant. To create a step, it is required to expend some energy, which can be treated as linear tension. Let $\varkappa$ denote the energy necessary to create a step of unit length at constant volume and temperature. As the surface energy, this energy is determined by the relative number of bonds broken on the steps. Then the surface tension of the vicinal surface determined as the function of the angle of deviation of this surface from the close-packed face can be represented as

$$\sigma(\theta) = \frac{\varkappa}{a}\sin\theta + \sigma(0)\cos\theta, \qquad (1.23)$$

where $\sigma(0)$ is the surface tension of the close-packed faces. In this formula, the ratio $\sin\theta/a$ is the number of steps per unit length of the vicinal surface, and $\cos\theta$ determines the overall step area. Note that the expression (1.23) is approximate, it does not take into account the interaction between the atomic steps. This interaction can be neglected at small angles θ, however, when the step density becomes large enough, their interaction makes a significant contribution to the energy surface.

If the vicinal surface deviates in the opposite direction, i.e. towards negative angles, then the tension is determined by a formula similar to (1.23):

$$\sigma(\theta) = -\frac{\varkappa}{a}\sin\theta + \sigma(0)\cos\theta, \qquad (1.24)$$

where for simplicity it is assumed that the energy levels of formation of steps do not depend on the direction of deviation of the vicinal surface from the close-packed surface.

Differentiating equation (1.23) and (1.24) over the angle θ and comparing values of derivatives at $\theta = 0$, we find that at this point the derivative $d\sigma/d\theta$ has a discontinuity with the value

$$\Delta\left(\frac{d\sigma}{d\theta}\right)_{\theta=0} = \frac{2\varkappa}{a}. \tag{1.25}$$

This means that when $\theta = 0$, the function $\sigma(\theta)$ has a sharp minimum. For this reason, $\sigma(\theta)$ can be expanded into a series with respect to θ in the neighborhood of the angles corresponding to the close-packed faces. For this reason, these phases are often called *singular*.

Performed in [234], the general analysis of the dependence $\sigma(\theta)$ has shown that $\sigma(\theta)$ is symmetric with respect to reflections in the horizontal $(\theta = \pi/2)$ and vertical $(\theta = 0)$ planes, i.e. invariant under the substitution θ to $-\theta$ and θ by $\pi \pm \theta$. The resulting dependence of surface tension on the orientation of the surface for a cubic crystal is characterized by sharp minima at $\theta = \pm\pi m/2$, where m are integers, and is presented in Fig. 1.4 in rectangular and polar coordinates. Note that the two-dimensional polar diagram of surface tension in Fig. 1.4b, as well as three-dimensional polar diagram

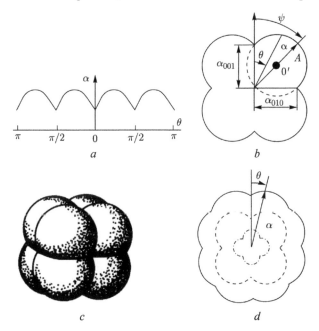

Fig. 1.4. The anisotropy of surface energy $\sigma(\theta)$ for a cubic crystal at $T = 0$ [234]: a – in rectangular coordinates, b – in the polar crystals in the two-dimensional cross-section, and c – the spatial polar diagram in the approximation of the nearest-neighbor interaction, d – a two-dimensional diagram in the approximation of the interaction of next nearest neighbors. The dashed line shows the contribution of the bonds with its nearest neighbors.

in Fig.1.4c, correspond to taking into account the bonds only between nearest neighbors. Taking into account the bonds between the next nearest neighbors gives a two-dimensional polar diagram in Fig. 1.4d. It should be emphasized that the diagrams shown in Fig. 1.4 are plotted for the absolute zero temperature, and the break points (cusps) on the diagrams form there for all directions, described by rational Miller indices. The break in the derivative function $\sigma(\theta)$ at the point of return decreases rapidly with increasing Miller index: $\Delta(d\sigma/d\theta) \sim -1/n^4$, where n is the width of the steps in the lattice constants. The more complex the crystal structure of the material, the greater is the number of the return points in the diagram $\sigma(\theta)$.

With increasing temperature the diagram of the surface tension undergoes the following changes: the minima on it, corresponding to singular orientations, become less acute, and at some diagrams specific to a given crystal, the temperature part of them disappears, i.e. with increasing temperature the relief of the surface tension diagrams becomes smoother.

Since the surface tension of the crystals depends on the orientation of the faces, the question arises about the shape of the crystal, which provides the minimum value of the total surface energy (for liquids is the sphere).

Obviously, the crystal must take the form in which minimizes the quantity

$$\Omega_s = \oint \sigma(\theta)d\Lambda, \tag{1.26}$$

calculated at a fixed volume of the crystal.

The value of this integral can be assessed by analyzing the form of polar diagrams $\sigma(\theta)$, the number of cusps on these diagrams and the change of surface tension in them [49]. If the minima corresponding to the singular planes on the diagram are sufficiently deep, the equilibrium crystal shape is a polyhedron with sharp edges.

The reduction of the depth of the minima and the disappearance of some of them, for example, at higher temperatures, lead to the situation where the singular plane are joined by rounded areas.

As was first pointed out in [298], in practice, the equilibrium shape can be observed only in crystals of very small sizes in which minimization of Ω_s in (1.26) significantly affects the reduction of the free energy of the system. In addition, this is also affected by the fact that the relaxation of the non-equilibrium crystal shape can be carried out at the time $\tau \sim L^k$ with index $k > 1$, where L – the linear dimension of the crystal.

1.2. The equilibrium shape of crystals. The Gibbs–Curie principle and Wulff theorem

The question of the equilibrium crystal shape is the problem of geometric crystallography, which was first solved by Wulff. At the equilibrium of the crystal with its saturated vapour or the melt, its shape is determined by the Wulff theorem, which he proved in 1885. This theorem expresses the specific condition of equilibrium of the crystal and can be established on the basis of the general conditions of equilibrium at constant volume V and temperature T:

$$\delta F = 0, \quad \delta^2 F > 0. \tag{1.27}$$

Before proving this theorem, we note that for each crystal there exists a point (the centre of the crystal), satisfying the condition

$$\frac{\sigma_1}{h_1} = \frac{\sigma_2}{h_2} = \frac{\sigma_3}{h_3} = \frac{\sigma_4}{h_4} = c, \tag{1.28}$$

where σ_i is the specific free energy (surface tension) of the faces, h_i are their respective distances to this point. Indeed, consider two adjacent faces of the crystal OM and ON (Fig. 1.5a) with surface tensions σ_1 and σ_2, respectively, and draw parallel to these faces two planes at distances h_1 and h_2 from the faces to satisfy the condition

$$\frac{\sigma_1}{h_1} = \frac{\sigma_2}{h_2}. \tag{1.29}$$

Line A of the intersection of the planes will obviously be parallel to the edge O and with an increase in h_1 and h_2 by the same number of times it will move along the plane OA. Thus, the plane OA is a locus defined by (1.29). Considering of the third and the fourth face, we see in the same way that on the plane OA there is a direct line, satisfying the condition

$$\frac{\sigma_1}{h_1} = \frac{\sigma_2}{h_2} = \frac{\sigma_3}{h_3}, \tag{1.30}$$

and on the line there is a point that satisfies the condition (1.28).

The existence of such point is not connected with the assumption of equilibrium of the crystal and, hence, in general, these points are different for every given four faces (of the total number N). In the case of the equilibrium crystal, this point is unique. Otherwise speaking, the equilibrium

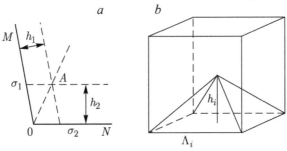

Fig. 1.5. Graphical explanation of the geometrical constructions in the proof of Wulff theorem [9].

crystal shape is characterized by the fact that its faces are situated away from a certain point (Wulff points) at distances proportional to the surface tension of faces:

$$\frac{\sigma_1}{h_1} = \frac{\sigma_2}{h_2} = \frac{\sigma_3}{h_3} = \cdots = \frac{\sigma_N}{h_N}. \tag{1.31}$$

This statement is called the *Wulff theorem*. Let us prove it.

At $T = $ const, the differential of the free energy of a system consisting of a crystal, the melt (') and the interface between them, is given by

$$dF = -pdV - p'dV' + \sum_{i=1}^{N} \sigma_i d\Lambda_i = \sum_{i=1}^{N} \sigma_i d\Lambda_i - (p - p')dV, \tag{1.32}$$

where Λ_i is the area of the i-th face of the crystal.

At equilibrium, $dF = 0$, so the crystal that is in equilibrium with the melt takes for a given volume V the faceting (form) in which its surface free energy has a minimum value (the Gibbs–Curie principle)

$$\sum_{i=1}^{N} \sigma_{i-1} d\Lambda_i = 0, \quad dV = 0. \tag{1.33}$$

The volume of the crystal can be considered as the sum of volumes built on the faces Λ_i of pyramids with a common vertex at an arbitrary point inside the crystal (Fig. 1.5b). It is obvious that

$$V = \frac{1}{3} \sum_{i=1}^{N} \Lambda_i h_i \tag{1.34}$$

and

$$dV = \frac{1}{3} \sum_{i=1}^{N} (\Lambda_i dh_i + h_i d\Lambda_i). \tag{1.35}$$

On the other hand, any change in the volume with an accuracy of the second order is equal to the displacement of the surface Λ_i by the change of height dh_i:

$$dV = \sum_{i=1}^{N} \Lambda_i dh_i. \tag{1.36}$$

Therefore:

$$\sum_{i=1}^{N} \Lambda_i dh_i = \frac{1}{2} \sum_{i=1}^{N} h_i d\Lambda_i,$$

$$dV = \sum_{i=1}^{N} \Lambda_i dh_i = \frac{1}{2} \sum_{i=1}^{N} h_i d\Lambda_i,$$

and the Gibbs–Curie principle (1.33) that defines the relative minimum surface free energy $\sum_i \sigma_i \Lambda_i$, can be written as

$$\sum_{i=1}^{N} \sigma_i d\Lambda_i = 0 \qquad (1.37)$$

provided

$$\sum_{i=1}^{N} h_i d\Lambda_i = 0. \qquad (1.38)$$

But the condition (1.38) does not exhaust all the bonds between changes in the areas for possible displacements. Some of these displacements include the displacement of the crystal in space without changes in shape. There are only three such independent displacements of the crystal in the space. Therefore, possible changes in the areas of the crystal faces must satisfy the three equations of bonds:

$$\sum_{i=1}^{N} a_i d\Lambda_i = 0, \qquad (1.39)$$

$$\sum_{i=1}^{N} b_i d\Lambda_i = 0, \qquad (1.40)$$

$$\sum_{i=1}^{N} c_i d\Lambda_i = 0, \qquad (1.41)$$

where a_i, b_i, c_i are the projections of the displacement vector of the crystal in the space on the direction of the normals to the crystal faces.

To determine the minimum of $\sum_i \sigma_i \Lambda_i$ under the conditions (1.38)–(1.41), we use the Lagrange method – multiply the equation (1.38)–(1.41), respectively, by the factors λ_1, λ_2, λ_3, λ_4, add them with equation (1.37) and obtain:

$$\sum_{i=1}^{N} (\sigma_i + \lambda_1 h_i + \lambda_2 a_i + \lambda_3 b_i + \lambda_4 c_i) d\Lambda_i = 0. \qquad (1.42)$$

In connection with the conditions (1.38)–(1.41), four variations of $d\Lambda_i$ are not independent. Let this be $d\Lambda_1$, $d\Lambda_2$, $d\Lambda_3$, $d\Lambda_4$. We choose λ_1, λ_2, λ_3, λ_4 so that changes in the coefficients of the dependent areas convert to zero:

$$\sigma_1 + \lambda_1 h_1 + \lambda_2 a_1 + \lambda_3 b_1 + \lambda_4 c_1 = 0,$$
$$\sigma_2 + \lambda_1 h_2 + \lambda_2 a_2 + \lambda_3 b_2 + \lambda_4 c_2 = 0,$$
$$\sigma_3 + \lambda_1 h_3 + \lambda_2 a_3 + \lambda_3 b_3 + \lambda_4 c_3 = 0, \qquad (1.43)$$
$$\sigma_4 + \lambda_1 h_4 + \lambda_2 a_4 + \lambda_3 b_4 + \lambda_4 c_4 = 0.$$

From these equations we find:

$$\lambda_1 = \frac{\Delta_1}{\Delta}, \ \lambda_2 = \frac{\Delta_2}{\Delta}, \ \lambda_3 = \frac{\Delta_3}{\Delta}, \ \lambda_4 = \frac{\Delta_4}{\Delta}, \qquad (1.44)$$

where

$$\Delta = \begin{vmatrix} h_1 & a_1 & b_1 & c_1 \\ h_2 & a_2 & b_2 & c_2 \\ h_3 & a_3 & b_3 & c_3 \\ h_4 & a_4 & b_4 & c_4 \end{vmatrix}, \quad \Delta_1 = -\begin{vmatrix} \sigma_1 & a_1 & b_1 & c_1 \\ \sigma_2 & a_2 & b_2 & c_2 \\ \sigma_3 & a_3 & b_3 & c_3 \\ \sigma_4 & a_4 & b_4 & c_4 \end{vmatrix},$$

$$\Delta_2 = -\begin{vmatrix} h_1 & \sigma_1 & b_1 & c_1 \\ h_2 & \sigma_2 & b_2 & c_2 \\ h_3 & \sigma_3 & b_3 & c_3 \\ h_4 & \sigma_4 & b_4 & c_4 \end{vmatrix} \quad \text{and so on.} \tag{1.45}$$

Given the expressions (1.44) and (1.45), from (1.42) for independent $d\Lambda_i$ ($i = 5, 6, ..., N$) we get:

$$\sigma_i + \frac{\Delta_1}{\Delta}h_i + \frac{\Delta_2}{\Delta}a_i + \frac{\Delta_3}{\Delta}b_i + \frac{\Delta_4}{\Delta}c_i = 0. \tag{1.46}$$

Equations (1.43) and (1.46) are geometric illustrations of the equilibrium conditions of the crystal and are valid for any point.

We apply equation (1.43) to the point at which the condition (1.28): $\sigma_i/h_i = c$, where $i = 1, 2, 3, 4$. Substituting these values of σ_i to the determinant (1.45), we find

$$\Delta_1 = -c\Delta, \quad \Delta_2 = \Delta_3 = \Delta_4 = 0. \tag{1.47}$$

From equation (1.46) we automatically receive:

$$\frac{\sigma_i}{h_i} = c, \quad i = 5, 6, ..., N, \tag{1.48}$$

which is the content of the Wulff theorem.

The tendency of the crystal to the equilibrium shape determined by Wulff theorem, becomes weaker with increasing crystal size, so that in practical terms, the conclusions of the theorem are relevant only to crystals of a sufficiently small size, not exceeding 10^{-6} m.

If for some face $\sigma_i/h_i > c$, then it will evaporate or melt, but if $\sigma_i/h_i < c$, then this will increase the face size, and the crystal growth rate u_i along the normals to various facets on approach to equilibrium will be proportional to the specific surface free energies of these facets:

$$u_i = \frac{h_i}{t}, \quad h_i = \frac{\sigma_i}{c}, \quad u_i \sim \sigma_i. \tag{1.49}$$

1.3. The role of surface tension in the formation of a new phase. Nuclei

It is well known that in equilibrium any system, depending on external conditions under consideration, is characterized by the minimum of one of their thermodynamic potentials and changes in these conditions result in the system moving from one stable equilibrium state to another. For example, if water receives heat at atmospheric pressure, then it is either heated or boils and partially transforms into steam as soon as its temperature reaches 100°C. However, by cleaning the liquid we can overheat it and the phase transition does not occur even at temperatures significantly above the boiling point at a given pressure. A similar behaviour is observed in other first-order phase transitions: in clean vapour the process of condensation in the liquid phase is delayed and such vapours are called *supercooled*; in a pure liquid or solution the phase transformation to the crystalline state is delayed and this state of a liquid or solution is called *supersaturated*.

The homogeneous system (phase) for a given volume can exist in a certain temperature range, having a free energy higher than the free energy of the inhomogeneous system of the same particles. This phase state is metastable. Over time, the system goes into a state with a minimum value of free energy, i.e. becomes inhomogeneous. However, this transition is hampered by a surface effect, i.e. the fact that in this phase small objects of another phase are formed having a surface free energy, which leads to an increase in the free energy of the system, so the transition is thermodynamically disadvantageous. For example, the beginning of condensation is difficult for the reason that when small (radius R) liquid droplets form in the vapour, their surface free energy F_s, which is proportional to R^2, is growing faster than the volume term of the free energy, proportional to R^3, decreases Therefore, the appearance of small droplets turns out to be thermodynamically unfavourable and condensation is delayed. At large droplets, beginning with some $R = R_c$, on the contrary, the volume term of the free energy decreases faster than the surface term, and condensation becomes possible: a droplet that occurs as a result of the fluctuations will grow.

For any metastable phase there is a certain minimum size which should be the size of a cluster of another phase, formed within the metastable phase due to fluctuations to ensure that this other phase is more stable than the original one. At smaller sizes the main phase is still stable than these fluctuations and they disappear. Such clusters of a new phase, with the smallest dimensions are called *nuclei*.

We calculate the critical radius of the droplet–nucleus for the onset of vapour condensation. We assume that fluctuations result in the formation in the initial phase of an area of the new phase, for example, in vapour – a droplet of a liquid with radius R. The change in the free energy system during the formation of droplets without surface effects (i.e. on a flat surface) at given temperature T and vapour pressure p can be written as

$$\Delta F = (\mu_2 - \mu_1)N + \sigma\Lambda, \tag{1.50}$$

where $\mu_1(T, p)$ is the chemical potential of the vapour, $\mu_2(T, p)$ is the chemical potential of the liquid, N is the number of particles in the droplet, Λ is the surface of the droplet, σ is surface tension.

The values of N and Λ are easily expressed in terms of the droplet radius:

$$\Lambda = 4\pi R^2, \quad N = \frac{4\pi R^3}{3v_2},$$

where v_2 is the specific volume per one molecule of the liquid. Thus,

$$\Delta F = \frac{4\pi R^3}{3v_2}(\mu_2 - \mu_1) + 4\pi R^2 \sigma. \tag{1.51}$$

From this formula we see that ΔF in many ways depend on R in two possible cases:

$$1)\, \mu_2 > \mu_1; \quad 2)\, \mu_2 < \mu_1.$$

In the first case, the chemical potential of a new phase under the given T and p is higher than that of the 'old' and the new phase is less stable. Appearance of liquid droplets in the vapour (or bubble in a liquid) for all its sizes R leads to increased ΔF (Fig. 1.6a), so the formation of the new phase is always thermodynamically unfavorable. If as a result of fluctuations a droplet forms in such vapour, then no matter what its size is, it must quickly disappear. In the second case, with increasing droplet size the value ΔF initially increases until $R < R_c$, so that small droplets of the new phase are unstable (Fig. 1.6b). This instability is due to the fact that at small R the second term in the expression for ΔF increases with increasing R at a rate which is greater than the rate of decrease of the first term. However, for

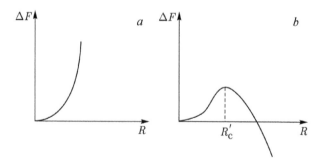

Fig. 1.6. Dependence of free energy ΔF on the radius of the droplet–nucleus [9].

large nuclei of the second phase, when $R > R_c$, the second phase becomes more stable.

The size of the nucleus is determined from the condition of maximum free energy change ΔF, which expresses the condition of unstable equilibrium:

$$\frac{\partial(\Delta F)}{\partial R}\Big|_{R=R_c} = 0$$

or

$$4\pi\frac{\mu_2 - \mu_1}{v_2}R_c^2 + 8\pi\sigma R_c = 0,$$

from which

$$R_c = \frac{2\sigma v_2}{\mu_2 - \mu_1}. \tag{1.52}$$

This shows that the critical radius of the nucleus is proportional to the surface tension coefficient. The term $\mu_2 - \mu_1$ in the denominator characterizes the degree of supersaturation and, consequently, the more it is supersaturated, the smaller the critical radius, and the sooner begins the process of vapour condensation.

The situation is similar for other first-order phase transitions: in a boiling liquid these nuclei are vapour bubbles, in crystallization – crystals. But the role of nuclei in these cases can be played not only by the bubbles or crystals of the substance but also by particle of any foreign substance (impurities).

The critical radius of the nucleus (vapour bubbles in a superheated liquid or droplets in a supersaturated vapour) can be found directly from the condition of mechanical equilibrium of the nucleus, i.e. the equality of pressure p_2 to the sum of the nucleus pressure p_1 in the main phase and pressure $2\sigma/R_c$, due to surface tension:

$$p_2 = p_1 + \frac{2\sigma}{R_c},$$

from which

$$R_c = \frac{2\sigma}{p_2 - p_1}, \tag{1.53}$$

which is equivalent to (1.52). Indeed, from the condition of equilibrium in the exchange of particles between the droplet and its vapour

$$\mu_2(T, p_2) = \mu_1(T, p_1). \tag{1.54}$$

Given the low compressibility of the liquid, i.e. the slight change in chemical potential with pressure, we find

$$\mu_2(T,p_2) = \mu_2(T,p_1) + \frac{\partial \mu_2}{\partial p_1}(p_2 - p_1).$$

Therefore, from (1.54) we obtain

$$p_2 - p_1 = \frac{\mu_1(T,p_1) - \mu_2(T,p_1)}{v_2}$$

and, hence, the expression (1.52).

The superheated liquid is heated above the boiling point (at the boiling point the vapour pressure of the liquid under external pressure p, is equal to this external pressure), but not boiling, i.e. not forming vapour bubbles under the surface but just evaporating from the surface. The saturated vapour pressure over a flat surface of such a liquid $p_\infty > p$. If a bubble forms in the liquid, its critical radius, obviously, can be found by setting in the expression (1.54) $p_2 = p_\infty$ and $p_1 = p$. As a result,

$$R_c = \frac{2\sigma}{p_\infty - p}.$$

1.4. Gibbs equation for adsorption of solids. Surface-active substances

At equilibrium of heterogeneous systems, a change in the concentrations of the components of the system takes place at the interface of different phases. Gases tend to condense on the surfaces of solids and dissolved substances also accumulate on these surfaces. This phenomenon is a consequence of the 'desire' of the free surface energy of solids to become smaller: the concentration in the surface layer is changed so as to decrease the surface free energy. Therefore, in the surface layer one (or more) of the components of a heterogeneous system can be in excess and we then talk about the adsorption of this component at the interface. The phenomenon of adsorption plays an important role in the processes that take place with the participation of interfacial surfaces. It is widely spread in nature and used in various technologies. For this reason, the studies of adsorption are give very serious consideration and it is one of the most studied surface phenomena.

We find the relationship between adsorption and surface stresses of a solid. To do this, we subtract from the equation (1.19) the differentials of the free energy related to the bulk phase. As a result, we obtain an expression for the differential of the surface free energy

$$dF_s = -S_s dT - \sum_i \mu_i dN_{si} + \sum_{\alpha,\beta} \sigma_{\alpha\beta}^s du_{\alpha\beta}^s, \qquad (1.55)$$

where S_s is surface entropy, N_{si} is an excess concentration of the i-th component in the surface layer. Using for the surface free energy of solids the expression (1.15), written through surface tension σ, differentiating it and subtracting from (1.55), we obtain

$$\Lambda d\sigma + \sigma d\Lambda + S_s dT + \sum_i N_{si} d\mu_i - \sum_{\alpha,\beta} \sigma_{\alpha\beta}^s du_{\alpha\beta}^s = 0, \qquad (1.56)$$

which is an analogue of the Gibbs–Duhem equation for the surface phenomena in solids. Using (1.20), we rewrite this equation in the form

$$d\sigma + \frac{S_s}{\Lambda} dT + \frac{1}{\Lambda} \sum_i N_{si} d\mu_i + \sum_{\alpha,\beta} (\sigma\delta_{\alpha,\beta} - \sigma_{\alpha\beta}^s) du_{\alpha\beta}^s = 0, \qquad (1.57)$$

or, introducing the values of adsorption $\Gamma_i = N_{si}/\Lambda$, each of which represents the excess number of particles of the i-th component in the surface layer on one surface, we write

$$d\sigma = -\frac{S_s}{\Lambda} dT - \sum_i \Gamma_i d\mu_i - \sum_{\alpha,\beta} (\sigma\delta_{\alpha,\beta} - \sigma_{\alpha\beta}^s) du_{\alpha\beta}^s. \qquad (1.58)$$

This equation is the Gibbs equation for the adsorption of solids. It shows that under the condition of equilibrium of two phases, taking surface phenomena into account, we arrive at (1.21), linking the surface stress with the surface tension of solids. We also note an important fact, following from the Gibbs adsorption equation: changes of the surface tension in solids may be due to adsorption of not only an excess number of particles of the components that determine the composition of the solid, but also of the particles of the liquid or gas phases bonded with the solid phase.

From the Gibbs equation (1.58) we immediately obtain the equalities

$$\Gamma_i = -\left(\frac{\partial\sigma}{\partial\mu_i}\right)_{T, u_{\alpha\beta}^s}, \qquad (1.59)$$

expressing the magnitude of adsorption of the i-th component of the material through the derivative of the surface tension with respect to the chemical potential this component. From these relations we see that the surface layer accumulates primarily those substances whose presence lowers the surface tension ($\Gamma_i > 0$ for $\partial\sigma/\partial\mu_i > 0$). Such substances by their presence lower the free energy of the surface layer and this is thermodynamically favorable for the equilibrium and also explains their positive adsorption (surface-active substances). Those substances, which cause increased surface tension ($\partial\sigma/\partial\mu_i > 0$) tend to leave the surface layer ($\Gamma_i < 0$, negative adsorption), as they increase the surface free energy by their presence.

It turns out that quite small additions of certain surface-active substances are sufficient for the body to dramatically alter its properties. In practice it is widely used in metallurgy and electronics. Small additions of surfactants increases the strength of metals. In addition, since the work function electrons from the metal is determined by the surface tension, adding a small amount of surface-active substances can greatly reduce the electron work function, which is especially important for the photoelectric effect.

Recently, surfactants have been widely used in oil production. Reducing the amount of oil, given by the oil well at a time when it is flooded, is due to the fact that the water droplets are trapped in the narrowing of capillary pore channels of rocks of oil deposits. The resistance of these droplets is proportional to the surface tension and reaches high values. This resistance must be overcome by the parallel flow of the surrounding oil phase. THe pressure in the stratum, however, is not enough for this. By reducing the surface tension at the water–oil interface the price of oil can be greatly reduced. Surface-active substances are also used for washing sand plugs in oil wells.

1.5. Surface phase transitions

The question of surface phase transitions is one of the most important aspects in the study of surface physics and two-dimensional structures that arise on the surface of the crystals in the processes of adsorption, as the number of different phases, even in the simplest experimental systems, is usually quite large. This is most clearly manifested in the case of adsorption systems for which it is easy to implement a huge range of densities – from a dilute lattice gas to an incommensurate crystal, with spacings smaller than the corresponding spacings of the three-dimensional crystal. In this case there are several modifications of two-dimensional crystals in which phase transitions occur when the temperature changes.

The reason for the interest in the surface phase transitions (as a rule, it is a second-order transition) is also connected with the fact that strong fluctuations in the system with low dimensionality (two-dimensional structure) substantially change the nature of the phase transition. The most striking example here is the possibility of melting of two-dimensional crystals by a second-order transition, while melting in the three-dimensional crystal is always a first-order transition.

1.5.1. Long-range order–disorder transition

The overall picture of phase transitions of both first and second kind from the phase with a long-range order to the disordered phase is similar for the two- and three-dimensional systems. Phase transitions of the second order in three-dimensional systems have been investigated in detail both theoretically and experimentally. The modern theory of phase transitions can be found in [128, 187]. Here is some information from the general theory of phase transitions necessary for future considerations.

Classification of phase transitions. Some of the phase transitions are accompanied by the release (or absorption) of heat and an abrupt change in density. In this case, a new phase appears in the form of nuclei, for example, vapour bubbles in water, and the phase transition occurs by a gradual increase in the volume of the new phase in the volume of the old one. In accordance with the well-known classification proposed by P. Ehrenfest (1933), such phase transitions are called *first-order phase transitions* (PTI).

Phase transformations in which the coexistence of two phases is possible and a new phase occurs immediately in the volume of the old place, completely replacing it, are called *second-order phase transitions* (PTII), *or continuous phase transitions*.

A more rigorous classification of phase transitions can be defined as follows:

– first-order phase transitions are characterized by breaks of the first derivatives of the chemical potential in transition through the curve of phase equilibrium

$$\left(\frac{\partial \mu}{\partial T}\right)_P = -S \rightarrow \Delta S = S_1 - S_2 \quad \text{– absorption or release of the latent heat of}$$

transition,

$$\left(\frac{\partial \mu}{\partial P}\right)_T = V \rightarrow \Delta V = V_1 - V_2 \quad \text{– a 'jump' in the density of matter, i.e. there}$$

is a clearly observable boundary between the phases;

– second-order phase transitions are characterized by the fact that the first derivatives of the chemical potential are continuous, and the second derivatives undergo a finite or infinite break

$$\left(\frac{\partial^2 \mu}{\partial T^2}\right)_P = -\frac{C_P}{T} \quad \text{– specific heat jump,}$$

$$\left(\frac{\partial^2 \mu}{\partial P^2}\right)_T = \left(\frac{\partial V}{\partial P}\right)_T \quad \text{– a jump of isothermal compressibility,}$$

$$\frac{\partial^2 \mu}{\partial P \partial T} = \left(\frac{\partial V}{\partial T}\right)_P \quad \text{– a jump in the thermal expansion coefficient.}$$

The second-order phase transitions are most often associated with spontaneous changes of any of the symmetry properties of the body.

For example, in the high-temperature phase the $BaTiO_3$ crystal has cubic lattice with the cell shown in Fig. 1.7 (Ba atoms at the vertices, O atoms at the centres of the faces and Ti atoms at the centers of cells). In the low-temperature phase, the Ti and O atoms are displaced in relation to the Ba atoms in the direction of one of the edges of the cube. As a result of this shift the symmetry of the lattice changes immediately, changing from cubic to tetragonal. This example is characterized by the fact that there is no abrupt change in the body condition. The location of the atoms in the crystal varies continuously.

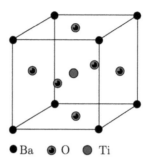

●Ba ◉O ●Ti

Fig. 1.7. Image of the unit cell of the BaTiO$_3$ crystal.

However, any arbitrarily small displacement of the atoms relative to their initial symmetrical arrangement is sufficient to change the lattice symmetry. Numerous studies of phase transitions of various physical nature clearly indicate a certain similarity between them and, most surprisingly, the quantitative coincidence of a number of their characteristics. To date, these observations and views have been realized in the construction of a sufficiently general universal theory of phase transitions [128, 187, 192].

The critical point. Critical phenomena. The second-order phase transitions are closely linked in a sense with the phase transitions occurring at the critical point.

The critical point is a point on the phase plane (for example, pressure-temperature for the liquid–vapour system) at which the phase equilibrium curve terminates. The temperature and pressure, corresponding to the critical point, are called *the critical temperature T_c* and *critical pressure P_c*. At temperatures above T_c, and at pressures higher than P_c there are no different phases and the solid is always homogeneous. We can say that at the critical point the difference between the two phases disappears. The concept of the critical point was first introduced by D.I. Mendeleev (1860).

If there is a critical point between any two phase states of matter a continuous transition can take place in which there is not separation into two phases at any moment of the process of changes in the state of matter – for this to take place we must change the state along any curve bending around the critical point and not intersecting the phase equilibrium curve. If there is a critical point, the very concept of the different phases becomes somewhat arbitrary, and it is not possible to show in all cases which state relates to one phase, and which to the other. Strictly speaking, one can talk about two phases only when they coexist simultaneously, touching each other, i.e. at points on the equilibrium curve.

The critical point can exist only for such phase states of matter where the difference between them is purely quantitative in nature. These are the liquids and gases which differ from each other only by a greater or lesser role of the interaction between molecules. Such phases as the liquid and solid (crystal) or different crystalline modifications of matter are characterized by qualitative differences among themselves, because their internal symmetry differs. It is clear that for any property (element) of symmetry we can say whether it is there or not; it may appear or disappear only once, abruptly, not gradually. In each state, the solid will have either one or the other symmetry and therefore one can always specify which of the two phases it belongs to. Consequently, the critical (terminal) point can not exist for such phases, and the equilibrium curve must either go to the infinity or intersect with the equilibrium curves of other phases.

It turns out that the first-order transitions close to the critical point, are very 'similar' to the second-order phase transitions. Namely, the jumps of the first derivatives (density, latent heat of phase transition) are small, but at the same time there is an anomalous behaviour of the second derivatives of the thermodynamic potential (specific heat, compressibility, etc.), as in the case of typical first-order phase transitions. This defines the physical similarity between the second-order phase transitions and critical phenomena.

Basic theoretical concepts for describing the second-order phase transitions. The first universal phenomenological theory of phase transitions and critical phenomena was proposed by L.D. Landau in 1937 [128, 187]. It was an important step in the creation of the modern theory of critical phenomena, since it allowed within a unified approach to describe the set of second-order phase transitions and critical phenomena in various systems. Landau managed to isolate the general feature which combines multiple phase transitions in materials with seemingly distant physical properties – the spontaneous breaking of symmetry for the description of which he introduced the fundamental concept of the modern theory of critical phenomena – *the order parameter*. The physical meaning of the order parameter can be different and depends on the nature of the phase transition.

Examples of the order parameter can be: magnetization at the ferromagnetic–paramagnetic transition, the difference between the densities of the liquid and the vapour near the critical point of the liquid–vapour system, the wave function of the superfluid component in the λ-transition He^4 to the superfluid state, etc. The common feature is that the order parameter is equal to zero in the high-temperature (disordered) phase with higher symmetry and is different from zero in the low-temperature (ordered) phase with lower symmetry.

Landau postulated the expandability of the thermodynamic potential of the free energy $F(\phi, T, ...)$ near the temperature of the phase transition to the powers of the order parameter ϕ, with the expansion coefficients being analytic functions of temperature T and external parameters. The explicit form of this series, as well as the number of components of the order parameter are determined by the symmetry group of the system at the phase transition point [128].

From the microscopic point of view, Landau's theory is a generalization of the self-consistent field method, used for describing the critical behaviour of specific microscopic models of real systems, such as the Ising model, the lattice gas model, etc. The most important and obvious disadvantage of this approach is that it does not take into account the correlation of microscopic variables. Landau's theory can be generalized taking into account the correlation effects, considering only the contribution from the degrees of freedom, corresponding to large spatial scales and being determining in the vicinity of the critical point. In this case, the order parameter is almost spatially uniform and can therefore be represented as a function in space $\phi(x)$ slowly varying in space and having small gradients. In the simplest case of symmetry $O(n)$ it leads to the thermodynamic potential

$$H_{GL}[\phi] = \int d^d x \left(\frac{1}{2}\left[r_0 \phi^2(\mathbf{x}) + (\nabla\phi)^2 \right] + \frac{u_0}{4!}(\phi^2)^2 - h(\mathbf{x})\phi(\mathbf{x}) \right), \qquad (1.60)$$

which is called the *Ginzburg–Landau–Wilson effective Hamiltonian* [41, 135]. Here: $\phi^2(\mathbf{x}) = \sum_{i=1}^{n} \phi_i^2(\mathbf{x})$, $(\nabla\phi)^2 = \sum_{i=1}^{n}(\nabla\phi_i)^2$; d is the dimension of the space; $h(x)$ is the external field conjugate to the order parameter; $r_0 = (T - T_c)/T_c$, T_c is the critical temperature.

The critical exponents (indices). In the study of phase transitions and critical phenomena, special attention is paid to the definition of values of a set of indicators, which are called critical exponents and which describe the effective power-law behaviour of various relevant thermodynamic and correlation functions near the phase transition temperature (critical point).

We give a general definition of the critical exponent describing the behaviour near the critical point of a function $f(\tau)$, where $\tau \sim (T - T_c)/T_c$ is the reduced temperature, which characterizes the difference between temperature and the critical temperature. Assume that the function $f(\tau)$ is positive and continuous for sufficiently small positive values of τ, and that the limit

$$\lambda \sim \lim_{\tau \to 0} \frac{\ln f(\tau)}{\ln \tau}. \qquad (1.61)$$

The limit λ is called the *critical exponent* associated with the function $f(\tau)$. For brevity, we can write

$$f(\tau) \sim \tau^{\lambda}, \qquad (1.62)$$

to emphasize the fact that λ is the critical exponent of the function $f(\tau)$.

It should be noted that the expression for the asymptotic behaviour of a thermodynamic function $f(\tau)$ can have a more complicated form characterized by the correction terms. The function (1.62) takes the form

$$f(\tau) = A\tau^x (1 + B\tau^y), \qquad y > 0. \qquad (1.63)$$

There may be a fair question: why the focus is drawn to such a quantity as the critical exponent which gives much less information than the form of a total function. The answer to this question is determined primarily by the experimental fact that the behaviour of the function (having the form of a polynomial) near the critical point is determined mainly by its leading terms. Therefore, the curves obtained in the experimental studies at temperatures sufficiently close to the critical point in the double logarithmic scale are given direct and the critical exponents are determined from the slopes of these lines. Thus, the critical exponents are always measurable which cannot be said of the complete function. The second reason for such attention to the critical indicators is that there are a large number of relations between critical exponents which are derived from general thermodynamic and statistical provisions and, therefore, valid for any particular system.

There is a simple unique relation between the critical exponent and the behaviour of the function near the critical point ($\tau \ll 1$). If the critical exponent λ, defined by equation (1.61) is negative, then the corresponding function $f(\tau)$ near the critical point diverges and tends to infinity; the positive values of λ correspond to the function $f(\tau)$ converting at this point to zero. The smaller λ, the more abrupt changes in temperature near the phase transition are shown by the behaviour of $f(\tau)$ – in the sense that for negative λ the divergence of $f(\tau)$ becomes stronger, and for the positive – the curve $f(\tau)$ tends 'faster' to zero.

Consider the behaviour of various physical quantities near the critical point T_c, which can be specified by a specific set of critical exponents (indices).

The critical index α for the specific heat:

$$C \sim |\tau|^{-\alpha}, \tag{1.64}$$

the critical index β for the order parameter:

$$\phi \sim |\tau|^{\beta}, \quad T < T_c, \tag{1.65}$$

the critical index γ for the susceptibility:

$$\chi \sim |\tau|^{-\gamma}, \tag{1.66}$$

the critical exponent δ for the critical isotherm:

$$\phi(h, T_c) \sim h^{1/\delta}. \tag{1.67}$$

To describe the fluctuations of the order parameter ϕ we introduce the critical index ν which determines the temperature dependence of the correlation length:

$$\xi \sim |T - T_c|^{-\nu}, \tag{1.68}$$

and the critical index η which determines the law of decay of the correlation function with distance x at $T = T_c$,

$$G^{(2)}(\mathbf{x}, T_c) \sim \frac{1}{|\mathbf{x}|^{d-2+\eta}}. \qquad (1.69)$$

The importance of the critical indices in the first place is determined by the fact that they can be most easily measured in the experiment.

The heat capacity (for some systems), susceptibility, and correlation length at $T = T_c$ are divergent values. The properties of systems at continuous phase transitions are determined by the strong and long-lived fluctuations of the order parameter. The measure of magnetic fluctuations is the linear size $\xi(T)$ of the characteristic magnetic domain – a region with strongly correlated spins. At $T \gg T_c$ the correlation length $\xi(T)$ in the order of magnitude is equal the lattice spacing. As with the approach of T to T_c from the top correlation effects in the spatial orientation of the spins are enhanced, $\xi(T)$ will increase when T approaches T_c.

Order–disorder transition in two-dimensional systems. The order–disorder transition in two-dimensional systems is described using the lattice gas model with nearest-neighbour interaction, analogous to the Ising model [128, 134, 187]. We introduce a square lattice, the nodes which may contain atoms, and each node contains more than one atom. Located in between the atoms of the neighbouring sites are repulsive forces, characterized by the energy $J > 0$. The situation when the number of atoms is equal to half number of nodes can be studied quite easily. The model Hamiltonian has the form:

$$H = \frac{J}{2} \sum_{\mathbf{r},\mathbf{b}} \sigma_\mathbf{r} \sigma_{\mathbf{r}+\mathbf{b}}, \qquad (1.70)$$

where $\sigma_\mathbf{r} = 0, 1$, vector $\mathbf{b}$ runs over nearest neighbors, $\mathbf{r}$ are the nodes of the lattice. This Hamiltonian as a simple model of the magnet was first used by Lenz. The ground state of the system is doubly degenerate, since the atoms occupy one of the two sublattices. We denote these sublattices by the indices A and B. Suppose that at $T = 0$ atoms occupy the sublattice A. At a non-zero temperature antiphase domains form by fluctuations in the system in which the atoms are on sublattice B. The domains have different sizes and are arranged randomly. The average domain size r_c increases with increasing temperature. The position, number, and the form of the domains fluctuate (Figure 1.8), i.e. the order parameter decreases with increasing temperature. In structural measurements the intensity of diffraction reflections decreases as a result of diffraction of waves by randomly distributed antiphase domains.

The change in energy and entropy in the formation of an antiphase domain is associated with the appearance of a domain wall. The energy of such a wall E_w is proportional to its length L, namely, $E_w \simeq LJ$. To estimate the entropy, we assume that in each node in a square lattice

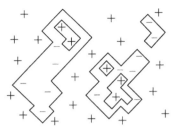

Fig. 1.8. Fluctuating domains of different sublattices in the vicinity of the phase transition point (The signs –, + denote different sublattices) [134].

the wall can, regardless of its state in other nodes, rotate on through the angles of $0°$, $±90°$, i.e. to take one of three possible states of a given node. Since rotations of the wall in different corners are assumed to be independent, the entropy is proportional to its length: $S_w \sim L\ln3$. The change of free energy F_w in the formation of a wall of length L, limiting the size of the antiphase domains of the order of L, is $F_w = F_w - TS_w$. So as the energy and entropy of the wall depend on its length by the same law, there is a temperature $T_c \sim J$, below which the formation of the antiphase domain with $L \to \infty$ is thermodynamically unfavorable, and above – lowers the free energy. Therefore, above T_c domains of both types are randomly mixed in the system, which corresponds to the disordered phase. At $T \gg T_c$ the atoms are evenly distributed throughout the lattice. With decreasing temperature, i.e. when $T \to T_c$ the disordered phase is characterized by the formation by fluctuations of ordered domains of both the first and second sublattice. As shown by the exact solution of the Ising model, obtained by Onsager, the characteristic size of these domains – the correlation radius r_c (the correlation length ξ in (1.68)) increases as the transition point is approached by the power law [128, 187]:

$$r_c \sim a \left| \frac{T - T_c}{T_c} \right|^{-1}, \tag{1.71}$$

where a is interatomic spacing.

1.5.2. Phase transitions associated with the reconstruction of the surface

When the temperature changes, the surfaces of solids have a wide variety of phase transitions characterized by surface reconstruction [95]. Unfortunately, the number of cases for which we can confidently construct a structural phase diagram (as is done for an identical bulk problem) is very limited. This problem has two aspects. First, in experimental studies it is difficult to determine the surface crystallography in order to establish the true parameters of emerging structures. Second, many of the surface phases are

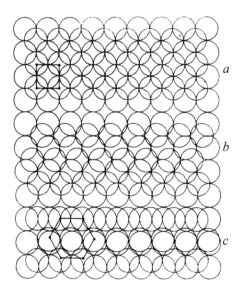

Fig. 1.9. The arrangement of atoms on the surface of Ir (100) in the phase transition from the structure 1×1 to 1×5 [333]: *a* – the ideal structure 1×1, *b* – a possible 'intermediate' structure, c – a reconstructed quasi-hexagonal structure 1×5.

actually metastable, i.e. the surface is not in true equilibrium. The latter is determined by simple reasons. With the splitting of the crystal due to dangling bonds we release only a certain part of energy, which may not be enough to move the surface atoms to the configuration with the minimum free energy. Therefore, the surface can easily be in a metastable state.

The establishment of the true thermodynamic equilibrium may require a significant thermal annealing of the sample. That is why the discussion of the phase diagram of the surface is often very similar to the history of its preliminary treatment. Consider, as an example, the reconstruction of the (100) surface of iridium, associated with the transition from the 1×1 structure to the structure 1×5 at temperatures above 800 K [333]. This transition is a first-order phase transition from the metastable to the stable state. The metastable structure 1×1 is typical for the metal surface (Fig. 1.9a). The structure of the ground state 1×5 (Fig. 1.9c) is best described as a close-packed distribution of the atoms in the ideal face-centered substrate with the orientation (100). It is likely that the energy barrier formed between the two surface configurations occurs when tight rows of atoms are displaced on the subsurface in the process of this transition (Fig. 1.9b).

This scheme of the changes in the surface structures is based on measurements performed with the use of high-quality equipment for low-energy electron diffraction (LEED), including a video camera, which allows the reconstruction to be recorded with a resolution of 20 ms in real time [333]. Measurements of the temporal and temperature dependences of the increase of the intensity of diffraction reflections, caused by the superlattice 1×5, shows that the activation energy of the transition is close to 0.9 eV/atom. The considered reconstruction is due to the fact that the formed close-packed metal surface has a lower energy than the open original surface. For iridium, this effect is very large (in absolute value), because this metal is characterized by the highest surface tension among all the elements. Reconstruction is also accompanied by the loss of energy as a result of the mismatch of a square structure of the substrate and the hexagonal structure of the top layer. The competition between these two effects plays a crucial role in the phenomenon of epitaxy.

In contrast to the example of the iridium surface (100), detailed information about the atomic geometry for the reconstruction of phase transitions is usually absent. Investigation of diffraction reflections in the LEED allows us to establish, as a rule, only the symmetry of the high- and low-temperature phases. However, according to the symmetry criteria of the Landau–Lifshitz theory [128], this information is sometimes enough to determine what type of phase transition we are dealing with – a continuous second-order or a first-order transition. Here are their suggestions.

Let $\rho(r)$ denote the surface density of atoms, corresponding to the crystal structure of a highly symmetric phase. This function is invariant with respect to the symmetry operations of the corresponding surface–spatial group, which we denote by G_0. After the phase transition the reconstructed crystal surface is described by the new density function $\rho'(r) = \rho(r) + \delta\rho(r)$, invariant with respect to the operations of a new spatial symmetry group G. According to the criteria of the Landau–Lifshitz theory, the transition may be a second-order phase transition, i.e. continuous, only if G is a subgroup of G_0 and the function $\delta\rho(r)$ is transformed in accordance with the unique irreducible representation of G_0, corresponding to the symmetry of the order parameter and responsible for the phase transition [99]. If this condition is not satisfied, we have a first-order phase transition.

As another phase transition with the reconstruction of the surface, we consider the tungsten surface W(100), characterized by weak structural changes of the 1×1 structure to the structure $\sqrt{2} \times \sqrt{2}$ $R45°$. In the low-temperature phase, the tungsten atoms on the surface are displaced from their ideal provisions to short distances and form a zigzag chain (Fig. 1.10). Symmetry analysis of surface reconstruction in accordance with the criterion of the Landau–Lifshitz theory shows that the considered phase transition can be smooth; such a conclusion has been confirmed by experiments [505]. Atomic displacements by which the ideal surface is transformed into a reconstructed one correspond to the longitudinal phonon branch with wave vectors $q = \pi/a$ $(1,1)$, where a is the lattice constant. The nature of this phase transition is associated with the effect of temperature 'softening' of the phonon branch, i.e. the tendency of its frequency to zero when approaching a certain temperature T_c. At T_c there is a phase transition accompanied by spontaneous deformation of the crystal. However, it is

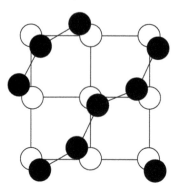

Fig. 1.10. The structure of the surface of W (100) [95]: the high-temperature phase 1×1 – open circles, low-temperature phase $\sqrt{2} \times \sqrt{2} - R45°$ – dark circles.

possible that the surface tungsten atoms are also displaced at a temperature above T_c, but in random directions.

In the case of sufficiently low temperatures, taking into account the surface interaction, the ordering displacement with the formation of the observed structure $\sqrt{2} \times \sqrt{2} - R45°$ is energetically favourable. The currently available experimental data do not conclusively give preference to any one of these mechanisms, and it is unclear which of them are really prevalent [358].

In any case, it is obvious that the reconstructed surface is characterized by a lower internal energy than the ideal surface. Exact calculations in the approximation of the local density of states for the model of the layered structure (sandwich) show that the zigzag reconstruction leads to splitting of the peak of the local density of states at E_F, characteristic of an ideal surface (see Fig. 1.11). As a result, the Fermi level is at the minimum between the two local maxima of the local density of states on the surface. The redistribution of the electron density of states reduces the total energy of the system, since the energy levels of some of the occupied states are shifted downward, while the levels of unoccupied states increase [314]. Note that the situation is reminiscent of the action of the driving forces of the reconstruction, which resulted in ejection of surface states outside the band gap on the surfaces of certain semiconductors.

Reconstructive phase transition from the 2×1 to 1×1 structure, which occurs on the surface of Au(110), may serve as a nice example that confirms the universality of the concept of the phase transitions.

Experiments using the method of scanning tunneling microscopy (STM) show [262] that at a temperature of around 650 K the high-temperature structure 1×1 reversibly transformed to the structure with the 'missing row' in Fig.1.12.

However, even without knowing the detailed structure, on the basis of only symmetry considerations it can be established that a continuous

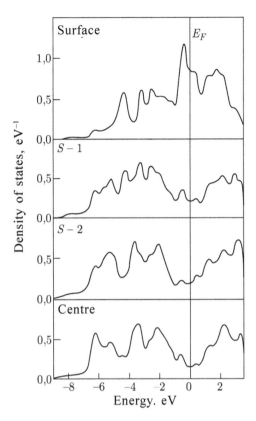

Fig. 1.11. The results of calculation of the local density of states for W (100) for the model of the layered structure. Surface states 'fill' the volume local density of states [441].

phase transition (if it occurs) must identify the critical properties inherent in the two-dimensional Ising model [248]. We predict that the temperature dependence of the order parameter (the intensity of diffraction reflections in the LEED method from the superlattice of the phase 2×1) will be characterized by the critical exponent β, corresponding to the known exact result of Onsager [135] ($\beta = 1/8$). The experiment gives $\beta = 0.13\pm0.2$ (Fig. 1.13).

1.5.3. Dislocation mechanism of melting of two-dimensional structures

Melting of two-dimensional adsorption structures by a second-order phase transition is based on the dislocation mechanism. We consider this mechanism, using the model of an uniaxial two-dimensional crystal, the atoms of which form a rectangular lattice with the primitive elementary unit cell compressed along one axis, for example, axis Ox [134]. In this model,

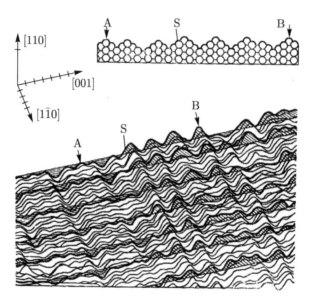

Fig. 1.12. STM image in real space surface of Au (110) 1 × 2. The axes correspond to 5 Å [264].

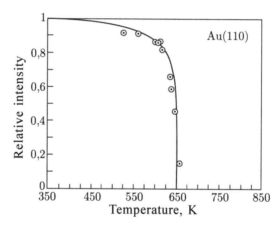

Fig. 1.13. Temperature dependence of the order parameter for continuous structural transition on the surface of Au (110). The intensity of the reflections in LEED for the structure 2×1 (circles), and the exact Onsager solution for the two-dimensional Ising model (solid curve) [280] are given.

it is assumed that the atoms are displaced only in one direction – Ox. The displacement is denoted by u. In the continuous medium approximation the expression for elastic energy displacement has the form

$$H_u = \frac{\lambda}{2} \iint_{S} \left[\left(\frac{\partial u}{\partial x} \right)^2 + \left(\frac{\partial u}{\partial y} \right)^2 \right] dx\, dy.$$

(1.72)

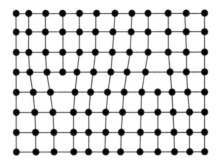

Fig. 1.14. A pair of dislocations with opposite Burgers vectors in a uniaxial crystal [134].

The first term in (1.72) describes the compression of the lattice along the axis Ox, the second – its rotation. The corresponding elastic constants should be, in general, different, yet they can be made identical by choosing different measurement units along the axes Ox and Oy.

The dislocation in the model of the crystal is nothing but the point of breaking a row of atoms located along the axis Oy. In the two-dimensional crystal, the dislocation is a point object (Fig. 1.14). The expression for the displacement field in a crystal with such a dislocation is [129]

$$u = \frac{a}{2\pi}\operatorname{arctg}(y/x). \tag{1.73}$$

Substituting (1.73) in the elastic Hamiltonian (1.72), we obtain the energy of the dislocation

$$E_d = \frac{\lambda a^2}{4\pi}\ln\frac{r}{a_0} + E_0, \tag{1.74}$$

where a_0 is the size, and E_0 the energy of the dislocation kernel. Thus, the energy of a single dislocation logarithmically increases with the size r of the system, and the birth of a single dislocation is energetically unfavourable. However, the increase of entropy, associated with the appearance of the dislocation, also occurs logarithmically with increasing system size, since the entropy can be estimated as the logarithm of a number of ways in which the dislocation kernel can be placed in the crystal:

$$S_d = 2\ln\frac{r}{a_0}. \tag{1.75}$$

It is clear that at a certain temperature T_m, considered as the melting point of the two-dimensional adsorption structure, the free energy change associated with the appearance of the dislocation

$$F_d = E_d - TS_d, \qquad (1.76)$$

will be negative. This means that dislocations can form spontaneously in the crystal. A more detailed analysis shows that connected pairs of dislocations with opposite Burgers vectors form at temperature below T_m. As temperatures rise, larger and larger pairs appear. Finally, at $T = T_m$ decay of pairs of individual dislocations begins. The order of the crystal is destroyed.

From the above estimates it is clear that $T_m \simeq a^2/(8\pi)$. From (1.74)–(1.76) it implies that

$$T_m = \frac{\lambda a^2}{8\pi}, \qquad (1.77)$$

confirmed by more rigorous calculations. Note that with the help of the Hamiltonian (1.72), corresponding to the Hamiltonian of a two-dimensional XY model [187], we can describe the behaviour of not only an uniaxial two-dimensional crystal but also of other two-dimensional systems, namely the two-dimensional planar magnet and two-dimensional superfluid films [187]. The picture of the destruction of order in these systems is similar to that described by the dislocation mechanism of melting. A role analogous to the role of dislocations in the destruction of the long-range order in planar magnets and superfluid films is played by vortex excitations. First their role in the destruction of the long-range order and the establishment of quasi-long-range order characterized by a power law decay with distance r of the correlation function $G(\mathbf{r}) \sim |\mathbf{r}|^{-\eta}$, was described in the works by Berezinskii [14]. Kosterlitz and Thouless developed a detailed picture of the transition [369, 370]. They found the exponent of the correlation function $\eta = 1/4$ at the phase transition.

The above phase transition is a second-order phase transition, as in this transition the correlation radius r_c tends to infinity. However, the growth of the correlation radius is exponential, in contrast to the power growth of r_c during the disorder–long-range order transition, as follows:

$$r_c \sim \exp\left[\text{const} \left(\frac{T - T_m}{T_m} \right)^{-1/2} \right]. \qquad (1.78)$$

The behaviour of the elasticity modulus λ is also unusual. The modulus does not tend to zero when approaching the point of transition from the ordered phase, while above the transition point, of course, it is equal zero. In the case of superfluid films, this effect should manifest itself in an abrupt change in the superfluid density, and this jump, as shown in [418], is expressed in terms of the transition temperature and the universal physical constants. This prediction was brilliantly confirmed in experimental studies [264, 449].

The dislocation mechanism of melting of an isotropic two-dimensional crystal, neglecting the influence of the potential relief of the substrate, was studied by Kosterlitz and Thouless [369, 370] and subsequently in the work of Yang [516] and Halperin, Nelson [329]. In these studies, a two-dimensional crystal is considered in the framework of the continuum theory of elasticity, where its elastic energy can be written as a Hamiltonian

$$H = \int \left[\frac{\lambda}{2}(u_{xx} + u_{yy})^2 + \mu(u_{xx}^2 + u_{yy}^2 + 2u_{xy}^2) \right] dxdy. \qquad (1.79)$$

As a result of detailed analysis in [329, 516], the following expression was obtained for the melting point of the dislocation

$$T_m = \frac{\mu(\lambda + \mu)}{4\pi(\lambda + 2\mu)}. \qquad (1.80)$$

Melting of a two-dimensional isotropic crystal was investigated experimentally in the xenon–graphite system [331]. There was an increase in the correlation radius r_c to about 50 nm, and although the experimental precision did not make it possible to make an unequivocal choice between the usual power-law dependence of the correlation length near the temperature of transition to the state, characterized by the establishment of the long-range order, and the exponential dependence of r_c (1.78), predicted on the basis of the two-dimensional XY model, the mere fact of observation r_c growth from 1 to 50 nm is a compelling evidence in favour of the dislocation theory of melting, which predicts a second-order phase transition. Note that the melting by the second-order transition is not compulsory for the two-dimensional crystal. For example, the studies have shown that the crystals in smectic films melt by the first-order transition [265, 408].

1.5.4. Phase transition of the smooth-textured surface type

At low temperatures the crystal surface consists of a number of flat faces with a small number of defects (Fig. 1.15a). As temperature increases the number of defects increases, and it is possible that at a temperature greater than T_R but lower than the melting temperature of the crystal T_m, a large number of defects form on the crystal face and it becomes atomically rough (Fig. 1.15b). The possibility of this phase transition was first pointed out by Burton, Cabrera and Frank [275, 276].

As shown below, the correlation properties of the atomically rough surface (i.e. the surface at $T > T_R$) are the same as in a free film. This means the disappearance at $T > T_R$ on the crystal surface of a flat region corresponding to a given face. Temperature T_R for the faces with different crystallographic orientations should be different, and lower for the faces with higher Miller indices. This means that, for example, decreasing the

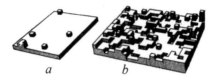

Fig. 1.15. The atomically smooth surface of the (100) face of the simple cubic lattice (*a*), the atomically rough surface (*b*) [134].

temperature from T_m on the uncut surface of the crystal results in gradual appearance of the faces with higher and higher Miller indices, so these transitions are also called *phase faceting transitions*. The faceting transitions investigated so far belong to the universality class of the XY-model.

Consider a simple model that describes such a transition, so called *SOS* (*Solid On Solid*), or a discrete Gaussian model. It is defined on a square lattice, where each node (*i, j*) has its discrete variable l_{ij} = 0, ±1, ±2, ... This variable describes the height of the column of atoms at the node (*i, j*). The model Hamiltonian has the form

$$H = \frac{J}{2}\sum_{i,j}\left[(l_{ij} - l_{i+1,j})^2 + (l_{ij} - l_{i,j+1})^2\right]. \qquad (1.81)$$

The quadratic (rather than linear corresponding to the number of dangling bonds) dependence of the energy of the face on the difference in the heights of the columns can be seen to be somewhat artificial. However, on the one hand, this simplifies the analysis and, on the other hand, as will be seen below – does not affect the long-wave correlation properties of the model. From (1.81) it can be seen that the ground state corresponds to l_{ij} = const, i.e. a smooth face. At very high temperatures $T \gg J$ the amplitude of fluctuations is $\langle l^2 \rangle \gg 1$. To study the transition, we substitute the variables in the partition function using the Poisson summation formula:

$$\sum_{l=-\infty}^{+\infty} f(l) = \sum_{m=-\infty}^{+\infty} \int_{-\infty}^{+\infty} d\varphi f(\varphi) e^{2\pi i m \varphi}. \qquad (1.82)$$

Then, the partition function Z is transformed as follows:

$$Z = \prod_{i,j}\sum_{l_{i,j}} \exp\left(-\frac{J}{2T}\sum_{i,j}\left[(l_{i,j} - l_{i+1,j})^2 + (l_{ij} - l_{i,j+1})^2\right]\right) =$$

$$= \prod_{i,j}\sum_{m_{i,j}} \int d\varphi_{i,j} \exp\left(-\frac{J}{2T}\sum_{i,j}\left[(\varphi_{i,j} - \varphi_{i+1,j})^2 + (\varphi_{i,j} - \varphi_{i,j+1})^2\right]\right) \times$$

$$\times \exp\left(\sum_{i,j} 2\pi i m_{i,j}\varphi_{i,j}\right). \qquad (1.83)$$

In the new expression for the partition function we perform integration over the variables φ_{ij}, since its integral is Gaussian. The formula convenient for integration can be used:

$$\prod_i \int d\varphi_i \exp\left(-\frac{1}{2}\sum_{i,j}\varphi_i G_{i,j}^{-1}\varphi_j + \sum_i \alpha_i \varphi_i\right) =$$

$$= \left(\det(2\pi G)\right)^{1/2} \exp\left(\frac{1}{2}\sum_{i,j}\alpha_i G_{i,j}\alpha_j\right). \qquad (1.84)$$

The matrix $G_{ij}^{-1} \equiv G^{-1}(\mathbf{r}_i, \mathbf{r}_j)$ corresponding to (1.83) is

$$G^{-1}(\mathbf{r}_i, \mathbf{r}_j) = \frac{J}{T}\left(4\delta_{\mathbf{r}_i,\mathbf{r}_j} - \sum_a \delta_{\mathbf{r}_i,\mathbf{r}_j+a}\right), \qquad (1.85)$$

where $\delta_{\mathbf{r}_i,\mathbf{r}_j}$ is the Kronecker symbol, vector $\mathbf{a}$ in the sum runs over the nearest neighbours. The matrix $G(\mathbf{r}, \mathbf{r}')$ is calculated using the Fourier transform. As a result, we have

$$G(\mathbf{r},\mathbf{r}') = G(0,0) - \frac{T}{2\pi J}V(\mathbf{r},\mathbf{r}'), \qquad (1.86)$$

where

$$G(0,0) = \frac{T}{2\pi J}\int_{-\pi}^{\pi}\frac{dq_x}{2\pi}\int_{-\pi}^{\pi}\frac{1}{4-2\cos q_x - 2\cos q_y}\frac{dq_y}{2\pi},$$

$$V(\mathbf{r},\mathbf{r}') = \int_{-\pi}^{\pi}\frac{dq_x}{2\pi}\int_{-\pi}^{\pi}\frac{1-e^{iq(\mathbf{r}-\mathbf{r}')}}{4-2\cos q_x - 2\cos q_y}dq_y. \qquad (1.87)$$

Hence we obtain the following expression for the partition function and Hamiltonian in the variables $m(r)$:

$$Z = \prod_r\sum_{m_r}\exp(-H), \quad H = \frac{\pi T}{J}\sum_{r,r'}m_r V(\mathbf{r},\mathbf{r}')m_{r'}. \qquad (1.88)$$

The constant $G(0, 0)$ imposes a condition on the partition function $\sum_r m_r = 0$. For large $|\mathbf{r} - \mathbf{r}'|$ the function

$$V(\mathbf{r},\mathbf{r}') = \ln|\mathbf{r}-\mathbf{r}'| + \text{const}. \qquad (1.89)$$

Thus, we arrive at the problem of two-dimensional Coulomb plasma [187, 516]. The role of the charges here is played by the variables m_r. However, unlike the two-dimensional XY model, in the high-temperature phase of the SOS-model charges m_r are bonded in pairs, and the low-temperature phase, on the contrary, corresponds to free (and with large values) m_r. The transition temperature $T_R = 2J/\pi$. For $T > T_R$ the number of non-zero m_r is small, and neglecting the term $2\pi i m_r \varphi_r$ in the partition function (1.83), we obtain the problem of a freely fluctuating two-dimensional film where the displacements φ_r take arbitrary values. In this extreme case the partition functions for l_{ij} and φ_{ij} according to (1.84) actually coincide, i.e. l_{ij} can be considered as a continuous variable. This means that at $T > T_R$

$$< (l_r - l'_r)^2 > \sim \frac{T}{J} \ln |\mathbf{r} - \mathbf{r'}|. \tag{1.90}$$

EXPERIMENTAL METHODS OF STUDY OF SURFACE PROPERTIES OF SOLIDS

2.1. Methods for studying the atomic structure

2.1.1. Slow electron diffraction

In studies of the atomic structure of surfaces and surface phase transitions currently the most widely used method is the slow electron diffraction (SED) method. Its high sensitivity to the atomic structure of surfaces is due to the fact that at an electron energy of about $10 \div 100$ eV, their mean free path is $0.5 \div 1$ nm. The low-voltage electronograph used in this method is so constructed (Fig. 2.1) that due to the delaying potential difference applied between the shield and filter grids, only the electrons which have undergone elastic scattering in the opposite direction from the crystal surface (about 1% of full output) are separated. The screen is attached to a large positive potential that defines the acceleration of electrons and determines the excitation energy or luminophor of the screen upon impact. Photo or video cameras are used record the resulting diffraction reflexes. Due to the selection of electrons which have undergone elastic scattering on atoms of the surface and near-surface ionic planes, the diffraction pattern displayed on the screen, is formed only by a few atomic planes closest to the surface. The basics of the SED method were discussed in detail in several monographs and reviews [169, 170, 349, 430]. Here we will focus on its most important of features and characteristics.

In any diffraction method, an important characteristic of the apparatus is the degree of coherence of the probe beam. In the case of SED, unless special measures are taken to reduce the energy and angular velocity scatter of the electrons in the beam, the average size of the electron wave packet (coherence width) is typically about 10 nm. This determines the maximum size of the area on the surface, within which we can explore the correlation in the mutual arrangement of particles. Advanced low-voltage electronographs with a reduced energy spread of electrons and improved focusing, in which the width of the coherence of the electrons was increased to about 100 nm [127, 323], have recently been developed.

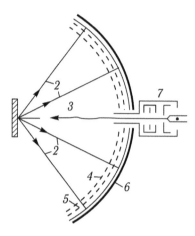

Fig. 2.1. The scheme of low-voltage electronograph with visual observation of diffraction patterns: *1* – crystal, *2* – scattered beams, *3* – the primary electron beam, *4* – shielding grid, *5* – filtering screen, *6* – screen, *7* – electron gun [170].

The resulting SED pattern is the image of the reciprocal surface lattice when observed from a large distance from the crystal along the normal to the surface. Since the distance between neighbouring points in the reciprocal lattice is inversely proportional to the distance between points in the corresponding direction in the direct spacial lattice, then for purely planar cell with the perfectly flat surface the spatial period in the direction normal to the surface is infinite. Consequently, the reciprocal lattice points along the normal to the surface are infinitely tight, and we can talk about *rods* in reciprocal space. Nevertheless, translational invariance in two dimensions guarantees the existence of diffraction, subject to the two-dimensional Laue conditions:

$$(\mathbf{k}_i - \mathbf{k}_f)\mathbf{a}_s = 2\pi m, \tag{2.1}$$

or

$$(\mathbf{k}_i - \mathbf{k}_f)\mathbf{b}_s = 2\pi n, \tag{2.2}$$

where $\mathbf{k}_i$ and $\mathbf{k}_f$ are the wave vectors of incident and scattered electrons respectively, m and n are integers.

The Laue conditions are best illustrated by the well-known Ewald scheme (Fig. 2.2). The reciprocal lattice rod passes through each point of the surface reciprocal lattice $\mathbf{g}_s = h\mathbf{A}_s + k\mathbf{B}_s$. The amplitude of the wave vector of the electron determines the radius of the sphere. The diffraction condition is satisfied for each beam propagating in the direction from the centre of the sphere to the point of its intersection with the rod the reciprocal lattice. As in the three-dimensional case, the beams are denoted indices of the reciprocal lattice vector, which determines diffraction.

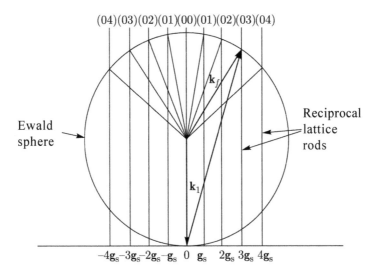

Fig. 2.2. Ewald scheme for the electrons incident normal to the surface. Nine beams that have experienced back-scattering are shown [95].

The crystal structure in the direction perpendicular to the surface, breaking the reciprocal lattice rods, leads to the formation of one-dimensional lattice of points and to fulfillment of the customary Laue conditions. Since the SED method analyses only a few surface atomic planes, the outgoing beams are visible at all electron energies, while the corresponding reverse lattice rod is within the scope of the Ewald sphere. The very existence of the point-diffraction pattern indicates the presence of an ordered surface and gives direct information about the symmetry of the substrate [346]. Note that the symmetry of the arrangement of surface atoms can not be higher than that set with the SED. The true surface structure may have a lower symmetry. Such a situation occurs when the surface contains regions (domains) oriented relative to each other according to the rules of symmetry. For example, in the study of the surface we can detect two systems of reflections, which have the symmetry of the third order and are rotated at an angle of $60°$ with respect to one another. As a result of averaging over the actual size of the electron beam the surface will have an apparent sixfold symmetry.

Further information can be obtained by analyzing the change in the intensity of diffraction within the width of a single spot – the so called profile of the spot [376]. For example, any deviation from the perfect two-dimensional periodicity of the reciprocal lattice rods can not be any longer described by the delta function. If the beam with the wave vector $k_{||}$, parallel to the surface, is investigated, we will observe the expansion and splitting of the diffraction pattern. Similarly, small variations in surface topography will break the bars in the direction normal to the surface. Expected point

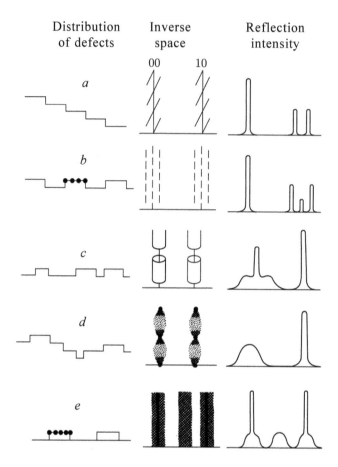

Fig. 2.3. Possible defect structures corresponding to the modifications of the reciprocal lattice rods, and the resulting profile of the SED point reflections (*a* – a regular monotonic sequence of steps, *b* – regularly spaced steps of two levels, *c* – a random arrangement of two levels of steps, *d* – a random arrangement of multi-level steps, *e* – adsorbate islands) [334].

reflections can be found by superposing the Ewald sphere on the modified 'rods' (Fig. 2.3). Note that this analysis does not answer the question the location of atoms within a single cell of the surface lattice.

The SED method in principle allows to produce the so-called complete structural analysis of the surface where the task is the definition of the positions of all the atoms in the unit cell of the surface lattice with respect to each other and relative to the atoms of the substrate. To do this, the experimental dependences of the intensity of diffraction reflections of the electron energy (voltage), the so-called $I(V)$-curves, are compared with those calculated on the basis of the dynamic SED theory. Quantum mechanical calculations are performed for various hypothetical structures of the surface lattice. True is the one for which the best agreement is achieved between the calculated and experimental data. It should be noted that calculations are very complex and achievable only for the simplest lattices. In most cases one is restricted to the geometric analysis of the SED patterns, based on kinematic diffraction theory, which ignores the multiple scattering of electrons. The latter effect is evaluated only qualitatively, since it leads to the appearance of the diffraction pattern of reflections, prohibited in the kinematic approximation. The symmetry and the surface lattice periods are determined with sufficient reliability, especially if the SED patterns are interpreted using also all sorts of additional information, such as the surface concentration of adsorbed particles, the mechanism of film growth etc. However, the positions of the surface atoms relative to the atoms of the substrate remain unknown.

As an example, we consider the use of structural analysis using SED to study the surface relaxation in metals when there is a shift of the surface atomic planes relative to their position in the bulk crystal. Figure 2.4 compares the experimental curves $I(V)$, obtained from the surface of (100) single crystal copper, and the calculated curves plotted according to the dynamical theory of SED [288]. From the data presented in Fig. 2.4 we can see that for all four beams there is good consent. The surface structure, as shown by these calculations, is characterized by oscillation relaxation of the interlayer distance, i.e. the distance between the upper layers is smaller than

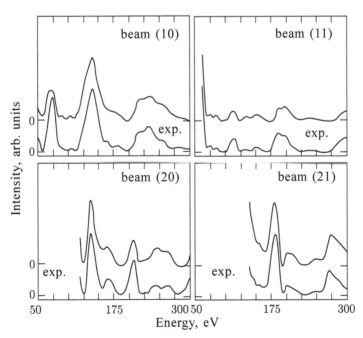

Fig. 2.4. Comparison of SED theoretical results with experimental data for the surface of Cu (100) [288].

the interplanar distances in the crystal ($\Delta d_{12}/d_{12}$ = −1.45%), While the distance between the following deeply located planes is greater than in the crystal ($\Delta d_{23}/d_{23}$ = +2.25%). Extensive SED experiments revealed that the effect of compression the upper layer is systematic. This result is consistent with the above notions of smoothing of the distribution of the charge density [468]. The largest smoothing and hence the largest compression take place on the surface of the crystal faces, characterized by low charge density and significant undulations (Fig. 2.5).

The SED method has been successfully applied to study phase transitions on the surface. At first order phase transitions, when various phases coexist on the surface, the observed diffraction pattern is a superposition of patterns from phases, because the diameter of the primary beam, as a rule, is greater than the characteristic size of the coexisting phases. At the same time, if the size of these islands is less than the coherence width of the electrons, then there is a noticeable broadening of the diffraction reflections, which allows to estimate its size.

For second order phase transitions, the SED method makes it possible to determine the critical exponents. Note that in this case, as a rule, it is not necessary ro use the dynamic theory of SED and we can confine ourselves to the kinematic approximation. The reason for this is the fact

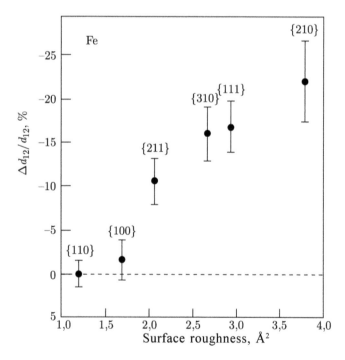

Fig. 2.5. The dependence of the relaxation of the upper layer of iron on surface roughness (roughness − the reciprocal value of the surface density of ions) [469].

that in the usual experimental setup the multiple scattering in the surface layer may experience only electrons scattered in the primary act by 90°, and the number of these electrons is relatively small [448]. In the kinematic approximation, the intensity of the reflections is proportional to the degree of long-range order [52, 123], so it can be fairly easy to identify the critical index of the order parameter β. Under certain conditions, the temperature dependence of the integrated intensity can also be used to find the critical index of change of the heat capacity α [251]. By measuring not only integrated intensity, but also its angular profile at various temperatures, we can find indices γ, ν [448].

2.1.2. Diffraction of fast electrons

The diffraction method of fast electrons (FED) is usually applied to the study of thin films when we observe diffraction patterns produced by the electrons passing through the film. In this case, the obtained information relates more to the volume than to the surface properties of the system. To study the structure of the surfaces the FED (electron energy $10 \div 100$ keV) is used in the reflective mode. In the experiments the primary and scattered beams propagate relative to the surface under sliding angles. This reduces the depth of the probed surface layer to about 1 nm [60]. Modern equipment provides fast electron beams with a spatial coherence of $10^3 \div 10^4$ nm, which provides high resolution. The advantage of the FED is also the fact that it is conveniently combined with electron microscopic studies. Great opportunities of the FED for study of mechanisms of fine surface phenomena have been demonstrated, for example, in [484], where the non-wetting–wetting transition in solid CF_4 films on graphite was studied.

The FED method is very convenient for use in molecular beam epitaxy systems, because thanks to design features of FED equipment there is larger space in front of the sample for evaporators than when using the SED method. Following the change in the intensity of specular reflection, we can simply determine the number of monolayers deposited on a substrate (Fig. 2.6) [295]. The point is that at nucleation of a new monolayer the surface becomes rough, and this leads to a marked decrease in the intensity of reflections. After passing through a minimum at half-filling the layer, the intensity again reaches its maximum at the completion of its 'development', so that the number of oscillations of the intensity of specular reflection is equal to the number of deposited monolayers.

2.1.3. X-ray diffraction

The scattering cross section by atoms of X-rays is several orders of magnitude smaller than that by the electrons. Therefore, the effect of x-ray scattering on the surface atoms, whose number for compact solids is only a small fraction of their total number, is usually negligibly small. However, at objects with

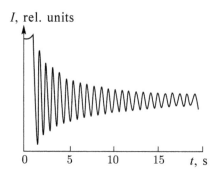

I, rel. units

0   5   10   15 t, s

Fig. 2.6. The dependence on time t, the intensity of specular reflection I of the FED pattern in molecular beam epitaxy of gallium arsenide on the surface of GaAs (001) (2 × 4). The oscillation period corresponds to the deposition time of a layer of Ga + As. Electron energy of 10 keV [295]

the highly developed surface this is not the case and in principle we can study the surface lattice.

X-ray structural analysis of surfaces has been used quite widely in recent years, particularly through the development of a special high-porosity material based on graphite. In the process of making this material the graphite is initially saturated with some substance, introduced into the space between the atomic planes, and is then rapidly heated, which leads to phase separation of graphite into thin plates along the basal plane. As a result, the graphite acquires a highly developed specific surface area of the order of tens of m^2/g. However, despite the high porosity, it has a fairly uniform surface, and in this respect is similar to the single crystal [300]. The use of this material as an adsorbent makes it possible to carry out efficient X-ray analysis of adsorbed films on the basal face of graphite.

An important advantage of the X-ray method is the high coherence of the probing beams. Thus, by using X-ray tubes with a rotating anode the spatial coherence of the order of 100 nm was achieved, and when using synchrotron radiation it was about 1000 nm (see, e.g. [472]). This allows to obtain information about the structure of the films and the relationships governing phase transitions in them with a very high accuracy. At the same time, it should be noted that the applicability of the classical techniques of X-ray analysis in the study of surface lattices is still limited to a rather narrow range of systems with a developed and at the same time quite homogeneous surface. In an attempt to overcome this limitation we can use other variants of the method. One possibility to overcome these limitations is to perform analysis at sliding incidence of the beam [302, 400]. In another variant, a standing X-ray wave is generated on the studied surface in the Bragg reflection mode [285, 292]. In this mode, the phase of the reflected beam is strongly dependent on the angle, so that the spatial position of nodes and antinodes of the standing wave varies strongly with rotation of the sample.

In particular, it is possible to ensure that centres of the adsorbed atoms are in the antinodes of the standing X-ray wave. This moment is recorded on the basis of the maximum of X-ray fluorescence [285] or photoelectron emission [7, 292] of the adatoms. After the appropriate calculations, we can determine the coordinates of the surface atoms to within a thousandth of a nanometer.

Another method for studying the structure of surfaces based on analysis of the so-called long-distance (Kronig) fine-structure of the X-ray absorption spectra (in the literature, this method usually abbreviated SEXAFS (Surface Extended X-ray Absorption Fine Structure)). For a review of the basics of this method see [328]. The non-monotonic variation of the absorption of X-rays with the energy change is due to the fact that the wave function of the excited state of the photoelectrons is formed as a result of interference of electron waves – primary and scattered by the atoms of the nearest coordination spheres. Multiple scattering of the photoelectrons can be neglected, because their energy is hundreds of electron volts, so the calculation of the absorption coefficient is relatively simple. Experimentally, the absorption is recorded on the basis of the yield of slow secondary electrons, which makes the method sensitive to the surface condition [258, 359, 482]. Comparison of calculations with the results of measurements allows to determine the interatomic distances from the surface up to 10^{-3} nm. The advantage of the method is that is it possible to determine the interatomic distances in systems without the long-range order. The experiments were performed, as a rule, in synchrotrons, since in the use of X-ray tubes the data acquisition time is too long.

2.1.4. The scattering of neutrons

We know that the study of the structure and dynamic characteristics of the three-dimensional matter by neutron scattering is a powerful method. For a long time it seemed that it cannot be used to study surfaces because of the smallness of the neutron scattering cross section. However, using laminated graphite as an adsorbent allowed to effectively apply neutron diffraction also in the physics of surface phenomena [282, 365, 421].

The neutron scattering method has several important advantages. In particular, there are practically no restrictions on the atomic number of the investigated substances: the method can be successfully used to study films of hydrogen and other light adsorbates. The coherent elastic scattering (diffraction) of neutrons provides information about the atomic structure of the adsorbed films (using thermal neutrons with a wavelength of the order of 10^{-8} cm), coherent inelastic scattering – the dynamics of the surface lattice. Incoherent inelastic scattering provides information about the mobility of adsorbed particles. Finally, the presence of the spin in the neutron allows one to study the magnetic ordering in two-dimensional crystals.

2.1.5. The scattering of ions and neutral atoms

This method has been well developed for studying the structure of the surfaces by scattering spectroscopy of slow ions (ion energy of the order of $10^2 \div 10^4$ eV) and by the Rutherford backscattering method for fast ions (the energy of the order of $10^5 \div 10^6$ eV) [10, 122, 188, 233]. It should be noted that, from a technical point of view, these methods are generally more complex than the methods based on the use of electron probing beams. Therefore, the use of ion structural analysis is most justified and effective in clarifying the issues that are difficult to solved by the SED or FED methods. This includes, for example, questions about the structure of clean and coated films of the reconstructed surfaces, surface relaxation (the change of the interplanar spacing at the surface), the surface melting of crystals, etc.

Consider a beam of light ions (H$^+$, He$^+$), incident on a solid surface. For these ions, the crystal represents a target in the form of columns or 'strands' of atoms, which lie parallel to the crystallographic directions with low indices. Coulomb scattering from the edge of each strand in the first atomic layer depends the collision parameter. The distribution of scattered ions forms a characteristic shadow cone behind the surface atom (Fig. 2.7). The atoms within the shadow cone do not contribute to the backscatter signal. If we neglect the effects of screening of the Coulomb interaction, the radius of the shadow cone is defined by expression

$$R = 2\sqrt{Z_1 Z_2 e^2 d / E_0},\qquad(2.3)$$

where Z_1 and Z_2 are the atomic numbers of the incident ion and target atom respectively, E_0 is the initial energy of the ions in the beam, d is distance along the strand.

The ion scattered directly back undergoes a simple binary collision with a surface atom. The final ion energy is determined by the laws of conservation of energy and momentum:

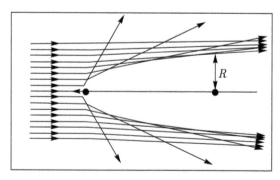

Fig. 2.7. Shadow cone formed by the trajectories of the ions which have undergone Rutherford scattering at the atom strand [481].

$$E = E_0 \left(\frac{M_1 - M_2}{M_1 + M_2} \right)^2. \qquad (2.4)$$

Formula (2.4) shows that the ions scattered in the opposite direction (M_1) are experiencing a shift in energy, which depends quite strongly on the mass of the surface atom (M_2). Consequently, the analysis of the energy spectrum of scattered ions can be regarded as another method for elemental surface analysis (currents in the beams must be substantially less than the value corresponding to the threshold of the sputtering defect). For structural studies it should be borne in mind that the radius of the shadow cone, and the Rutherford scattering cross section $d\sigma_R / d\Omega \approx E_0^2$ are strongly dependent on the initial energy of the ions in the beam. Therefore, for a complete analysis of data on the scattering of the ions it is more or less natural to use three power modes: scattering of low-energy ions (SLEI) (1 ÷ 20 keV) scattering of medium-energy ions (SMEI) (20 ÷ 200 keV) and scattering of high-energy ions (SHEI) (200 keV ÷ 2 MeV).

The SLEI method is suitable for laboratory studies of the surface. Due to the large values of the scattering cross section ($\sigma_R \sim 1$ Å^2) and the shadow cone radius ($R \sim 1$ Å), the vast majority of the ions never penetrate deeper into the surface layer. Thus, the ions are rapidly neutralized by electron

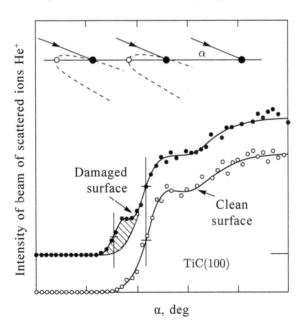

Fig. 2.8. SLEI from the surface of TiC (100). The vertical lines indicate calculated angles at which the shading of Ti atoms by nearest neighbouring C atoms and the nearest neighbouring Ti atoms takes place. The inset shows the geometry of the scattering on the C atom with the critical value of the angle. A single carbon vacancy is shown [246].

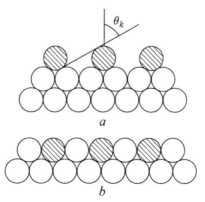

Fig. 2.9. Adsorption in the absence (*a*) and presence (*b*) of reconstruction of the substrate (θ_k – the critical angle of incidence of the ion beam at which the adatoms begin to obscure the substrate atoms) [338].

capture and do not contribute to pilot signal if the detector detects only charged particles [269].

To illustrate the effectiveness and simplicity of the concept of the shadow cone, Fig. 2.8 shows the data obtained in the scattering of He$^+$ ions with the energy of 1 keV on the surface of the TiC (100) single crystal [246]. The azimuthal angle of incidence was chosen so that the scattering plane contains alternating rows of surface atoms of titanium and carbon. Only the ions scattered from the atoms titanium in the opposite direction, for which the relation (2.4) is valid, were recorded. Significant scattering was observed at high polar angles of incidence α. However, the signal fell rapidly at angles smaller than the critical angle. In this mode, the back-scattering on Ti atoms is terminated due to the influence of the shadow cone of the carbon atoms. Since the radius of the shadow cone is given by (2.3), we can accurately calculate this critical angle (in this case is 22.1°).

If the surface is specially damaged by preferential sputtering of carbon atoms, even at angles less than critical we can observe scattering, which is caused by 'unshaded' atoms. Measurement of the angular dependence of the current of scattered slow ions in some cases leads to an unambiguous conclusion about the presence or absence of the reconstruction of the substrate during adsorption [338]. An example of such an experiment is shown in Fig. 2.9. If adsorption is not accompanied by reconstruction (displacement of the atoms of the substrate from the 'normal' positions), and in the surface monolayer there are only adsorbate atoms, the energy spectrum of elastically scattered ions depends on the angle of incidence to the surface. Indeed, at normal incidence of the beam the specified range will contain peaks corresponding to elastic scattering of the ions on both the adatoms and the substrate atoms (at ion energies greater than 10 eV, elastic scattering can be considered as a binary collision of the atoms). But at a certain critical incidence angle the adatoms start to screen the substrate atoms and only one peak remains in the spectrum. At the same time, in the presence of reconstruction the atoms of the adsorbate and of the substrate are mixed and independently of the angle of incidence the spectrum contains peaks due to scattering on the atoms on both varieties. A detailed analysis of the scattering spectra taking into account the effects of shading, multiple-scattering processes and channeling, provides a fairly complete information on the status of the adatoms on the substrate and the structure of their nearest environment.

Especially high-precision data on the structure of surfaces provides the method of Rutherford backscattering [287, 410]. The surface is bombarded by fast ions, for which the scattering centres are the core, and the probability of scattering is described by the Rutherford formula. In the SHEI method, only a small number of ions undergoes back-scattering from the surface. A large part of the ions, incident along the atomic rows with small indices, penetrates deep inside, and undergoes a series of correlated collisions with the neighbouring rows. This behaviour is known as channeling. Channeled ions collide with the loosely coupled electrons and lose energy in accordance with the inhibitory of solids. In the end, they are scattered back from the crystal, but with energies less than that of the ions which have undergone

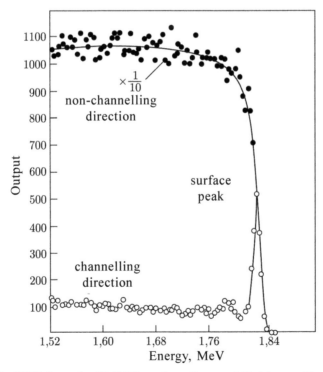

Fig. 2.10. SHEI from the W (100) surface. Scattered He$^+$ ions with energy of 2 MeV. The spectra are shown for incidence in the direction of channeling (100) and undirected incidence [307].

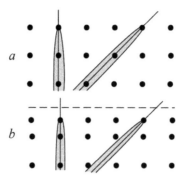

Fig. 2.11. Shadow cones at SHEI for the two directions of channeling in ideal case (*a*) and relaxed (*b*) surface [95].

elastic scattering with the surface. Therefore, analysis of the ion energy at a fixed angle of recording reveals a 'surface peak' (Figure 2.10) [307]. The area under the peak surface is proportional to the number of atoms related to the number of atomic rows in the beam path. This number is equal to unity for normal incidence on the ideal surface (100) of the face-centered cubic crystal at $T = 0$. Note that if the incident ion beam is not oriented along the direction of channeling, the probability of scattering in the reverse direction does not depend on depth.

The scattering of high-energy ions is particularly sensitive to the interplanar surface lattice relaxation, whereas application of the SED method for this purpose requires complex calculations and unambiguous conclusions cannot be always made. If the ions fall along the channeling direction, then in the case of the ideal surface there is full shading of the subsurface atoms. However, at the direction of incidence of the ions, different from normal, shading would be incomplete if the uppermost atomic plane is shifted (Fig. 2.11). If the crystal rotates back and forth about this axis, the ion yield does not will be symmetrical (Fig. 2.12). The height of the surface peak in Fig. 2.12 is greater than unity, reflecting the contribution of thermal vibrations of the surface atoms. Due to the movement of the atoms, the ion beam reaches the deeper layers of atoms, which are shaded at $T = 0$.

In [313], using Rutherford backscattering of the ions, the results convincingly demonstrated the existence of reversible surface melting, which occurs below the bulk melting point. In the interaction of ions with an ordered surface the main contribution to scattering comes from the surface atomic plane, since at the appropriate choice of the angle the lower planes are shaded, and on the energy distribution of scattered ions there is a narrow peak. At disordering of the surface layer the ions are scattered at different depths and in the end have different energy losses, resulting in significant energy distribution broadening.

Data on the surface structure can also be obtained by examining scattering of thermal beams of neutral atoms and molecules (mainly of

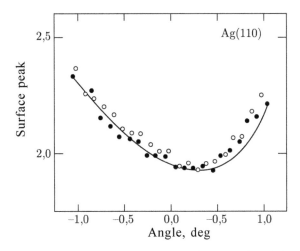

Fig. 2.12. The asymmetric angular dependence of the height of the surface peak (expressed in the number of atoms per row) in the scattering of He^+ ions with an energy of 400 keV incident along the (101) axis on the Ag (110) surface [375].

helium and hydrogen). Although the technique of these experiments is complicated, the method has important advantages: it is virtually non-destructive and has an exceptionally high surface selectivity. For the review of this method, see [445].

2.1.6. Methods of electron spectroscopy

As noted above, the diffraction methods of structural analysis of surfaces are effective in the study of symmetry and the surface lattice periods. However, the determination with their help of the positions of the adatoms relative to the substrate atoms is associated with considerable difficulties. However, this information is essential for understanding the nature of adsorption bonds, the interaction of adsorbed particles with each other, as well as the dynamic characteristics of the films (lattice oscillations, diffusion, etc.). This information is obtained by the methods of electron spectroscopy. This is based on two experimental facts. First, electrons with kinetic energy in the range $15 \div 1000$ eV has very small mean free paths in matter ($\lambda < 10$ Å). Second, the binding energy of the electrons of the deep shells of an atom is sensitive to the nature of the element. Therefore, the measurement of the kinetic energy of electrons emitted from the solid under the action of photon or electron bombardment, provides information on the elemental composition of the surface.

The most complete picture of the sensitivity of the electron to the properties of the surface gives a plot of the dependence of the mean free path for inelastic scattering of electrons on kinetic energy (Fig. 2.13). The experimental points are distributed around some 'universal curve', which

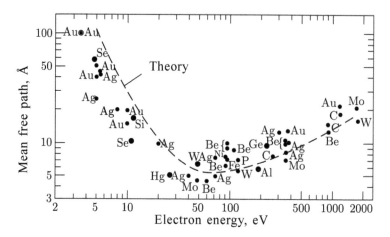

Fig. 2.13. A universal curve of the mean free path of electrons in solids (experimental data from [444, 470], the theoretical curve from [431].

has a broad minimum near the energy of 50 eV. The universal nature of the curve is easy to understand. Recall that the dominant mechanism of energy losses by electrons in solids is the excitation of electrons in the valence bands. It should also be noted that the density of electrons in the valence band is approximately constant for most materials and is about 0.25 electron/$Å^3$. If the electrons of the solid are considered as a gas of free electrons with a specified density, the results of calculating the mean free path for inelastic scattering [431] are in good agreement with experimental data (dashed curve in Fig. 2.13). The available data suggest that electrons with energies in the corresponding band, which left the solid body without a consequent loss of energy, must have been emitted from the surface region.

The most common method of elemental analysis is Auger electron spectroscopy (AES). It is based on the use of an electron beam of sufficiently high energy ($E > 1$ keV). The beam is directed onto the sample and the spectrum of backscattered electrons $N(E)$ is recorded. The dependence of $N(E)$ is characterized by the peak of elastically scattered electrons (which exited the solid without scattering) and a long, seemingly structureless 'tail' caused by electrons that have lost energy in the solid state (Fig. 2.14a). The contribution to the 'tail' is provided by two types of electrons. The electrons of he first type leave the sample, losing energy in a single act of inelastic scattering of a known type. The others so-called secondary electrons, lose energy in a number of inelastic collisions. The pilot signal from the latter is completely structureless. However, the signal from the electrons of the first type is characterized by minor fluctuations depending on $N(E)$, the origin of which allows the analysis of the derivative $dN(E)/dE$ (inset in Fig. 2.14a).

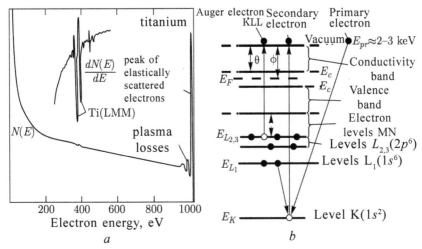

Fig. 2.14. *a* – depending on $N(E)$ and $dN(E)/dE$ for the electrons which have undergone backscattering from the titanium target. The energy of the electrons in the incident beam of 1 keV [428], *b* – diagram of the Auger transition $KL_1L_{2,3}$

The exact energy position of a distinct structure in the spectrum of the derivative $dN(E)/dE$ in Fig. 2.14a is determined by the nature of the elements that make up the surface. To see this, suppose that some electron in the incident beam encounters an atom in a solid and ionizes the 1s-electron whose binding energy is equal to E_{1s}. If E_{1s} is less than 2000 eV, then the remaining hole in the 1s-shell is filled predominantly by the non-radiative Auger transition, i.e. a 2s-electron drops to the level of the hole, but at the expense of the energy of this transition from the 2p-level the second Auger electron is emitted. The scheme of the Auger transition is shown in Fig. 2.14, *b*.

In accordance with the law of conservation of energy, the kinetic energy of the emitted electron is

$$E_{\text{kin}} = E_{1s} - E_{2s} - E_{2p}, \qquad (2.5)$$

where E_{2s} and E_{2p} is the binding energy for the 2s-and 2p-atomic levels, respectively, in the presence of a hole on the 1s-level. Note that the kinetic energy of the emitted electron depends only on the properties of the atom. Similarly, the energy relaxation process of the hole in a deep atomic shell occurs in all elements of the periodic table, except hydrogen and helium. The characteristic energy of the Auger electrons are well known and tabulated. Fundamentally, it is important that each element has a Auger transition for which the kinetic energy of the emitted electron is in the range corresponding to the maximum sensitivity of the method to the state of the surface. Auger spectroscopy is ideally suited for elemental analysis of the surface, for each surface atom leave their 'fingerprints' on the energy spectrum $N(E)$.

X-ray photoemission spectroscopy (XPS) is a method of surface analysis which also uses the advantage of the small mean free path of electrons in matter, and the value of the binding energy of a hole in deep atomic electron shells strictly defined for each element. Here, the photoelectric effect is observed when exposed to monochromatic X-rays. As a rule, MgKα (1254 eV) or AlKα (1487 eV) lines are used. In the spectrum of emitted electrons, called the *energy-distribution curve*, we record peaks at kinetic energies corresponding to the 'surface-sensitive' range (Fig. 2.15). The position of the peaks in the electron kinetic energy E_{kin} depends on the binding energy of holes E_{bind} as follows:

$$E_{kin} = \hbar\omega - E_{bind}. \tag{2.6}$$

The atoms that are present on the surface are identified by comparing the measured values of E_{bind} with the tabulated values of the binding energies of the atomic shells of different elements. Note that the energy-distribution curve also shows additional peaks (see Fig. 2.15), corresponding to electrons emitted by means of the Auger process, which follows the primary act of photoemission.

The sensitivity of XPS to surface contamination is close to the sensitivity of Auger spectroscopy, and is about 1% of the monolayer. However, in the XPS method, the sample does not acquire an electric charge and, more importantly, small changes in the observed binding energies E_{bind} allow us to identify the same element in different chemical environments [487].

Investigation of the symmetry of the angular distributions of photoelectrons emitted by irradiation of the surface with ultraviolet light, makes it possible determine the symmetry of the centres occupied by

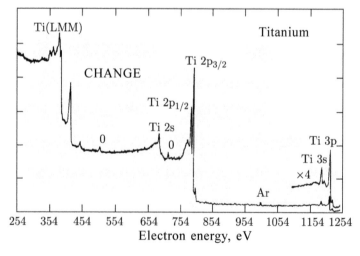

Fig. 2.15. XPS spectrum of the energy of the electrons from the titanium target under the action of MgK$_\alpha$-radiation [500].

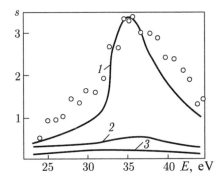

Fig. 2.16. The energy spectrum of photoelectrons emitted at an angle 45° to the surface of the (100) face of nickel coated with a film of carbon monoxide [440] (s – the relative peak area; o – experimental data); curves show calculated results (1 – CO axis normal to the surface, the atom C is the bottom; 2 – same as above but with the C atom at the, and 3 – the CO axis parallel to the surface).

adsorbed particles on a substrate, as well as the orientation of the adsorbed molecules (Fig. 2.16) [57, 439, 440].

Spectroscopy of the characteristic electron energy loss (SCEEL) at a resolution of about 10 MeV can effectively investigate the vibrational spectra of adsorbed particles. Analysis of these spectra also gives information on what positions are occupied by adsorbed particles on the substrate [356, 439].

Finally, it is important to note the method based on measuring the angular distributions of ions emitted from the surface in electron-stimulated desorption [1, 396, 397]. In this method, the surface is irradiated with electrons with an energy of 100 eV.

Depending on the nature of the adsorbate and the nature of the adsorption bond, the effective cross section of electron-stimulated desorption in the form of ions is $10^{-24} \div 10^{-18}$ cm^2. In [396] it was observed that the angular distribution of desorbed ions is anisotropic and can be used to determine the symmetry of the adsorption centre. Despite the low intensity of the ion fluxes, the spatial distribution of the ions can be observed visually on the screen through the use of microchannel multiplier plates, which makes this method quite convenient [458].

2.1.7. The methods of field ion and electron microscopy. Scanning tunneling microscopy

Modern microscopy techniques make it possible to obtain extensive information about the structure of two-dimensional systems – from direct observation (with atomic resolution) of the lattice and forms of growing two-dimensional crystals to obtaining the quantitative data on surface phase transitions.

   The projection of the image of the arrangement of atoms on a metal surface can be obtained using a field ion microscope (FIM), developed by Erwin Muller [411]. In this device, the thin edge of the sample material receives a high positive potential, so that the strength of the electric field on the surface is close to 10^9 V/cm. Then, a gas of neutral atoms, usually helium or a mixture of helium and hydrogen, is supplied to the chamber with the sample. These atoms are attracted to the solid surface and lose their kinetic energy by multiple collisions with the surface (Fig. 2.17 a). In the end, they remain in the vicinity of the surface long enough for the external field to ionize the atoms. The image of the faceted surface of the sharp edge is formed by ions that are accelerated rapidly when moving from metal to the fluorescent screen (Fig. 2.17 b). The image of the tungsten tip shows that the (100) and (111) planes are well resolved.

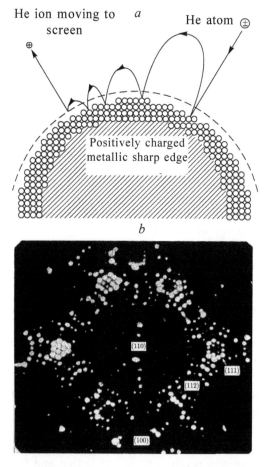

Fig. 2.17. Field ion microscope: a – scheme of the mechanism of formation of the image, b – the image of a tungsten tip with a curvature radius of 120 Å [495].

Unfortunately, the use of FIM is limited to studies of transient metals and their alloys, as the very tip should be stable at electric field strengths required to ionize the gas forming the image. At sufficiently strong fields metal atoms can be removed from the surface also in the absence of gas. This process (called *field evaporation*) on the one hand, leads to the restriction noted above for the studied materials and, on the other – can be used for chemical analysis of alloys, if the FIM is fitted with a mass spectrometer.

The FIM method has the resolution of the order of tenths of a nanometer, so that it makes it possible to observe phenomena on the surface on the atomic scale [167, 168]. In particular, the screen immediately displays the faces of the lattice with a relatively loose packing. One can observe the order–disorder phase transition in alloys, as well as the reconstruction of clean surfaces. There is a unique possibility of chemical identification of individual atoms (atom probe mode). The atom, selected by an observer on the screen of the microscope, is desorbed by the electric field and sent to the input of the time-of-flight mass spectrometer. Field-ion microscopy has been used to obtain a lot of data on migration (random wandering) of individual adatoms and clusters consisting of 2÷3 adatoms [168, 252, 301, 311], as well as on the interaction of adatoms with each other.

Electron microscopy has been used to obtain images of the surface after the recent progress in studies of volume, with which it was possible to reach a resolution of the order of the atomic spacing. Data acquisition can

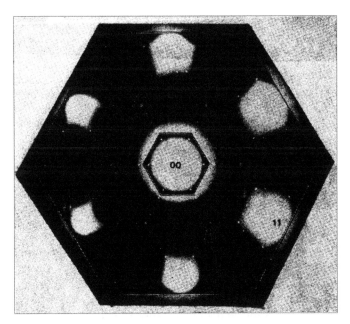

Fig. 2.18. Reflections in electron diffraction on the surface of Si (111) 7 × 7 in transmission geometry [486].

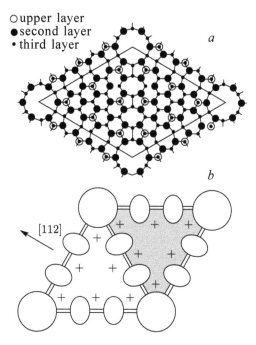

Fig. 2.19. The structure of the Si (111) 7 × 7 surface: *a* – top view of the first three layers [447], *b* – schematic drawing, which shows the grooves on the surface (circles and ovals), dimers (double line) and stacking faults (shaded area) [405].

be used either in 'transmission' (dark field and bright field) or reflection geometry. This yielded excellent images (see Fig. 1.1). Moreover, it is possible to analyze the intensities of the diffracted rays (used for image formation) in the framework of simple kinematics. At high electron energies ($E > 100$ keV) multiple scattering is not longer a problem. Figure 2.18 shows point reflections derived from the Si (111) 7 × 7 surface in the study of electron diffraction in transmission geometry. Based on similar results, and also microscopic images, the authors of [486] have suggested a significant reconstruction of this surface (Fig. 2.19), which was later confirmed by other methods such as surface scattering of X-rays, ion scattering, photoemission, etc.

The first direct observations of phase transitions of the type of two-dimensional condensation and sublimation of submonolayer adsorbed films were performed using a field electron microscope – projector [55, 236]. Its screen displays the distribution of the work function on the surface of the monocrystalline tip of a conductive material. The resolution of the microscope is 2÷3 nm. The formation of islands of a dense phase in two-dimensional condensation and their disappearance during sublimation is recorded directly on the magnified electronic image of the surface of the tip (Fig. 2.20). In determining the rate of growth and sublimation of two-

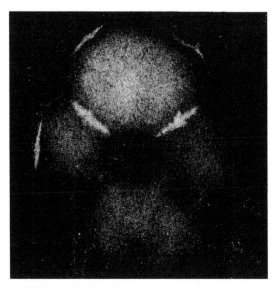

Fig. 2.20. Picture of a W tip of the Zr film adsorbed on it, obtained in the field electron microscope (bright spots – two-dimensional islets of Zr) [55].

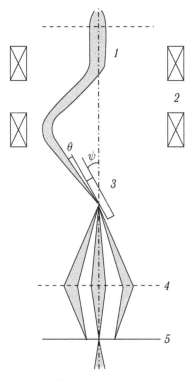

Fig. 2.21. The path of rays in the reflection electron microscope: *1* – electron beam, *2* – deflection system, *3* – sample *4* – lens, *5* – aperture diaphragm [6].

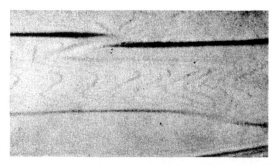

Fig. 2.22. The image in the electron microscope the surface of a silicon crystal with steps with monatomic height [6].

dimensional crystals at different temperatures, one can find the binding energy of adsorption particles in crystals and the activation energy of surface diffusion.

In recent years, great progress has been made in the study of the structure of surfaces using transmission and reflection electron microscopy. In particular, in the transmission mode we can observe with atomic resolution the structure of the surface lattice, consistent building of the atomic planes in the layered growth of thin films, and study the shape of the growing two-dimensional crystals [347, 399, 513]. Modern reflective electron microscopes can see on the surface steps with the height of one lattice constant [6, 350, 424, 487, 489]. The electron beam formed in these devices is deflected by an angle of $\theta + \psi$ to the optical axis of the microscope (see Fig. 2.21), where the angle ψ is close to the Bragg angle, and the angle θ is $0.001 \div 0.01$ rad, and the surface of the single crystal is oriented along the crystallographic plane (hkl). The image is formed from a single or several diffracted beams which due to the smallness of the angle θ are the result of reflection from the atomic planes closest to the surface. This ensures a high sensitivity of the method to the surface structure, so that even a small (of the order of 10^{-4}) deformation of the lattice near the atomic level is sufficient to create a noticeable contrast on the image (see Fig. 2.22). An important prerequisite for these studies is to maintain the ultrahigh vacuum around the sample. The effective illustration of the possibilities of this method was to investigate the phase transition leading to the reconstruction of the (111) surface of silicon when the temperature is lowered [6, 487, 489]. It was established that the emergence of superlattice domains occurs on the outer dihedral angles of monatomic steps. It is also shown that the traditional view that the steps are the preferred sites of adsorption and nucleation of a new two-dimensional phase is not always justified. Thus, when germanium is applied to the (111) face of silicon nuclei are formed in the gap between the steps, in the middle of the terraces [6].

The outstanding achievement in the diagnosis of the surfaces was the creation of a scanning tunneling microscope (STM) [255, 260, 261, 263,

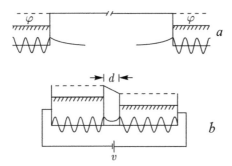

Fig. 2.23. Potential barriers and wave functions of electrons with energies equal to the Fermi energy, in the case of tunneling into vacuum: a – at the macroscopic gap between the barriers, and b – the microscopic gap between the barriers, to which external voltage is applied [95].

284, 429]. Scanning tunneling microscopy allows to obtain the direct image of the surface topology of the real space with atomic resolution. This non-destructive method does not require periodicity on the surface or ultrahigh vacuum; it provides information on both the chemical and electronic structure.

The principle of the STM is very simple, easy to understand when studying the interaction of two closely spaced solids by the model of a particle in a potential well. Figure 2.23 a shows two energy barriers of a finite size, separated by a macroscopic gap. According to quantum mechanics, the wave functions of the electrons with the energy equal to Fermi energy, 'flow' outside the limits defined by the potential barrier, and decay exponentially, and the decay constant is $\kappa = (2m\varphi)^{1/2}/h$, where m – electron mass, φ – the electron work function from the surface of the solid.

Now we reduce the distance between the barriers to the microscopic size and introduce a potential difference V between them (Fig. 2.23 b). The wave functions overlap which makes possible quantum-mechanical tunneling, and, therefore, an electric current can pass through the vacuum gap. The tunneling current I is a measure of the overlap of wave functions and is proportional to exp $(-2\kappa d)$, where d is the width of the vacuum gap. In a real microscope [262] the sensing element is a metallic tip (like the tip of the FIM), which at a distance of $d \sim 5$ Å above the surface. The tip scans the surface at a fixed bias voltage, and the piezoelectric systems, controlled by a feedback, adjust the vertical displacement of the tip so that the tunneling current ($I \sim 1$ nA) is maintained constant. Thus, the tip follows the contours of constant overlap of the wave functions, i.e. the surface topography (Fig. 2.24). The voltage on the piezoelectric plates has information about the surface topography and is shown on the display.

The resolution along the normal to the surface reaches 0.005 nm and along the surface – 0.2 nm.

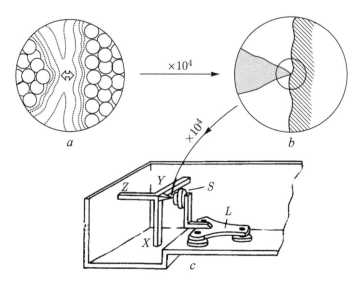

Fig. 2.24. Diagram showing the principle of the scanning tunneling microscope [260]: *a* and *b* – respectively the scanned surface and the tip on the atomic scale and the macroscale, c – the circuit of the microscope (*X*, *Y*, *Z* – piezoelectric rods, *S* – sample, *L* – piezoelectric plate for coarse setting of the distance from the sample to the tip).

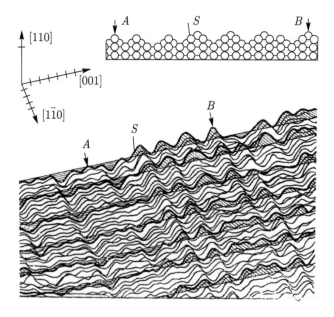

Fig. 2.25. STM image in real space of the surface of Au (110) 1 × 2. The divisions of axes correspond to 5 Å [264].

Fig. 2.26. STM image of Si (111) 7 × 7 near the atomic step [256].

In contrast to the method of field ion microscopy, the STM method is not associated with any sample size limitations and the electric field applied to the sample. It is only necessary to ensure that the material is conductive. The number of studies in this direction is rapidly growing. Although a full assessment of the possibilities of STM is a task for the future, it is not doubted that the prospects for its use are highly promising. In particular, the question of the structure of the reconstructed surfaces [94], discussed for many years, has been greatly clarified, and the possibility of removing with atomic-resolution the charts of the electronic density of states on the surface [95] has been indicated. It is expected that by using the STM on a qualitatively new level it will be possible to study a variety of surface phenomena.

For example, STM has been used to construct a model of 'missing rows' the surface of Au (110) (Fig. 2.25). Gold belongs to a very small group of metals where true reconstruction rather than simple relaxation can be observed. The image obtained using STM with atomic resolution shows that the surface is quite disordered in both the vertical direction and in directions along the surface. Despite this, the model of the hard spheres, used to describe the topography (see the inset in Fig. 2.25), not only detects the 'missing number' step along the [110] direction, but also shows that the monolayer steps S open close-packed (111) faces having a low surface tension.

The STM method is also a means of selecting the most plausible model in analyzing the complex structure Si (111) 7 × 7, which was discussed previously. Topographic images show clearly visible deep holes and an isolated top layer of atoms. Particularly striking results are obtained when the initial microscopic data are subjected to computer processing to produce images (Figure 2.26). In this example, the reconstruction of Si (111) is shown in the vicinity of surface steps.

The method of scanning tunneling microscopy has an exceptional potential. Recently, for example, researchers have used the fact that in tunneling electron transfer takes place from the filled states on one side of the vacuum gap to unoccupied states on the other side (see in Fig. 2.23 b). Therefore, by changing the amplitude and sign of the bias voltage, the electronic structure of the surface can be studied by STM.

2.2. Methods of investigating electronic properties

2.2.1. Methods of electron spectroscopy

In the study of the electronic structure of surfaces, the most informative are the methods of photoelectron and electron-photon spectroscopy. Depending on the frequency of exciting radiation we have X-ray and ultraviolet photoelectron spectroscopy (XPS and UVES) [57, 174, 202, 209]. The XPS method allows reliable recording of the shifts of the inner energy levels of atoms in the formation of chemical bonds and is currently the main source of information about the chemical state of matter at the surface. At the same time, UVES gives information about the energy structure of the outer electron shell and its restructuring as a result of adsorption. An example of such spectra is shown in Fig. 2.27. As noted in section 2.1.6, this method allows to determine also the symmetry of the occupied adsorption sites,

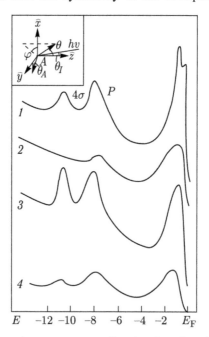

Fig. 2.27. Energy photoelectron spectra of molecules of carbon monoxide on the (100) nickel at different experimental conditions (scheme top left) (4σ and P – levels of the CO molecule corresponding to 4σ- and 1π-orbitals [209], E_F – the Fermi level): $1 - \theta = 45°$, $\theta_I = 0°$, $\varphi = 90°$; $2 - \theta = 45°$, $\theta_I = 0°$, $\varphi = 0°$; $3 - \theta_I = 45°$, $\theta = \varphi = 0°$; $4 - \theta = 63°$, $\theta_I = 45°$, $\varphi = 90°$.

which greatly complements the data obtained with the geometric analysis of the SED patterns.

Electron-photon spectroscopy (EPS) is based on an analysis of electromagnetic radiation produced by bombarding a solid by electrons (the phenomenon is called *electron–photon emission*, and sometimes the *reverse photoelectric effect*) [5, 19, 297, 467]. UVES provides information on the distribution of the density of occupied electron states, and EPS – the distribution of free states. However, EPS, apparently, is less sensitive to the surface state because the depth of the photon yield is greater than the mean free path of photoelectrons before the loss of energy.

Spectroscopy of the characteristic electron energy loss (SCEEL) can be realized experimentally with different resolution. At high resolution (about 10 meV), it is a powerful tool for studying the vibrational spectra of adsorbed particles (see sections 2.1.6 and 2.1.3), but measurements require a complex apparatus. The resolution of $0.1 \div 1$ eV is achieved more easily [57]. In this case it is possible to record energy loss due to excitation of plasma oscillations as well as due to one-electron transitions in adsorbed films. In particular, it is of great interest to observe the conditions under which surface plasmons can be excited in the adsorbed film as the latter becomes denser. This moment is usually correlated with a change in the surface work function.

Measurement of the electron energy loss in the one-electron excitation makes it possible to determine the energy structure of the valence electron shell of the adsorbed particle and the position of the levels of inner electrons. In the latter case, the action of the probing beam leads to ionization of these shells, in connection with which this kind of SCEEL is often defined as *ionisation spectroscopy*. Like the XPS method, it allows the chemical state of the surface to be evaluated [54, 57].

2.2.2. Measurements of work function and surface ionization

A number of surface diagnostic methods is based on the detection of changes in the work function caused by the adsorbed films. From the change in the work function we can determine (for coatings not exceeding a monolayer) the component of the dipole moment of the adsorption bond (dipole potential barrier), normal to the surface [84]. Obviously, these data are important for clarifying the nature of the interaction of adsorbed particles with both the substrate and with each other. In particular, in the presence of a significant dipole potential barrier the interaction of adsorbed particles with each other (lateral interaction) can be repulsive.

If the surface of a monocrystalline substrate during deposition the film becomes non-uniform as regards the work function, it means that different phases coexist in the film, i.e., the first-order phase transition takes place.

For the detection of surface contrast by the work function it has been proposed to use simultaneously the methods use of surface ionisation (SI)

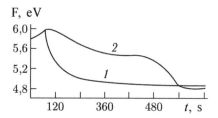

Fig. 2.28. Changes in the work function calculated from the thermionic current and the current of ions during the deposition of carbon on the (111) face of iridium [91] (with the deposition time $t < 100$ s and for $t > 600$ s the film has a single plase, and in the intermediate region two phases).

and thermionic emission (TE) [91–93]. A flux of atoms is directed on the investigated surface and the atoms ionised during desorption with a certain probability. According to the Saha–Langmuir equation, the probability of SI is proportional to exp $[(\Phi - V_i)/T]$, where Φ is the surface work function, V_i is the energy of ionisation of the desorbed atoms. The surface is probed using the atoms for which $V_i > \Phi$. Obviously, the ion current is determined mainly by the surface areas with the maximum work function Φ_{max}, so that the processing of measurement results of the ion current on the basis of the Saha–Langmuir formula gives a value of Φ very close to Φ_{max}. At the same time measuring the TE current gives a value $\Phi \approx \Phi_{max}$. Thus, the difference in values of Φ, determined by these two methods, indicates the phase inhomogeneities of the surface (Fig. 2.28). It is essential that both methods are applicable at high temperatures and thus can significantly increase temperature range in which we study the surface phase transitions. In particular, the combination of SI and TE methods is very effectively used in the study of phase transitions in adsorbed films of carbon on the metal.

It was found, for example, that the two-dimensional carbon phases coexisting on the surface – the gas phase and the condensed phase with a graphite structure – vary greatly not only in the work function, but also in catalytic activity in dissociation reactions [94]. If the composition of dissociated molecules contains an easily ionised component then the formation of the catalytically active phase on the surface can be easily detected on the basis of the appearance of ion current.

Using the method of contact potential difference (CPD), which has various variants [84], we can determine the mean value of the work function $\langle \Phi \rangle$ on the surface. In the first-order phase transition range at arrival of the adsorbate on the surface at a constant speed the area occupied by islets of a denser phase increases linearly with time due to reduction in the size of the less dense phase. Accordingly, the value of $\langle \Phi \rangle$ should depend linearly on the time of deposition of the film (and the average concentration of the adsorbate on the surface), varying between the values characteristic of the coexisting phases. Thus, the the presence of distinct linear sections on the dependence of the work function on the concentration of the adsorbate, measured by the CPD method, usually makes it quite safe to predict phase transitions the first kind in the corresponding intervals of coatings [17, 305]. On heating the sample the adsorbed substance is distributed between phases (heating temperature is chosen so that desorption does not occur, so that the total amount of the adsorbate on the surface remains unchanged). Since the dipole moments of adatoms in phases of different densities are different, this redistribution results in changes in $\langle \Phi \rangle$. By analyzing these changes, we can study the regularities of the phase transition [368].

In the regions of homogeneity of the film, where the changes of the concentration of the adatoms results in continuous changes of the structure, the work function is usually non-linearly dependent on the concentration, if only the interaction between the adatoms is not so weak that it does not affect the dipole potential barrier near the surface region.

The above are only those methods of studying the electronic surface properties that are most closely related to the topics discussed in this book. Of course, they make up only a small fraction of the modern experimental arsenal in this field. Especially numerous and varied are the methods of studying the electronic properties of surface and subsurface semiconductor layers, based on studies of surface conductivity, transport phenomena in magnetic fields of different photoelectric effects, and others (in this regard, see, e.g. [4, 105, 130, 173, 201], and the optical methods of investigation of surfaces – [3, 8, 86, 179]).

2.2.3. Methods for studying surface magnetic properties: diffraction of slow spin-polarized electrons

The vast majority of works in which the objective was to investigate the effect of dimension reduction on the properties of magnets, has been carried out for layered systems. Examples of such objects are crystals of dichalcogenides of transition metals ($NiCl_2$, etc.) in which the planes, formed by the metal atoms, are separated by two atomic planes of chalcogen. These systems are studied using the whole arsenal of methods developed for studying three-dimensional magnets. The magnetic characteristics of layered crystals are covered in an extensive literature [290, 480]. However, to study the magnetic properties of clean surfaces and adsorbed films it is essential to use more sensitive methods because the amount of the analyzed matter in this case is much less than in the case of layered systems.

One of them is the method of diffraction of slow spin-polarized electrons (DSSPE) [109, 303]. This method differs from the conventional SED method by the fact that the surface is probed by a polarized electron beam and the scattered beams are analyzed for both intensity and spin polarization. The sources of polarized electrons are GaAs photoemitters coated with a cesium–oxygen film. Another possibility is to use one of the diffracted beams, obtained after scattering of the unpolarized primary beam on the crystal.

The polarization of scattered electrons is due to spin-orbit interaction, and if there is magnetic ordering, it is then also due exchange interaction, whose energy depends on the orientation of the spin in relation to the magnetization axis. The theory of the DSSPE method has been developed in some detail [303]. The asymmetry of the polarization of the scattered beam is proportional to the magnetization and is most pronounced at small scattering angles. The sensitivity is sufficient to investigate the magnetic properties of the surface atomic plane, and observe how the magnetization varies with distance from the surface into the metal. So, with the help of the DSSPE method it has already been found that the critical magnetization exponent β of the surface of the faces (100) and (110) of nickel is 0.8, whereas for the volume $\beta \approx 1/3$ [244, 304].

It is believed that the method DSSPE will play in the study of magnetism of surfaces such vital role as that played by the method of neutron scattering in the study of the magnetic properties of three-dimensional objects. Significant opportunities for the study of the magnetic properties of surfaces are offered by the methods of photoelectron and electron–photon spectroscopy with spin analysis of the electrons [278]. Note also that positron can be used instead of electrons. It is convenient that positron beams have always natural polarization and that the calculations in this case are much easier because there is no need to consider the exchange-correlation interaction of the beam with the adatoms [303].

In [321] the authors implemented another highly sensitive method, which is the development of the classical method of torsional vibrations of a magnetic sample suspended by an elastic string in a magnetic field. The method allows to investigate the magnetic properties of extremely thin films down to monolayers. To ensure their cleanliness, the substrate is placed in a UHV chamber equipped with different means of diagnosis.

2.3. Methods for studying the dynamics of the lattice, diffusion and film growth mechanisms

The question of surface lattice dynamics is crucial for the physics, chemistry and mechanics of the surface. A number of methods to investigate the vibrational spectra of particles at the surface has been proposed. These are the methods of infrared spectroscopy, Raman spectroscopy, tunneling spectroscopy, as well as methods based on inelastic scattering of electrons, atoms and neutrons. Each of these methods has its advantages and limitations. The method of IR spectroscopy has a high resolution, but because of the smallness of the absorption of IR light special measures must be taken to increase sensitivity (the regime of multiple reflection of the probing beam from the surface under study [216]). Raman scattering can be applied to the rather limited number of objects, such as large organic molecules on the rough metal surface. The method of neutron scattering requires the use of samples with a highly developed surface of the fibre type graphite or powders, etc. A fairly complete comparison of features of these methods is given in [499].

2.3.1. The method of IR spectroscopy

The detailed analysis of the symmetry of the adsorption spectrum requires a local method, which is sensitive to the chemical specificity of the object. Unique advantages in this respect has a method of infrared spectroscopy (IRS). The sensitivity of IRS is about 0.005 monolayers. The energy resolution of IRS, as well as of any optical method, is excellent – a typical value is 0.05 meV. Useful information on the structure may be obtained due to the specific symmetry properties of vibrational excitations.

The N-atomic molecule has $3N$ degrees of freedom, of which 3 are translational, 3 (for a diatomic molecule 2) – rotational, and the rest – oscillating. If, however, a chemical bond forms between the molecule and the solid surface, translation and rotation cannot occur is pure form. In addition, the vibrational spectrum of the adsorption centre can be characterized by varying degrees of degeneration, so the number of non-degenerate vibrational modes is determined by the point group symmetry of the surface of the cluster and to find the number of separate modes we must apply the methods of group theory [356].

More importantly, infrared absorption does not necessarily lead to excitation of vibrations, required by the symmetry. Thus, the method of IRS provides the excitation of oscillations in accordance with the characteristic value proportional to matrix element $\left\|\langle B|\mu \cdot E|O\rangle\right\|^2$, where $|O\rangle$ is the ground state of this mode, and $\langle B|$ is the first excited state. Interaction operator $\mu \cdot E$ includes the local electric field E on the adsorption centre and the dipole moment μ, emerging due to the vibrational mode associated with the motion of nuclei within the adsorbate. From the group theory we have the selection rule according to which the given matrix element may be different from zero only if the considered vibrational mode belongs to the same irreducible representation of the point group that there is at least one of the Cartesian components of the dipole moment μ. These modes are called *dipole-active*.

If the specific conditions of adsorption 'fix' the direction of the field E, then additional 'pseudo-selection rules' come into effect. For example, the reflection of metals at frequencies below the plasma edge is explained by their good shielding properties. The same properties give rise to fields outside the solid which can be described by means of electric images. So, if some tension mode causes a dipole moment to appear in the adsorbate oriented perpendicular to the metal surface, then the dipole image induced in the metal enhances the local electric field. As a result, one can expect the appearance of strong vibrational excitations (see Fig. 2.29). In contrast, the mode, generating a dipole parallel to the surface, induces the dipole image, which largely compensates the initial local dipole field, so only a weak

Fig. 2.29. Electric images of dipoles at the metal surface [95].

quadrupole field remains. The resulting excitation can be experimentally undetectable. Although the materials transparent in the spectral region corresponding are not governed by similar 'pseudo-selection rules', the matrix of the excitation can always appear to be 'accidentally' small, even for those modes that are nominally dipole-active.

The adsorption of hydrogen on the surface of a semiconductor Si (100) can be used as an example, demonstrating very clearly the merits of the IRS method. The IR spectrum of the so-called monohydride phase Si (100) (2 × 1)–H clearly shows two local vibrational modes (see Fig. 2.30), which could not be resolved by other methods of vibrational spectroscopy. There is a high-frequency symmetric mode, creating a dipole moment in the direction normal to the surface, and an anti-symmetric mode (hydrogen atoms are moving out of phase), creating a dipole moment parallel to the surface. The resolution, inherent in the IRS method, is necessary to obtain quantitative data on the width of the vibrational lines and splitting of the modes. These values, combined with theoretical calculations, provide information on both the structural and dynamic properties of the system.

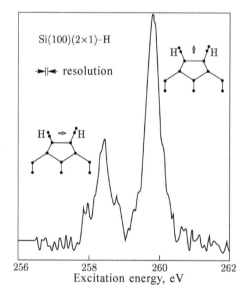

Fig. 2.30. IR spectrum of the structure of the Si (100) $c(2 \times 2)$–H in the monohydride phase. The insets give the identification of modes. Small black arrows indicate the direction of motion of atoms, light – induced dipole moment [493].

2.3.2. Methods of electron spectroscopy: the method of characteristic electron energy loss and the method of Auger spectroscopy

In the method of spectroscopy of the characteristic energy loss of electrons (SCEL) the resolution is approximately 10 meV [57, 356]. This is achieved by monokinetisation of the primary electron beam and the use of advanced energy analyzers of secondary electrons. Equipment was also developed which allows to analyze the inelastic scattering of beams of helium atoms with an average energy of about 20 meV with a resolution of about 0.5 meV [294, 501]. Because of this it has became possible to investigate in detail the dispersion curves for various vibrational modes of both clean surfaces and adsorbed films. The SCEL method is particularly highly sensitive in the study of localized vibrational modes in adsorbed films because these modes are typically characterized by the presence of dynamic dipole moments, which leads to a strong electron–phonon interaction.

As an example, Fig. 2.31 shows the dispersion curves for surface phonons in the oxygen–(001) face of nickel system obtained in [442]. In

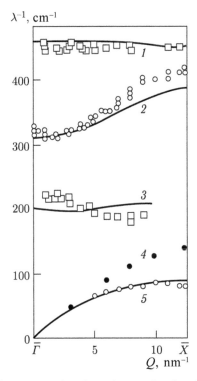

Fig. 2.31. The dispersion curves of surface phonons for the nickel (001) face – lattice $c(2 \times 2)$ of oxygen system [442] (●,○, – the experimental data, curves – calculated results): *1, 2* – longitudinal and transverse modes of vibrations of oxygen atoms, *3, 4* – longitudinal and transverse surface resonant modes of nickel, *5* – data for the clean nickel surface.

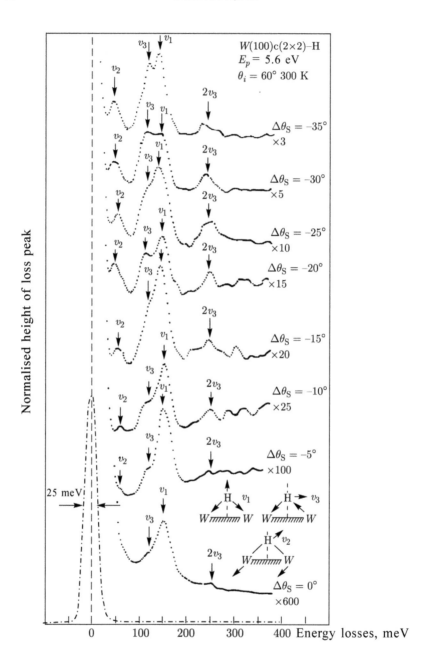

Fig. 2.32. The SCEL spectra of the surface structure W (100) $c(2 \times 2)$–H. The energy of the primary electron beam is 5.5 eV, the polar angle of incidence 60°, signal detection was carried out at different angles relative to the direction of specular reflection [250].

this case the oxygen atoms form a lattice of the type $c(2 \times 2)$. Curves *1* and *2* characterize the vibrations of oxygen atoms, respectively parallel and perpendicular to the surface, curves *3* and *4* – surface resonant modes of nickel. The solid curves are drawn according to the results of calculations performed with certain assumptions about the geometry of the surface layer.

The SCEL method has one major advantage: the short-term impact scattering of incident electrons on the local potential of the adsorbate can lead to the excitation of vibrational modes which are not of the dipole-active type. The scattering cross section for this excitation mechanism is very small, so to prevent the loss of the signal on the background of the dominant dipole scattering it is necessary to detect the electrons that are strongly deflected from the direct (specular) direction. Figure 2.32 illustrates the use of this technique to identify the centre of hydrogen adsorption on tungsten [250]. In the direction of specular reflection the dipole-active 'beat' mode for hydrogen (v_1) with an energy of 130 MeV is dominant in the spectrum. However, with rotation of the electron detector further away from the specular direction $\left(\Delta\theta_s \neq 0 \right)$ there are two more independent modes and their overtones, which indicates the local symmetry C_{2v}, characteristic for the bridge centre.

Similar results were obtained for other adsorption systems and clean surfaces by the scattering of atomic beams [283, 294, 445, 501]. These results indicate that the experimental conditions for successful research of the atomic dynamics of surfaces have been creates.

In a number of methods the starting material for the calculation of the dependence of the diffusion coefficient on the concentration of the adsorbate is the data on the change of the concentration profiles of the adatoms due to diffusion [40, 277, 392, 417]. To register these profiles, it is necessary to provide a sufficiently high resolution of the instrument both with respect to the concentration and the coordinates. The concentration of the adatoms can be estimated from, for example, changes of the work function and the latter is determined by the methods of the contact potential difference, photoelectronic or secondary electron emission. At the same time, spatial resolution is 20÷50 μm. When using scanning Auger microscopy, the achieved resolution in a typical case is about 1 μm, and in the most sophisticated instruments it is even better. With regard to the resolution ability of the device for the concentration, depending on the nature of adsorbate and the features of the method of registration, it is now one hundredth-tenth of a monolayer.

Auger spectroscopy is one of the main methods of elemental analysis of surfaces and, at the same time, it allows to study the mechanism of film growth. For layered growth of the film, corresponding to the case of ideal wetting, the dependence of the current of Auger electrons on the surface concentration of the adsorbate is usually linear within the fill of each next monolayer. But with the growth of the monolayer the slope of this dependence is reduced due to shielding of the layers that lie below

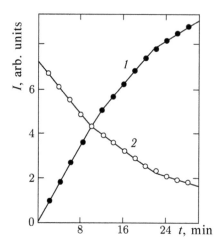

Fig. 2.33. The time t dependence of the amplitude $I(t)$ of the Auger signal of silver (*1*) and tungsten (*2*) at the adsorption of silver on the (110) face of tungsten [254].

by the upper growing layer of the. Therefore, in general, this dependence is described by the broken line on which the time of development of each next monolayer is clearly indicated by a break, and the almost constant level of the Auger signal is achieved at the film thickness corresponding to the exit depth of Auger electrons (Fig. 2.33) [254].

In the case where the adsorbate does not wet the substrate and forms three-dimensional crystallites or droplets, the dependence of the Auger signal on the amount of the adsorbate is non-linear. An alternative variant is also observed quite frequently when there is layer growth to a certain critical film thickness, and only then the nucleation and growth of three-dimensional structures start. In such a situation corresponding to incomplete wetting (the so-called Stransky–Krastanov growth mechanism, very popular during the adsorption on metals and semiconductors), the growth of the film is accompanied by a change the nature of this dependence of the Auger signal [253]. If temperature is changed, there can occur a transition from the wetting to non-wetting regime, or vice versa. Auger spectroscopy, applied parallel with the diffraction methods, is an effective method for studying such transitions.

As mentioned in section 2.1.2, a convenient way to monitor the layer growth of films is the registration of the periodic dependence of the intensity of diffraction reflections in the method of diffraction of fast electrons on the spraying time of the film (Fig. 2.6). It has been shown that the ellipsometry method may be used for the same purpose [197].

Note also that the information about the amplitudes of the oscillations of surface atoms can be obtained by examining the temperature dependence of the intensity of the reflections in slow electron diffraction (Debye–Waller factor) [170, 173].

2.3.3. The methods of field ion and electron microscopy. Scanning tunneling microscopy

The mobility of the surface particles is the most important characteristic, which depends on the kinetics of many processes on the surface. The ultimate sensitivity in the study of surface diffusion processes – the ability to follow random wandering of individual adatoms – is provided by the method of field ion microscopy [168, 252, 301]. At the same time, the migrating atom can be regarded as a probe, sensing changes in the potential of interaction of the atom with the surface. In addition, it is necessary to bear in mind that the observation of atomic diffusion do not give absolute values of the binding energy, but only provides insights on the strength of these bonds. A review of works on the application of field ion microscopy in the study of surface diffusion of atoms was presented in [238, 494].

High resolution (2÷3 nm) is also typical of the method of field electron microscopy, which is used in two variants [417]. In the first, the adsorbate is deposited on one half of the sample having the shape of a tip, and then its migration to another ('shaded') half is studied. Unfortunately, due to the fact that the surface of the tip is a mosaic of different crystal faces, this method is not suitable for obtaining with the required accuracy the dependence of the diffusion parameters on the concentration of adatoms on the individual faces. In the second variant, the field electron microscope is used to measure fluctuations in the number of particles in a small area of the tip having a linear size of about 10 nm [319]. The processing of these data allows to determine the diffusion parameters in a wide range of concentrations of adatoms on the individual faces of the crystals.

The dependences of the diffusion parameters on the concentration of the adsorbate are of interest not only as direct information about the mobility of particles in different coatings, but also due to the fact that they reflect the interaction of adsorbed particles and hence the phase transitions in the film.

The surface diffusion processes can be studied with a scanning tunneling microscope [259]. The observations of this type allow one to find the activation energy and the pre-exponential factor in the equation for the coefficient of diffusion of individual adatoms on the surface of the well-defined atomic structure. The diffusion of dimers and larger associates of atoms can be studied by the same method.

2.4. Methods for measuring the thermodynamic characteristics of adsorbed films

2.4.1. Measurements of adsorption energy

The interaction energy of the particles with the surface is conventionally divided into two components: the interaction with the substrate and the interaction of the adsorbed particles with each other (lateral interaction). Because these components are not independent, such a division, of course, can not be considered strict, but it is used for reasons of clarity. It is usually assumed that the energy of interaction with the substrate can be determined by considering the adsorption of a single particle (atom or molecule) on the surface. All changes to the adsorption energy, which take place when a large number of adsorbed particles is applied to the surface, are attributed to lateral interactions. Information about adsorption energy is important for understanding the physical nature of interactions on the surface and from a purely practical point of vision (thermal resistance of the films, catalytic reactions, etc.).

We first consider the methods of the lateral interactions as the main factor determining the properties of two-dimensional systems. The most direct information about the lateral interactions can be obtained by observing in the field ion microscope correlated random wandering of two adsorbed atoms on the surface [168, 252, 301, 311]. In this way, in particular, it has been possible to confirm theoretical predictions of the existence of interaction of adatoms through the substrate, oscillating with changes of the distance. The binding energy of adatoms in dimers and larger two-dimensional clusters can also be defined. Similar data for the size of islands of the order of nanometers are provided by the method of field electron microscopy, i.e. in this case we are already talking about the two-dimensional sublimation energy [55, 236].

Quite often, the lateral interaction energy is determined using the data on the order–disorder transitions in adsorbed films. The results of experiments usually performed using diffraction methods are processed on the basis of a theoretical model of the transition [271].

In all these methods, as well as the method based to record changes in the work function of the film during heating [368], the amount of material on the surface during the experiment remains unchanged, i.e. there are purely two-dimensional transitions. Another possibility is to explore the balance between the three-dimensional gas and adsorbed phases or the kinetics of desorption into the gas phase. At the same time, information is obtained about the differential heat of adsorption at different concentrations of the adsorbate on the surface. In the limit of small concentrations, it characterizes the adsorbate-substrate interaction, and its change with increasing concentration (if the surface is adsorption-homogeneous) occurs only due to lateral interaction of adsorbed particles.

A classic example of this type of method is the method of adsorption isotherms. In the three-dimensional gas–adsorbed film equilibrium the dependence of the concentration of adatoms on the gas pressure at constant temperature was recorded (Fig. 2.34). With the family of the isotherms, we can calculate the heat of adsorption by the Clausius–Clapeyron equation [471, 483]. The presence of jumps in the isotherms indicates the process of two-dimensional condensation in the film, so that the adsorption isotherms

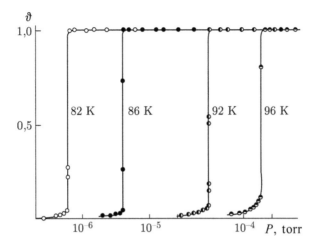

Fig. 2.34. Adsorption isotherms of xenon on the basal face of graphite – the degree of coverage, as measured by Auger spectroscopy, P – pressure of xenon; jumps correspond to the transition from two-dimensional gas to the two-dimensional crystal with structure $\sqrt{3} \times \sqrt{3} - R30°$ [483].

can in principle be used to construct a phase diagram of the film in the same way as is done for three-dimensional systems.

The method of thermal desorption spectroscopy (TDS) has been efficiently developed to investigate the adsorption energy [241]. It consists of registration of changes in the flow of desorbed particles in the process of raising the substrate temperature. In general, the obtained dependence has the shape of the curve with a number of peaks corresponding to the consecutive evaporation of particles from the adsorption states with increasing binding energy. Data processing is based on the desorption rate equation. When interpreting the data on the binding energy obtained by TDS, there are some difficulties caused by the fact that, as a rule, the structural state of the film at the desorption temperature is not known. As we know, this temperature is quite high, while structural studies are carried out in most cases at low temperatures. Apparently, only the short-range order is usually retained in the films at the desorption temperature

2.4.2. Measurements of the heat capacity of adsorbed films

The use of exfoliated graphite as a substrate allows absorb such a large amount of material that even with submonolayer coatings, this amount is sufficient to measure the heat capacity by the calorimetric method [268, 286].

Reference 268 describes the design of the calorimeters, designed to measure the specific heat of helium films on graphite. The heat capacity of the film is defined as the difference between the heat capacities measured for the substrate with the film deposited on it and without it.

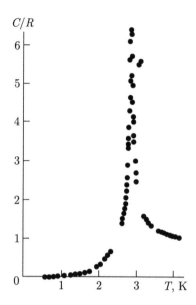

Fig. 2.35. Temperature dependence of heat capacity C of the He film with the $\sqrt{3} \times \sqrt{3} - R30°$ structure on the basal face of graphite [286].

Figure 2.35 shows an example of the data. The temperature dependence of the specific heat shows a clearly defined peak which, as shown by comparison of the phase diagram of the film, is related to the order–disorder transition. This method was used in detailed investigations of the characteristics of the process of disordering of the films with the commensurate and incommensurate structures.

In [279] measurements were taken with record sensitivity (about 10^{-9} J/K) of the heat capacity of monolayer films of helium adsorbed on the surface of a cleaved graphite crystal. This work showed the possibility of measuring the heat capacity of submonolayer films adsorbed on single crystal substrates.

As can be seen from this review of experimental methods, at present there are ample opportunities to study the surface properties of solids. A number of effects have already been studied on the atomic level.

3

LOCALIZED STATES AND SURFACE ELEMENTARY EXCITATIONS

Introduction

The surface of the crystal is a two-dimensional periodic arrangement of atoms. For the surface, as for the three-dimensional periodicity of the bulk crystal, it is possible to determine the elementary excitations that propagate along the surface, but, unlike in the bulk crystal, are localized in a narrow surface region in the direction normal to the surface.

These elementary excitations include quasi-particles and collective excitations. Quasi-particles are electrons localized in the surface states. As the collective excitations of the surface layer of atoms we can introduce surface phonons and surface polaritons and also surface plasmons as collective excitations of the electronic gas near the surface and surface magnons as excitations of spin density in the surface layer of a ferromagnetic or antiferromagnetic. A review of the general properties and features of these elementary surface excitations are subject of this chapter.

In this case, the periodic structure of the surface can be taken as coinciding with the structure within the solid, and then we are concerned with an ideal surface, or it is a superstructure, resulting from the rearrangement of surface atoms. If the crystal surface is covered with an adsorbed layer, it can also have a structure different from the crystal structure inside the solid. If the adsorbed layer covers the surface incompletely or if the surface is locally deformed, then local surface states may appear. However, in this chapter, we restrict ourselves to surface excitations for an idealized surface of the crystal.

3.1. Electronic surface states

The possibility of formation of surface states was first predicted in 1932 by I.E. Tamm in [211], where it was shown that the boundary of an ideal crystal lattice can itself serve as a source of special electronic states localized near this boundary. These electronic surface states, called subsequently 'Tamm' states, split off from the allowed region spectrum and are located inside the forbidden zone. By their nature, they are very similar to the usual bound states studied in quantum mechanics courses. Both are characterized by an energy lying in the region of the spectrum not available for the 'free' particle, and are described by an exponentially decaying wave function. Both correspond to the poles of the scattering matrix (for Tamm states – the scattering of Bloch waves by the crystal surface).

With the emergence of the concept of surface states, it became clear that the surface of the crystal plays the role of its independent subsystem, and its owns electrons move in a two-dimensional periodic field. It became possible to talk about these more complex mixed structures like the metal with a dielectric surface or, alternatively, an insulator with a two-dimensional metal on its surface.

The Tamm surface states arise in a situation where the surface makes a strong perturbation in the crystal structure. The surface in [211] was introduced as a stepped potential jump and the wave function and its derivative were 'sewn' together at the point of discontinuity. A.V. Maue in [403] considered the model of a crystal in the approximation of nearly free electrons. In this paper, the surface is also introduced as a stepped jump in a periodic potential at the point of its high or low, and the conditions of existence of surface states were determined. In 1939, three studies of E.T. Goodwin [320] on the surface states were published. In the first paper the results of A.V. Maue were summarized, but the other two contained a brand new theoretical approach for that time, which was based on the method of linear combinations of atomic orbitals (LCAO). It is essential that here the discontinuities of the physical quantities on the surface are described in another way: it was assumed that the surface atoms correspond to the values of the Coulomb and resonance integrals different from the corresponding values for the atoms in space.

Also in 1939 W. Schockley in [459] considered a model of the electronic surface states on the basis of the study of the properties of the one-dimensional periodic potential of a general form, limited by a step at the maximum point. His approach also used 'stitching' of the wave functions at the discontinuity. Schockley showed that if the areas of bulk electronic states overlap, the surface states also if the surface perturbation is sufficiently small. This was contrary to previous findings, according to which the surface states arise only when the surface perturbation is sufficiently large. Later these two types of surface states were identified, respectively, as *Shockley and Tamm states*. Schockley suggested that the overlapping zones and the negligible surface perturbations are characteristic of semiconductor crystals such as silicon and germanium.

So, if in the 1940s the theory of surface states had already been to some extent developed, the experimental evidence for their existence began to appear only at the end of the 1940s with the development of semiconductor technology and purification technology of crystals. In 1947

G. Bardin in [249] explained the 'anomalous' results obtained in [407] for contacts of metals with silicon, with the involvement of surface electronic states. In 1948 Schockley and G. Pearson in [460] validated the Bardin's assumptions. Imposing a uniform electric field, they changed the potential of the free surface of the semiconductor (the so-called experiment with the 'field effect'). It turned out at the imposition of an external field parallel to the surface part of the surface charge begins to move, while its larger part remains stationary. This fixed part of the charge was attributed to the electrons trapped in surface states. Later it was found that there are both 'slow' and 'fast' traps associated with the formation on the surface of silicon and germanium when exposed to air.

These experimental results again increased the interest in the theory of surface states. In [247, 344, 390, 478] attention was given to the issues associated with the Tamm and Shockley states. The next step towards a more detailed theory of localized surface states, which allows to describe the real system, was made in [371, 372], where the authors were able to consider all types of states localized at the surface of an ideal crystal. In their method, called the one-electron resolvent method, most results were obtained in the LCAO representation. At the beginning of 1960s, experiments with the diffraction of slow electrons (DSE) showed [377, 404] that some of the free surfaces of crystals undergo restructuring (e.g. (100) surface in silicon, cleaved in ultrahigh vacuum). This restructuring is unstable and depends on temperature. All this showed that the silicon surface experiences a strong perturbation which can not be neglected, as suggested by W. Schockley and others. On the contrary, as shown by similar experiments, electrically neutral surfaces of partially ionic compounds of elements of the III–V, II–VI and I–VII groups are not reconstructed [98, 306, 412]. Theoretical studies, devoted to the study of surface states in crystals of compounds of this type, are presented in [289, 353, 386, 387].

Description of the appearance of surface Tamm states, which follows the original paper [211], was presented in a recent book by V.I. Roldugin 'Physical chemistry of surfaces' [203] as well as in a more specialized monograph by S. Davison and G. Levin [85], in a large amount of the material devoted to the theoretical description of surface electronic states was systematized and generalized. We propose in this book, in our view, a more intuitive approach to the description of surface electronic states, considering the model of a crystal in the approximation of almost free electrons.

Consider a simplified model of an idealized surface. Let the periodic potential of a semi-infinite crystal occupies the half-space with $z < 0$. Suppose that in the half-space with $z > 0$, corresponding to vacuum, the potential is constant and equal to V_0. The surface is thus a sharp transition between the strictly periodic lattice and vacuum. Let us, in addition, make the following assumptions: we consider the lattice potential as a weak perturbation, using thus the approximation of almost free electrons, we reduce the problem to a one-dimensional model by introducing a periodic potential $V(z) = V(z + na)$ for $z < 0$, where a is the lattice constant, $V(z) = V_0$ for $z > 0$.

First, we solve the Schrödinger equation for the one-dimensional problem

$$\left(-\frac{\hbar^2}{2m}\frac{d^2}{dz^2} + V(z)\right)\psi(z) = E\psi(z) \tag{3.1}$$

separately for $z > 0$ and for $z < 0$, and then 'sew' the solution for $\psi(z)$ at $z = 0$ from the continuity condition of the wave function and its derivative.

The vacuum with $V(z) = V_0$ for $z > 0$ corresponds only to the solutions of the Schrödinger equation, decreasing with increasing z:

$$\psi = A\exp\left(-z\sqrt{\frac{2m}{\hbar^2}(V_0 - E)}\right). \tag{3.2}$$

For $z < 0$, we use the approximation of nearly free electrons for movement of the electron in a one-dimensional periodic potential $V(z) = V(z + na)$. Since the potential is periodic, it can be decomposed into a Fourier series: $V(z) = \sum_n V_n \exp(i2\pi nz/a)$. A term with $m = 0$ is the average value of the potential and we believe it to be zero. The energy of the electron in the zeroth approximation is equal to $E^{(0)} = \frac{\hbar^2 k^2}{2m}$ with the wave function $\psi_k(z) = \frac{1}{\sqrt{a}}\exp(ikz)$. The initial energy state is twice degenerate: for k and $-k$. Therefore, the wave function of the zero-order approximation should be sought as a linear combination of plane waves: $\psi^{(0)}(k, z) = \alpha\psi_k(z) + \beta\psi_{-k}(z)$. The matrix elements of the perturbation operator, whose role is played by $V(z)$, is characterized by $V_{k,k'} = \langle k'|V(z)|k\rangle = V_n\delta_{k,k'+g}$, where $g = 2\pi n/a$ is the reciprocal lattice vector. Given the fact that $|k| = |k' + g|$, the wave function of the zeroth approximation is sought in the form of

$$\psi^{(0)}(k,z) = \alpha\exp(ikz) + \beta\exp\left[i\left(k - \frac{2\pi}{z}\right)z\right]. \tag{3.3}$$

Then the system of equations for the undetermined multipliers α and β, as well as the eigenvalues of the electron energy in the first order perturbation theory will be the following:

$$\left(\frac{\hbar^2 k^2}{2m} - E(k)\right)\alpha + \beta V_1 = 0,$$

$$\alpha V_1^* + \beta\left[-\frac{\hbar^2}{2m}\left(k - \frac{2\pi}{a}\right)^2 - E(k)\right] = 0. \tag{3.4}$$

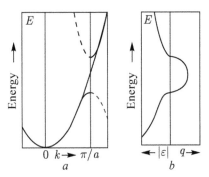

Fig. 3.1. The plot of the band structure for the one-dimensional periodic potential with lattice constant a in the approximation of nearly free electrons [137].

We introduce the parameters $\varepsilon = k - \dfrac{\pi}{a}$ and $\gamma = \dfrac{\hbar^2 \pi \varepsilon}{ma|V_1|}$. Then from (3.4) we get:

$$\psi(k,z) = B\left[\exp\left(i\frac{\pi}{a}z\right) + \frac{|V_1|}{V_1}\left(-\gamma \pm \sqrt{1+\gamma^2}\right)\exp\left(-i\frac{\pi}{a}z\right)\right]\exp(i\varepsilon z), \quad (3.5)$$

$$E(k) = \frac{\hbar^2 k^2}{2m} + |V_1|\left(-\gamma \pm \sqrt{1+\gamma^2}\right). \quad (3.6)$$

For real ε the expression (3.6) gives the energy bands of the electron states (Fig. 3.1a), typical for the approximation of almost free electrons using a one-dimensional periodic potential: the parabola $E(k)$ for free electrons is distorted near the Brillouin zone boundary, and the zones are separated by an energy gap. The wave functions (3.2) and (3.5) can be 'stitched' to each other to an arbitrary value of E. In the half-space $z < 0$ for this we require two solutions $\psi(k, z)$ and $\psi(-k, z)$, whose linear combination is 'crosslinked' with the vacuum solutions. The energy zones of the infinite lattice in such a case, except for minor amendments, remain unchanged.

In addition to the zones, there are also solutions that are localized on the surface. Since (3.5) is the solution only in the half-space $z < 0$, the parameter ε can also be apparent. For $\varepsilon = -iq$ with the real and positive q there are solutions that decay exponentially in the crystal. Thus, at $\gamma = i\sin(2\delta) = -i\hbar^2 \pi q/(ma|V_1|)$ we find from (3.5)

$$\psi(k,z) = C\left\{\exp\left[i\left(\frac{\pi}{a}z \pm \delta\right)\right] \pm \frac{|V_1|}{V_1}\exp\left[-i\left(\frac{\pi}{a}z \pm \delta\right)\right]\right\}\exp(qz). \quad (3.7)$$

The energy corresponding to this solution has the form

$$E(q) = \frac{\hbar^2}{2m}\left[\left(\frac{\pi}{a}\right)^2 - q^2\right] \pm |V_1|\left[1 - \left(\frac{\hbar^2 \pi q}{ma|V_1|}\right)^2\right]^{1/2}. \quad (3.8)$$

It is valid for $0 \leq q \leq q_{max} = ma|V_1|/\pi\hbar^2$. For $q = 0$, we obtain solutions

$$\psi(k,z) = C\left[\exp\left(i\frac{\pi}{a}z\right) \pm \frac{|V_1|}{V_1}\exp\left(-i\frac{\pi}{a}z\right)\right],$$

$$E(q) = \frac{\hbar^2}{2m}\left(\frac{\pi}{a}\right)^2 \pm |V_1| \tag{3.9}$$

with $\psi(k,z) \sim \cos(\pi z/a)$ for $V_1 > 0$ and $\psi(k,z) \sim \sin(\pi z/a)$ for $V_1 < 0$, which correspond to the energies of the two band edges, i.e. the lowest eigenvalue is associated with the sine function, and the top one – with the cosine function.

To ensure the 'stitching' of the solutions (3.2) for $z > 0$ and (3.7) for $z < 0$ at our disposal we have two free parameters – the ratio of the coefficients A and C and the energy E. Both parameters are determined by the continuity of the wave function and its derivative at $z = 0$.

Thus, we obtain the following result: while the solutions of the Schrödinger equation with a real k correspond conventional solutions for the zones, solutions are possible for the imaginary k, which decreases with distance from the surface. The corresponding values of energy, according to (3.8), lie in the energy gap between the bands (Fig. 3.1b). One of these solutions can be 'sewn' with the solution for the outer space. It is a condition in which the electron is localized in a narrow region near the surface. This is the desired surface condition.

A deeper analysis shows that the 'stitching' is possible only for $V_1 > 0$. Thus, in this model the surface states may exist, but not in every case.

In the one-dimensional model the surface state has a discrete level in the energy gap. Extending this model to three dimensions, we can consider the results as relating to the component of the wave vector $\mathbf{k}$, perpendicular to the surface. For each fixed component $\mathbf{k}$, parallel to the surface, one can expect different levels of energy of the surface state. Thus, instead of individual levels we obtain the energy bands for the surface states. Since the energy gap, which should contain each surface level, differs for each value of $\mathbf{k}$, the band of the surface states can overlap with the bands of bulk-states (Fig. 3.2).

However, the one-dimensional model considered here, and its qualitative extension to three dimensions, are not realistic in some cases. In particular, the surface is not an abrupt transition from the unperturbed periodic potential of the lattice to the external space.

Despite this, the states of the type under consideration are available for many surfaces. Amendments are needed primarily to explain the three features of the real surface.

1. In the surface atomic layer of the lattice the forces acting in it are directed only perpendicular to the surface; as a result, deformation of the periodic potential takes place and changes the distance between atomic

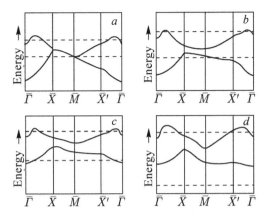

Fig. 3.2. Surface area for the surface (110): *a* – Ge, *b* – GaAs, *a* – InP, *i* – ZnSe. The dashed lines indicate the upper edge of the valence band and the lower edge of the conduction band of the semiconductor [137].

planes in the surface field. The phenomenon of displacement of a few (2 ÷ 3) atomic planes in the surface region relative to their position in the bulk crystal became known as *lattice relaxation of the surface*. In Chapter 4 we will describe the technique of calculating the displacement of the subsurface ionic planes and their influence on surface energy and the electron work function from the surface.

2. The free valence bonds on the surface may be connected otherwise than within the crystal, which leads to the appearance of a surface superstructure, i.e. a change of symmetry in the surface layer.

3. The surface may be covered with an ordered adsorbed layer.

All of these amendments change the position of the surface areas and, therefore, change the density of states in them. However, these factors do not eliminate the possibility of the appearance of surface states.

In addition to these bands of delocalized states (the states belonging to the entire surface and extending along it) there are also possible localized states as discrete levels associated with local distortions of the surface (individual adsorbed atoms, an incomplete adsorbed layer, etc.). Such localized states are observed when the conditions for 'stitching' of the wave functions (3.2) and (3.3) on the surface can not satisfied, i.e. when the surface areas are not available.

In this section we only illustrate the basic principles of formation of surface states. For a more detailed analysis the reader should refer to the monograph [85].

3.2. Surface plasmons

Surface plasmons are introduced as the quanta of oscillations of the density of electrons localized near the surface. These high-energy elementary excitations are now coming under scrutiny because they are one of the main characteristics for the metallic nanoparticles and thin films. As a result, the theoretical understanding of surface plasmons has become an important tool for studying the processes occurring at the interphase boundaries [26, 27, 190, 374]. Before proceeding to the consideration of surface plasmons we discuss the nature of the bulk plasma oscillations.

3.2.1. Three-dimensional plasma waves and plasmons

Plasmons are the elementary excitations associated with the collective motion of electrons relative to ions in solids. These elementary excitations are due to Coulomb interaction between electrons and positive ions. They correspond to longitudinal waves, which are called *plasma waves*. The quanta of plasma waves are called *plasmons*.

Plasma oscillations occur in metals and semiconductors, i.e. in solids with electrons, weakly bound to the ions. In the ground state, the electrons fully compensate the positive charge of the ions, and each unit cell of the crystal is neutral. Let n_0 be the mean number of electrons per unit volume of the crystal, corresponding to such a neutral state. The deviation of the electron density n from the average value n_0 leads to a violation of electrical neutrality and the appearance of electrical forces restoring the balance. This leads to variations in the density of the electrons relative to the mean value n_0.

In the simplest theory of plasma oscillations in solids the discrete distribution of positive ions is replaced by a homogeneously distributed positive charge with a density equal to the average electron charge density n_0. Such a model of the solid received the name of the *'jelly model'*. The valence electrons and the conduction electrons are considered as an electronic gas, and the dilution and compression of this gas with respect to the average value lead to longitudinal vibrations. In the approximation of the 'jelly' model for shear wave there are restoring force as such waves do not change the local electroneutrality system. Therefore, the intrinsic transverse oscillations are absent. The longitudinal waves cause compression and rarefaction in the electron gas, breaking the neutrality and thereby inducing strong Coulomb forces restoring violations. The intrinsic frequencies of longitudinal vibrations, called *plasma frequencies*, are relatively high frequencies. The density of electrons in metals is of the order of $10^{22} \div 10^{23}$ cm^{-3}. At high electron densities, the electron kinetic energy of their 'zero' ($T = 0$) motion is much higher than the energy of thermal motion, so the latter is not taken into attention.

Consider the long-wavelength plasma oscillations in an isotropic crystal. For long-wave oscillations, the electrons can be considered as a continuous medium. The change of the electron density in relation to the mean value n_0 can be written as

$$\delta n(\mathbf{r},t) = n(\mathbf{r},t) - n_0 = n_0 div \mathbf{R}(\mathbf{r},t), \qquad (3.10)$$

where $\mathbf{R}(\mathbf{r}, t)$ is the vector of small displacements of the electron gas from its normal position.

Changes of the electron density lead to local disruption of electrical neutrality of the crystal. The result is the electrostatic potential $\varphi(\mathbf{r}, t)$, satisfying the Poisson equation

$$\nabla^2 \varphi(\mathbf{r},t) = 4\pi e n_0 div \mathbf{R}(\mathbf{r},t). \qquad (3.11)$$

The potential energy that occurs when the electrons are displaced will consist from the changes of the elastic and electrostatic energies:

$$U = \frac{1}{2} \int \left[\gamma (div\mathbf{R})^2 - e\delta n\varphi \right] d^3\mathbf{r}, \qquad (3.12)$$

where γ is the elastic modulus of the electron gas without taking charges into account. We consider only the longitudinal displacements, i.e. we assume that rot $\mathbf{R}(\mathbf{r}, t) = 0$.

If m is the mass of the electron, the kinetic energy of the displacements of the electrons is

$$K = \frac{mn_0}{2} \int \left(\frac{d\mathbf{R}(\mathbf{r},t)}{dt} \right)^2 d^3\mathbf{r}. \qquad (3.13)$$

Let the crystal has the shape of a cube with side L and volume $V = L^3$. We introduce cyclic boundary conditions. Then, the wave functions are

$$\psi_k(\mathbf{r}) = \frac{1}{\sqrt{V}} \exp(i\mathbf{kr}), \qquad (3.14)$$

where the components of the wave vector k_i take values $2\pi l_i / L (l_i = 0, \pm 1, ...)$ and form a complete orthonormal system of functions. We expand the displacements $\mathbf{R}(\mathbf{r}, t)$ in this system of orthonormal functions:

$$\mathbf{R}(\mathbf{r},t) = \frac{1}{\sqrt{V}} \sum_k A_k(t)\mathbf{e}(\mathbf{k})\exp(i\mathbf{kr}), \qquad (3.15)$$

Here $\mathbf{e}(\mathbf{k})$ is the unit vector of the longitudinal polarization satisfying the conditions: $\mathbf{e}^2(\mathbf{k}) = 1$, $\mathbf{e}(\mathbf{k}) = \mathbf{e}(-\mathbf{k})$, $\mathbf{k}||\mathbf{e}(\mathbf{k})$. The condition of the reality of displacement implies that $A_k = A_{-k}^*$.

From (3.15)

$$div\,\mathbf{R}(\mathbf{r},t) = \frac{i}{\sqrt{V}} \sum_k (\mathbf{ke}(\mathbf{k})) A_k \exp(i\mathbf{kr}). \qquad (3.16)$$

We expand the electrostatic potential on the orthonormal system of functions (3.14):

$$\varphi(\mathbf{r},t) = \frac{1}{\sqrt{V}}\sum_{k}\varphi_{k}(t)\exp(i\mathbf{kr}). \tag{3.17}$$

From the Poisson equation (3.11) with (3.16) taken into account we obtain:

$$\varphi_0 = 0, \quad \varphi_k = -\frac{4\pi i n_0 e}{k^2}(\mathbf{ke(k)})A_k, \quad \mathbf{k} \neq 0. \tag{3.18}$$

Taking into account these relations the potential energy is converted to the form

$$U = \frac{1}{2}\sum_{k}\left(\gamma k^2 + 4\pi e^2 n_0^2\right)A_k A_{-k}, \tag{3.19}$$

and the kinetic energy of the form

$$K = \frac{mn_0}{2}\sum_{k}\dot{A}_k\dot{A}_{-k}. \tag{3.20}$$

Then the Lagrange equations for the system takes the form

$$mn_0\ddot{A}_k + \left(\gamma k^2 + 4\pi e^2 n_0^2\right)A_k = 0. \tag{3.21}$$

Putting $\ddot{A}_k = -\omega^2(\mathbf{k})A_k$, we obtain the dispersion law of plasma oscillations tions for small values of wave vectors:

$$\omega^2(\mathbf{k}) = \omega_p^2 + \frac{\gamma}{mn_0}k^2, \tag{3.22}$$

where $\omega_p^2 = 4\pi e^2 n_0/m$ is the square of the plasma frequency.

When $e \to 0$, electrostatic effects disappear and $\omega(k) \approx k\sqrt{\gamma/(mn_0)} = \omega_{ac}(k)$. This dependence coincides with the dispersion law of the frequency of the sound waves propagating in a gas at the speed $\sqrt{\gamma/(mn_0)}$. THe value $\sqrt{\gamma/(mn_0)} \sim 5\cdot10^5\,\text{cm/s}$, $k_{max} \approx 10^8\,\text{cm}^{-1}$. Therefore $\omega_{ac} \approx 5\cdot10^{13}\,\text{s}^{-1}$. Evaluation of plasma frequency for $n_0 \approx 10^{23}\,\text{cm}^{-3}$ gives

$$\omega_p = \sqrt{\frac{4\pi e^2 n_0}{m}} \approx 2\cdot10^{16}\,\text{s}^{-1},$$

where $\hbar\omega_p \approx 12\,\text{eV}$. Consequently, $\omega_p \gg \omega_{ac}$ and the variance of plasma waves is very small. The relative change $\omega(k)$ within the first Brillouin zone is less than 10^{-3}. The value of $\gamma/(mn_0)$ in (3.22) has the dimension of the velocity square, so for the metals it is natural to assume that $\gamma/(mn_0) \sim v_F^2$, where v_F is the electron velocity at the Fermi surface. A more accurate analysis shows [149] that $\gamma/(mn_0) = 3v_F^2/5$, and, therefore, $\omega_{ac}^2(k) = 3v_F^2k^2/5$.

We introduce the generalized momentum conjugate to the collective coordinate A_k:

$$P_k = \frac{\partial(K-U)}{\partial \dot{A}_k} = mn_0 \dot{A}_{-k}.$$

Therefore, the classical Hamiltonian function of plasma oscillations in terms of the generalized coordinates and momenta is of the form

$$H = \frac{1}{2mn_0} \sum_k P_k P_{-k} + \frac{mn_0}{2} \sum_k \omega^2(\mathbf{k}) A_k A_{-k}. \tag{3.23}$$

We apply the procedure for the quantization of plasma oscillations in the system, passing to the Hamiltonian operator in the representation of occupation numbers of the plasmons. For this we carry out the following transformations in (3.23):

$$A_k \rightarrow \hat{A}_k = \sqrt{\frac{\hbar}{2mn_0\omega(k)}} \left(\hat{a}_k + \hat{a}^+{}_{-k}\right), \tag{3.24}$$

$$P_k \rightarrow \hat{P}_k = i\sqrt{\frac{\hbar\omega(k)mn_0}{2}} \left(\hat{a}^+{}_k - \hat{a}_{-k}\right),$$

where $\hat{a}^+{}_k$, $\hat{a}_k$ are the Bose operators of formation and annihilation of the plasmons, respectively, with the wave vectors $\mathbf{k}$.

As a result of these transformations, we obtain the Hamiltonian operator for the plasmons as elementary excitations of the system:

$$\hat{H} = \sum_k \hbar\omega(\mathbf{k})\left(\hat{a}^+{}_k \hat{a}_k + \frac{1}{2}\right), \tag{3.25}$$

where the energy of the plasmons in the states defined by wave vectors $\mathbf{k}$ is defined as $\varepsilon(\mathbf{k}) = \hbar\omega(\mathbf{k})$ with $\omega(\mathbf{k})$ from (3.22). In this approximation the plasmons behave as non-interacting Bose quasi-particles.

The conducted long-wave description of plasma oscillations, when the idea of the distribution of the electrons in a crystal as a continuous medium is justified, is characterized by the range of applicability of the following dynamic equations for the Fourier components n_k of the electron density $n(\mathbf{r}) = \sum_i \delta(\mathbf{r}-\mathbf{r}_i) = \sum_k n_k \exp[(i\mathbf{k}\mathbf{r}_i)]$ [77]:

$$\ddot{n}_k + \omega^2(\mathbf{k})n_k + \sum_i (\mathbf{k}\mathbf{v}_i)^2 \exp(-i\mathbf{k}\mathbf{r}_i) + \frac{4\pi e^2}{m} \sum_{q\neq k, q\neq 0} \frac{(\mathbf{k}\mathbf{q})}{\mathbf{q}^2} n_q n_{k-q} = 0, \tag{3.26}$$

where $\mathbf{r}_i$ and $\mathbf{v}_i$ are the radius vector and velocity of the i-th electron, considered as a point particle.

From equation (3.26) it can be seen that the harmonic law of plasma fluctuations for n_k is justified if we can neglect the contributions of the third and fourth terms in (3.26). Neglecting the contribution of the fourth term is well known in the solid state theory as the random approximation (RPA), when neglecting the effects of correlations of the electron density, i.e. it is assumed that the sum over $\mathbf{q} \neq \mathbf{k}$ in this term contains a large number of small alternating terms for $\mathbf{q} \neq 0$. The third term in (3.26) depends on the velocity of the electrons. This movement has a disordering effect on the collective plasma oscillations. Its influence becomes weaker with the reduction of $\mathbf{k}$. To estimate the values of $\mathbf{k}$ at which we can neglect the contribution this term (3.26), $\mathbf{v}_i$ can be replaced by the maximum value of the electron velocity on the Fermi surface $v_F = \hbar(3\pi^2 n_0)^{1/3}/m$. Then

$$\sum_i (\mathbf{k}\mathbf{v}_i)^2 \exp[-i(\mathbf{k}\mathbf{r}_i)] < v_F^2 k^2 \sum_i \exp[-i(\mathbf{k}\mathbf{r}_i)] = v_F^2 k^2 n_k.$$

Consequently, if the inequality

$$k^2 < k_c^2 = \frac{\omega^2(k)}{v_F^2} \approx \frac{\omega_p^2}{v_F^2} = \frac{2\pi e^2 n_0}{E_F}, \tag{3.27}$$

is fulfilled, where E_F is Fermi energy, and in the random phase approximation, equation (3.26) of the changes in the electron density changes to the equation of collective plasma oscillations. The value of the critical wave vector k_c can be taken as an upper bound for the wave vectors of the plasmons. It may be clarified in the light of the kinetic energy of single-particle excitations – electron – and is determined by the condition

$$\hbar\omega(k_c) = \frac{\hbar^2 k_c^2}{2m} + \hbar k_c v_F. \tag{3.28}$$

Critical wave number k_c corresponds to a point in the space of the wave vectors of the first Brillouin zone, in which the dispersion curve of the plasmons first reaches the upper limit of existence of the single-particle electron excitations. Thus, the wave vectors of the plasmons occupy the central region of the Brillouin zone with the volume $4\pi k_c^3/3$. Since the proportion of one value of the wave vector in the space of wave vectors has the volume $(2\pi)^3/V$, then in the crystal there can be $k_c^3 V/(6\pi^2)$ plasmons.

When the inequality (3.27) is satisfied, the phase velocity of the plasmons equal to ω_p/k, is greater than the maximum velocity of the electrons and, therefore, the damping effects associated with the transfer of the energy of the plasmons to the energy of motion of the individual quasi-particles–electrons, do not exist. The damping of plasmons in solids is due to the interaction with the lattice vibrations, impurities and other lattice defects.

The elementary excitations with the wave vectors $k > k_c$ do not have a collective character, and the plasmons decay into single-particle excitations – electrons. Under these conditions, the electron gas should be viewed as a system of individual quasi-particles and their electrostatic interaction is characterized by the screened Coulomb potential

$$V_{scr}(\mathbf{r}) = \sum_{k>k_c} \frac{4\pi e^2}{k^2} \exp(ikr) \approx \frac{e^2}{r} \exp(-k_c r). \tag{3.29}$$

Thus, the interaction between the electrons occurs only at small distances $r < k_c^{-1}$. The effective radius of interaction $r_{TF} = k_c^{-1} = v_F / \omega_p$ is the *Thomas-Fermi screening radius*. In this case, we assume that the density of electrons in metals n_0 is large and they fill all the states with energy $E < E_F$, i.e. we consider a degenerate electron gas.

In semiconductors, the electron density n_0 is generally much lower and there is no degeneration, with the exception of heavily doped semiconductors. The distribution of the electrons in the energy states is determined by the Boltzmann distribution. The interaction of electrons in semiconductors is also screened. Thus, the Coulomb interaction $e^2/(\varepsilon_0 r)$ is replaced by a screened interaction:

$$V_{scr}(\mathbf{r}) = \frac{e^2}{\varepsilon_0 r} \exp(-r / r_D), \quad r_D^2 = \frac{\varepsilon k_B T}{4\pi e^2 n_0}, \tag{3.30}$$

where r_D is the Debye screening radius.

The energy of the plasmons is high so they are not excited by heating. The plasmons are excited by fast electrons with energies of the order of kilovolts, passing through thin (about 100 Å) films. With the passage of fast electrons through films of beryllium, magnesium and aluminium, they lose energy $\hbar\omega_p$, $2\hbar\omega_p$, ... according to the number of plasmons which they excited. At the same time, the observed plasma frequency is in good agreement with that calculated with the valence electrons taken into account (two in Be and Mg, and three in Al). In some metals and semiconductors (C, Si, Ge) the electrons excite one plasmon. In carbon, silicon and germanium the plasma frequency is also determined by the valence electrons – four per atom. In the metals Cu, Ag, Au, and many other transition metals with other electrons take part together with the valence electrons in plasma oscillations. Table 3.1 shows the values of the plasmon energy for some solids.

Table 3.1. The values of the plasmon energy for a number of metals and semiconductors

Materials	Be	Mg	Al	C	Cu	Ag	Pb	Ni	W	Si	Ge	ZnS
$\hbar\omega_p$, eV	18.9	10.6	15.3	22.0	17.8	23.1	14.0	22.1	23.5	16.9	16.0	17.0

Plasma oscillations also appear in the interaction of the electromagnetic waves with solids.

We consider an electromagnetic wave incident perpendicularly on a flat metal sheet

$$E_x = E_x^{(0)} \exp[i(kz - \omega t)].$$

If the sheet thickness is small compared with the wavelength, inside the sheet $\exp(ikz) \approx 1$. In the transverse vibrations of the electrons induced by an electromagnetic wave there are no electrostatic forces and, therefore, taking into account the elastic forces, the induced oscillations electron gas are determined by the equation

$$m\left(\ddot{x}+\omega_{ac}^{2}(k)x\right)=-eE_{x}, \quad \omega_{ac}^{2}(k)=\frac{3}{5}v_{F}^{2}k^{2}. \tag{3.31}$$

Its solution is $x=eE_{x}/\left\{m\left[\omega^{2}-\omega_{ac}^{2}(k)\right]\right\}$. Consequently, the specific electrical polarization formed in the crystal under the action of the electromagnetic field, caused by the displacement of electrons, is proportional to the field

$$P_{x}=-en_{0}x=-\frac{e^{2}n_{0}E_{x}}{m\left[\omega^{2}-\omega_{ac}^{2}(k)\right]}. \tag{3.32}$$

On the other hand,

$$P_{x}=\frac{\varepsilon(\omega)-1}{4\pi}E_{x}. \tag{3.33}$$

Comparing (3.32) and (3.33), we obtain

$$\varepsilon(\omega)=1+\frac{\omega_{p}^{2}}{\omega_{ac}^{2}(k)-\omega^{2}}. \tag{3.34}$$

From (3.34) it follows that the frequency of plasma oscillations, determined by the relation $\omega^{2}(k)=\omega_{p}^{2}+\omega_{ac}^{2}(k)$, is the zero of the dielectric permeability and, consequently, the ratio

$$\varepsilon(\omega)=0 \tag{3.35}$$

can serve as a condition specifying the intrinsic frequencies of the bulk plasma oscillations.

Due to the interaction of electrons with phonons, impurities and other crystal defects, the free movement of electrons is possible only for a mean time τ between collisions. Given these factors, equation (3.31) must be replaced by the equation

$$m\left(\ddot{x}+\frac{\dot{x}}{\tau}+\omega_{ac}^{2}(k)x\right)=-eE_{x}. \tag{3.36}$$

In this case, the expression (3.34) takes the form

$$\varepsilon(\omega)=1+\frac{\omega_{p}^{2}}{\omega_{ac}^{2}(k)-\omega(\omega+i/\tau)}. \tag{3.37}$$

The imaginary part of dielectric permittivity characterizes the damping of plasma oscillations. In pure metals at low temperatures the value of τ is of the order of 10^{-9}, and the condition $\omega_{p}\tau\gg1$ of weak attenuation is easily satisfied.

3.2.2. Surface plasmons

We now consider the actual surface plasmons. *Surface plasmons* are called the plasmons localized in the surface layer of a metal or a semiconductor. From the above analysis of bulk plasma oscillations it is clear that the frequency of plasma oscillations is the intrinsic frequency of oscillations of the electron density, which, as shown above for the three-dimensional case, corresponds to the zero dielectric constant of the crystal, which depends on the frequency (3.35). However, in order to find the intrinsic frequency of the surface plasmons it is necessary to study the oscillations of the electron density in the bounded systems. These studies were conducted in a number of works [26, 27, 315, 357, 398], where various approximations have been used for identifying the characteristics of the surface plasmons. In this section, to determine the intrinsic frequency of the surface plasmons, we use an approach based on the dielectric formalism subject to the restrictions imposed by the existence of a plane boundary between, for example, a metal and vacuum.

The oscillations of the electron density are accompanied by oscillations of the electromagnetic field. Therefore, the propagation of surface plasmons is equivalent to the propagation of surface electromagnetic waves localized in the surface layer and defined by the dielectric permittivity of the medium taking into account the influence of the metal–vacuum interface. Consider the conditions for the existence of surface electromagnetic waves. Let the metal occupies the half-space with $z < 0$. The interface is taken as the plane (x, y) and we denote vectors in this plane by $\mathbf{X}$.

The problem consists in solving Maxwell's equations for a two-phase metal–vacuum system. Neglecting retardation effects in the interaction, in Maxwell's equations we can formally set $c \to \infty$, and thus use the electrostatics equations. In both environments, the electrostatic potential φ can be determined from Laplace's equation

$$\nabla^2 \varphi = 0. \tag{3.38}$$

The solution of equation (3.38) will be sought in the form

$$\varphi = \varphi(z) \exp[(i\mathbf{KX})]. \tag{3.39}$$

Substitution of (3.39) into (3.38) leads to the equation for $\varphi(z)$

$$\frac{d^2 \varphi(z)}{dz^2} - K^2 \varphi(z) = 0, \quad K^2 = K_x^2 + K_y^2. \tag{3.40}$$

We are interested in solutions having the character of collective oscillations and localized at the surface (vanishing at $z \to \infty$). As a result of (3.40) we have:

$$\phi(z) = A e^{Kz}, \quad z < 0,$$
$$\phi(z) = B e^{-Kz}, \quad z > 0. \tag{3.41}$$

We 'sew' together the solution of equation (3.41) at the interface $z = 0$ from the conditions of continuity of the tangential component of the strength of the electric field (equivalent to the continuity of $\varphi(z)$) and the normal component of electric induction (equivalent to continuity $\varepsilon(\omega)\dfrac{d\varphi(z)}{dz}$). From the first condition we have $\phi(z) = Ae^{-K|z|}$. From the second condition, given the fact that the permittivity of vacuum is unity, and the metal is characterized by the frequency-dependent permittivity $\varepsilon(\omega)$, we obtain the condition for the existence of collective oscillations of the surface in the form

$$\varepsilon(\omega) + 1 = 0. \tag{3.42}$$

To find from the condition (3.42) the natural frequencies of surface plasma oscillations, we must specify a model form of $\varepsilon(\omega)$. For this we use the expression (3.34). We then obtain the frequency of the surface plasmons in the form

$$\omega_s(K) = \frac{1}{\sqrt{2}}\left[\omega_p^2 + 2\omega_{ac}^2(K)\right]^{1/2} = \frac{1}{\sqrt{2}}\left(\omega_p^2 + \alpha v_F^2 K^2\right)^{1/2}. \tag{3.43}$$

The main feature of the expression (3.43) is that neglecting the dependence on the wave vector leads to an expression for the surface plasmon frequency, which is $\sqrt{2}$ times smaller than the bulk plasmon frequency: $\omega_s(K) \approx \omega_p/\sqrt{2}$. Note that the law of dispersion of the surface plasmon is also different from the bulk plasmon. Thus, in (3.43), the coefficient $\alpha = 89/90$ (this question is discussed in section 7.2). In Chapter 7 we also discuss the connection of surface plasmons with the surface energy and adhesion interaction of metals and semiconductors and describe the method of their calculation.

The theory of dispersion of surface plasmons is quite complex. Construction of the corresponding theory requires the use of the microscopic approach (see [24]). When $K/k_F \ll 1$, it was found that in the frequency of surface plasmons along with the quadratic term K/k_F there is a linear term K/k_F, which begins to determine the dispersion of plasmons in the long-wave approximation

$$\omega_s(K) \approx \frac{\omega_p}{\sqrt{2}}\left(1 + (A_1 + iA_2)\frac{K}{k_F}\right). \tag{3.44}$$

The numerical values of A_1 and A_2, and even the sign of A_1 are strongly dependent on the microscopic model of the surface (the type of potential at the surface and electron density).

Knowing the number of valence electrons per atom, we can find the concentration of free electrons n_0, and then, using the values n_0, e, m, calculate ω_p and estimate the frequency $\omega_s \approx \omega_p/\sqrt{2}$ and the surface plasmon energy. The results of calculations for some simple metals are presented in

Table 3.2. The energy values of the surface plasmon $\hbar\omega_s$, eV, for a number of metals

Metals	Mg	Al	In	Ga
Experiment	7.1	10.6	8.7	10.3
Theory	7.7	11.2	8.9	10.3

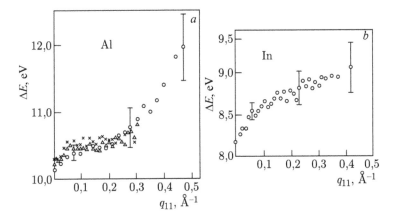

Fig. 3.3. Data obtained by SELE for the plasmon dispersion on the surface of the polycrystalline films: a – aluminium, b – indium [373].

Table. 3.2, where for the same metals there are measurements of the energy of surface plasmons which are in fairly good agreement with the results of theoretical estimates, even without taking into account the dispersion law of surface plasmons.

The spectroscopy of the energy loss of electrons (SELE) allows us to perfectly identify the dispersion of plasmons (see Fig. 3.3) for the approximation of the excitation wavelength to the atomic dimensions. In this experiment [373], high-energy (50 keV) electrons passing through a thin (~100 Å) metallic film lose energy and transferring it to the plasmons.

The component of the wave vector parallel to the surface is varied by adjusting the angle of incidence – the same as in experiments with angle-resolved photoemission.

The surface plasmons and their energy properties can be also found in experiments on optical absorption on the surface specially prepared for this purpose. Thus, for the results of measurements of the dispersion of surface plasmons in InSb [401], shown in Fig. 3.4, a dashed lattice with lattice constant d was cut out on the surface. In this case, we can separate the component of the wave vector **k**, tangent to the surface with values $k_x = (\omega/c)\sin\alpha + 2\pi m/d$, where α is the angle of incidence, and m assumes the integer values. Only such 'tricks' have made it possible to utilize the effects of plasmon excitation by light, as for the flat unperturbed surfaces the plasmons are excited by light as little as surface polaritons.

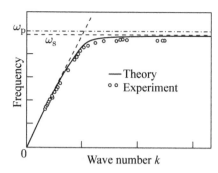

Fig. 3.4. The dispersion curve of surface plasmons in InSb and comparison with the theory [401].

3.3. Surface phonons and polaritons

Consider the elementary excitations of a semi-infinite crystal associated with the collective vibrational motion of atoms and ions relative to the crystal lattice sites, the amplitude of which decays exponentially with the distance from the surface deep into the crystal. In the description of such surface density waves there are three main types:

   a) long-wavelength elastic waves
   b) the long-wavelength optical wave
   c) short-acoustic and optical waves.

The first discussion of the phenomena associated with the peculiarities of propagation of surface elastic waves was carried out by Rayleigh.

3.3.1. Surface phonons

Long-wave elastic (acoustic) surface waves will be considered in the classical theory of elasticity for continuum [129]. We denote by $\mathbf{u}$ the displacement vector of an element of an isotropic semi-infinite continuum from the equilibrium state. In a general case, this vector can be represented as the sum of the transverse (div $\mathbf{u}_t = 0$) and longitudinal (rot $\mathbf{u}_l = 0$) displacements. The components of the vectors $\mathbf{u}_t$ and $\mathbf{u}_l$ satisfy the conventional wave equation taking into account the corresponding sound velocities c_t and c_l:

$$\frac{\partial^2 \mathbf{u}_t}{\partial t^2} - c_t^2 \nabla^2 \mathbf{u}_t = 0,$$

$$\frac{\partial^2 \mathbf{u}_l}{\partial t^2} - c_l^2 \nabla^2 \mathbf{u}_l = 0. \tag{3.45}$$

Spatially, these mutually perpendicular displacements are not related with each other. In the case of surface waves this division into two independent parts is not possible due to the influence of the boundary conditions. Near the surface we seek the solution of equations (3.45) in the form (the regions of the medium correspond to $z < 0$)

$$\mathbf{u}_i = \mathbf{A}_i \exp[i(\mathbf{Q}\mathbf{X} - \omega t)]\exp(\varkappa z), \quad \varkappa = \left(Q^2 - \frac{\omega^2}{c^2}\right)^{1/2}, \quad (3.46)$$

where $\mathbf{X}$ and $\mathbf{Q}$ determine the spatial and wave vectors in the plane of the surface (x, y). Note that these waves are macroscopic and in the limit $\mathbf{Q} \to 0$ the displacement continues to exist inside the crystal at a sufficiently great depth.

For surface waves, the displacement vector $\mathbf{u}$ must be determined by the linear combination of the vectors $\mathbf{u}_l$ and $\mathbf{u}_t$. In order to determine this combination we need to use the surface boundary conditions. In an infinite medium equilibrium stresses correspond to the forces that balance each other on each side of an infinitesimal volume element. On the surface, all components of the forces that intersect boundary plane must vanish, i.e.

$$dF_i = \sigma_{iz} df_z = 0, \quad (3.47)$$

where df_z is the component of the vector $d\mathbf{f}$ of the surface element directed along the outward normal to the surface. This implies the conditions

$$\sigma_{xz} = \sigma_{yz} = \sigma_{zz} = 0. \quad (3.48)$$

We assume that the stress tensor components σ_{ij} are related to components of the strain tensor $\varepsilon_{ij} = (\partial u_i/\partial x_j + \partial u_j/\partial x_i)/2$ by Hooke's law. For an isotropic medium, the elastic constants, which play the role of the coefficient of proportionality in Hooke's law, can be written in terms of Young's modulus E and Poisson's ratio σ. Therefore, from conditions (3.47) we obtain:

$$u_{xz} = u_{yz} = 0, \quad \sigma(u_{xx} + u_{yy}) + (1 - \sigma)u_{zz} = 0. \quad (3.49)$$

Since all quantities are independent of coordinate y, then the condition $u_{yz} = 0$ yields

$$u_{yz} = \frac{1}{2}\left(\frac{\partial u_y}{\partial z} + \frac{\partial u_z}{\partial y}\right) = \frac{1}{2}\frac{\partial u_y}{\partial z} = 0.$$

Consequently, $u_y = 0$. This boundary condition reflects the requirement that the displacement of surface modes occurs only in the sagittal plane, i.e. the plane containing the direction of propagation $\mathbf{Q}$, parallel to the axis x, and the normal to the surface. Since the definition

$$\operatorname{div}\mathbf{u}_t = 0 = \frac{\partial u_{tx}}{\partial x} + \frac{\partial u_{tz}}{\partial z},$$

$$\operatorname{rot}\mathbf{u}_l = 0 = \frac{\partial u_{lx}}{\partial z} - \frac{\partial u_{lz}}{\partial x}, \quad (3.50)$$

then, in view of (3.46) and the fact that $\mathbf{Q} = \mathbf{Q}_x$, we obtain the relations

$$iQu_{tx} + \varkappa_t u_{tz} = 0,$$
$$iQu_{1z} - \varkappa_1 u_{1x} = 0,$$

(3.51)

where the expressions for the components of the displacement vector:

$$u_x = \left(a\varkappa_t e^{\varkappa_t z} + bQe^{\varkappa_1 z}\right)e^{i(Qx - \omega t)},$$
$$u_z = -i\left(aQe^{\varkappa_t z} + b\varkappa_1 e^{\varkappa_1 z}\right)e^{i(Qx - \omega t)},$$

(3.52)

where $\varkappa_1$ and $\varkappa_t$ are determined by the expression for $\varkappa$ in (3.46) with velocities $c = c_1$ and $c = c_t$.

Now we use the first and third of the conditions (3.49). Expressing u_{ij} by the derivatives of u_i and introducing velocities c_1 and c_t, we can write these conditions as

$$\frac{\partial u_x}{\partial z} + \frac{\partial u_z}{\partial x} = 0,$$
$$c_1^2 \frac{\partial u_z}{\partial z} + (c_1^2 - 2c_t^2)\frac{\partial u_x}{\partial x} = 0.$$

(3.53)

Substituting into (3.53) the expressions (3.52) for the components of the displacement vector, we obtain the following equation:

$$a(Q^2 + \varkappa_t^2) + 2bQ\varkappa_1 = 0,$$
$$2aQc_t^2\varkappa_t + b[c_1^2(\varkappa_1^2 - Q^2) + 2c_t^2 Q^2] = 0.$$

(3.54)

The condition of consistency of the these homogeneous equations imply the characteristic equation for determining the dispersion of surface waves

$$(Q^2 + \varkappa_t^2)^2 = 4Q^2 \varkappa_t \varkappa_1.$$

(3.55)

Substituting into (3.55) the expressions for $\varkappa_1$ and $\varkappa_t$, we can define the relationship of ω with Q. From (3.55) it follows that for surface waves, as well as for bulk elastic acoustic waves the frequency is proportional to the wave vector. To determine the velocity of surface acoustic waves we will search for connection between ω and Q in the form

$$\omega = c_t Q \zeta.$$

(3.56)

Substituting (3.56) into equation (3.55), we obtain the equation for ζ

$$\zeta^6 - 8\zeta^4 + 8\zeta^2\left(3 - 2\frac{c_t^2}{c_1^2}\right) - 16\left(1 - 2\frac{c_t^2}{c_1^2}\right) = 0.$$

(3.57)

From (3.57) it can be seen that ζ is a number that depends only on the ratio c_t/c_1, which is a constant characteristic of each substance and depends only on Poisson's ratio: $\dfrac{c_t}{c_1} = \sqrt{\dfrac{1-2\sigma}{2(1-\sigma)}}$. For $\varkappa_1$ and $\varkappa_t$ to be real, the value of ζ should be real and positive, with $\zeta < 1$. Equation (3.57) has only one root satisfying these conditions.

The ratio of the constants a and b, appearing in (3.54), gives the ratio of the amplitudes of the transverse and longitudinal components of the wave and is given by

$$\frac{a}{b} = -\frac{2-\zeta^2}{2\sqrt{1-\zeta^2}}. \tag{3.58}$$

The ratio c_t/c_1 varies for different substances actually in the range from 0 to $\sqrt{2}/2$, which corresponds to a change of σ in the interval $0 \le \sigma \le 1/2$. In this case the numerical values of ζ change from 0.874 to 0.955 (Fig. 3.5). As a result, the surface acoustic wave velocity $c_R = c_t\zeta$ is less than c_t and c_1. Such a 'slow' surface acoustic wave is called a *Rayleigh wave*.

For a more complete examination we should consider that the real substances usually have anisotropic elastic constants. In this case, either we obtained exactly the same monotonically decreasing solution as (3.52), or there is a generalized Rayleigh wave with subcritical damping whose amplitude decreases with the distance from the surface into the interior of the crystal, but is sinusoidally modulated. Figure 3.6 shows a typical phase diagram of Rayleigh waves generated at the (100) surface for a number of known substances [322]. The area enclosed between the dashed lines is the area of elastic stability.

Another class of long-wavelength surface modes exists in ionic compounds, where the oscillations of the ions, corresponding to the optical branches of vibrations (transverse (TO) and longitudinal (LO)), cause the appearance of an oscillating dipole moment in each unit cell of the crystal. In the ionic crystals, along with the elastic forces proportional to the displacement, an important role is played by long-range electrical interactions between the ions. The displacements of ions from their equilibrium positions create electric polarization, causing the appearance of the field, which in turn interacts with the ions. Specific polarization **P**

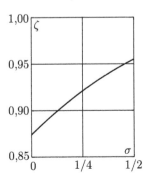

Fig. 3.5. Dependence of ζ on σ [129].

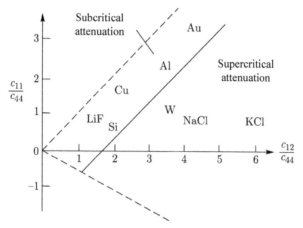

Fig. 3.6. Phase diagram of Rayleigh waves. The solid line separates waves with subcritical and supercritical damping [322].

is due to both the displacement of the ions and the internal polarization of the ions due to the displacement of electrons relative to the nuclei under the influence of an electric field. Therefore, in general terms we can write

$$\mathbf{P} = \gamma_1 \boldsymbol{\xi} + \gamma_2 \mathbf{E}, \tag{3.59}$$

where the parameters γ_1 and γ_2 can be expressed via directly measurable quantities, and in accordance with [77], if we ignore retardation effects of interaction carried by the transverse component of the electric field $\mathbf{E}_t$,

$$\gamma_1 = \omega_{TO} \sqrt{\frac{\varepsilon_0 - \varepsilon_\infty}{4\pi}}, \quad \gamma_2 = \frac{\varepsilon_\infty - 1}{4\pi}, \tag{3.60}$$

where ε_0 and ε_∞ are the static and high-frequency dielectric permeabilities of an ionic crystal, ω_{TO} is the frequency of the bulk transverse optical vibrations.

The frequencies of the bulk long-wavelength longitudinal and transverse intrinsic ion oscillations are related by the well-known Lindeman–Sachs-Teller relationship

$$\omega_{LO} = \omega_{TO} \sqrt{\frac{\varepsilon_0}{\varepsilon_\infty}}. \tag{3.61}$$

In this case the electrical properties of crystals are characterized by the following frequency dependence of the dielectric constant:

$$\varepsilon(\omega) = \varepsilon_\infty + \frac{(\varepsilon_0 - \varepsilon_\infty)\omega_{TO}^2}{\omega_{TO}^2 - \omega^2} = \frac{\varepsilon_\infty (\omega_{LO}^2 - \omega^2)}{\omega_{TO}^2 - \omega^2}. \tag{3.62}$$

From (3.62) it follows that the natural frequencies of the bulk longitudinal fluctuations correspond to the zero dielectric constant of the crystal, and the intrinsic frequencies of the transverse bulk oscillations to the poles of the dielectric constant.

Macroscopic electrostatic fields induced by their intrinsic long-wavelength optical vibrations, taking into account the influence of the crystal surface must, as is the case with surface plasmons, be defined by the electrostatic potential, defined by (3.39)–(3.41). Carrying out the same arguments as in section 3.2, we find that the surface mode of long-wavelength optical vibrations must satisfy the condition $\varepsilon(\omega) + 1 = 0$, where $\varepsilon(\omega)$ is given by (3.62). As a result, we have

$$\omega_{SO} = \omega_{TO}\sqrt{\frac{1+\varepsilon_0}{1+\varepsilon_\infty}}. \qquad (3.63)$$

For direct observation of surface phonons corresponding to such long-wavelength surface optical vibrations, it is ideal to use the energy-loss spectroscopy of electrons (ELSE). The physical phenomenon here is the inelastic Coulomb scattering of electrons on the long-range dipole potential of the crystal. This scattering is characterized by a sharp maximum in the forward direction. To see this, consider an incident electron with wave vector $\mathbf{k}$ and the kinetic energy E. Since we are talking only about the long-wavelength modes, i.e. $\mathbf{Q} \rightarrow 0$, the modulus of the wave vector after scattering is not very different from $|\mathbf{k}|$.

Consequently, if the electron is actually dissipated by a small angle θ, then we have $Q = k\theta$. Since the radius of the potential is of order Q^{-1}, the time during which the electron is situated in the dipole field is equal to $t \sim 2m/(\hbar kQ) = h/(E\theta)$. Furthermore, the mode with energy $\hbar\omega_{SO}$ is excited most strongly by the electron transit time of the order of ω_{SO}^{-1}. Thus, the strongest scattering will occur at an angle $\theta = \hbar\omega_{SO}/(2\pi E)$. The initial assumption of small angle scattering is justified in the case of excitation of phonons ($\hbar\omega_{SO} \sim 0.1$ eV) by low-energy electrons ($E \sim 1 \div 100$ eV). Since the low-energy electrons are diffracted on the crystal surface, the strongest ELSE signal will be observed in the specular reflected beam.

The spectrum of the energy losses of electrons, scattered on the surface of ZnO (1100), contains a significant peak at about 69 meV (Fig. 3.7) [355]. The peak is in excellent agreement with those calculated by the formula (3.63) if we use the bulk dielectric characteristics of zinc oxide. The signal of inelastic scattering is particularly intense in this case, since the contribution to the field is provided by the dipole moments up to the crystal cells to the atomic layer with the number $Q\lambda_{SO}$, measured from the surface, where λ_{SO} is the wavelength of the surface of the optical vibrations. The spectrum also has a small peak at 69 meV, due to the absorption of the pre-existing surface phonon by an electron beam. Using the ratio of the amplitudes of the peaks, corresponding to the increment and loss of energy, we can calculate the surface temperature.

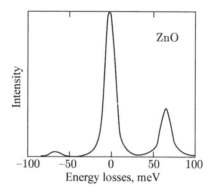

Fig. 3.7. The macroscopic surface optical mode at the surface ZnO (1100), discovered through ELSE [355].

The observation of macroscopic surface optical modes at the central point of the surface Brillouin zone by ELSE is crucially related to the small momentum transfer characteristic of dipole Coulomb scattering. Therefore, this mechanism can not be used to study excitations with a finite wave vector, i.e. the dispersion of surface phonons. But electrons can also be scattered directly on the short-range atomic potential. In this case, the phonons are excited as a direct result of collision, and in the experiment we can achieve the momentum transfer, corresponding to the transition through the whole Brillouin zone. The resolution which is provided by the data obtained in this way is limited mainly to energy smearing of the primary electron beam. Using the best modern experimental technique, by the ELSE method we can determine the dispersion of the phonons with a resolution of about 5 meV.

3.3.2. Bulk and surface polaritons

If in the description of the transverse optical vibrations we take into account the delayed electromagnetic interaction, carried by electromagnetic waves induced by these fluctuations, then in the conditions when the photon energy as quanta of the electromagnetic field is equal to the energy of the transverse optical phonons, the stationary ionic states correspond to the elementary excitations called *polaritons*.

The term 'polariton' was introduced in 1957 by D. Hapfield [348] to denote the normal wave polarization in the crystal. Later V.M. Agranovich [2] used it as the short equivalent of the term 'Normal electromagnetic wave in a medium' (a plane monochromatic electromagnetic wave in an infinite crystal, satisfying the macroscopic Maxwell's equations). This interpretation proved to be successful and has been almost generally accepted.

Most often, the term 'polariton' is used when the frequency of an electromagnetic wave enters the neighbourhood of the frequency of the dipole active transition in the crystal. In this case, the interaction of the electromagnetic field with a specified transition is expressed particularly clearly and leads to a strong 'mixing' of interacting subsystems – the electromagnetic and the 'mechanical' (oscillations of ions). The term 'polariton' is used here to emphasize the mixed electromagnetic–mechanical nature of the normal waves system satisfying the Maxwell macroequations. It is mixing of electromagnetic radiation with oscillations corresponding to a particular dipole active (polar) transition, for example, the polar optical vibrations of the crystal. It should also be kept in mind that in the quantum description, the term 'polariton' is used not for the electromagnetic wave, but for an appropriate quasiparticle – the quantum of the electromagnetic field in a medium ('a photon in a medium').

The macroscopic theory of polaritons in isotropic media can be developed [77], if in the study of transverse vibrations of the ions we take into account the transverse component of the electric field:

$$\ddot{\xi}_t + \omega_{TO}^2 \xi_t = \gamma_1 \mathbf{E}_t,$$

$$\mathbf{P}_t = \gamma_1 \xi_t + \gamma_2 \mathbf{E}_t. \tag{3.64}$$

Equations (3.64) must be supplemented by Maxwell's equations

$$\text{rot}\mathbf{H} = \frac{1}{c}\left(\dot{\mathbf{E}}_t + 4\pi \dot{\mathbf{P}}_t\right), \quad \text{rot}\mathbf{E}_t = -\frac{1}{c}\dot{\mathbf{H}},$$

$$\text{div}\mathbf{H} = 0, \quad \text{div}\left(\mathbf{E}_t + 4\pi\mathbf{P}_t\right) = 0, \tag{3.65}$$

connecting the transverse fields $\mathbf{H}$, $\mathbf{E}_t$ with transverse specific polarizability. The system of equations will be sought in the form of

$$\frac{E_x}{E_{x0}} = \frac{P_x}{P_{x0}} = \frac{\xi_x}{\xi_{x0}} = \frac{H_y}{H_{y0}} = \exp[i(kz - \omega t)], \tag{3.66}$$

and we then obtain a system of equations

$$(\omega_{TO}^2 - \omega^2)\xi_x - \gamma_1 E_x = 0, \quad kE_x = \frac{\omega}{c}H_y,$$

$$P_x = \gamma_1 \xi_x + \gamma_2 E_x, \quad kH_y = \frac{\omega}{c}(E_x + 4\pi P_x). \tag{3.67}$$

The condition of non-trivial solvability of this system is reduced to the equality

$$\frac{c^2 k^2}{\omega^2} = 4\pi\left(\gamma_2 + \frac{\gamma_1^2}{\omega_{TO}^2 - \omega^2}\right). \tag{3.68}$$

Substituting the expressions for the parameters γ_1, γ_2, from (3.60), we obtain

$$\frac{c^2 k^2}{\omega^2} = \frac{\varepsilon_\infty (\omega_{LO}^2 - \omega^2)}{\omega_{TO}^2 - \omega^2}.$$ (3.69)

Equation (3.60) can be used to determine the dispersion of elementary excitations – polaritons. Thus, the equation (3.60) with respect to ω takes the following form:

$$\varepsilon_\infty \omega^4 - \omega^2 \left(\varepsilon_\infty \omega_{LO}^2 + k^2 c^2 \right) + k^2 c^2 \omega_{TO}^2 = 0.$$ (3.70)

Taking into account the equality (3.61), the solutions of (3.70) are written in the form

$$\omega_{1,2}^2 = \frac{1}{2\varepsilon_\infty} \left(\varepsilon_0 \omega_{TO}^2 + c^2 k^2 \pm \sqrt{\left(\varepsilon_0 \omega_{TO}^2 + c^2 k^2 \right)^2 - 4\varepsilon_\infty c^2 k^2 \omega_{TO}^2} \right).$$ (3.71)

For small values of k solutions (3.71) take the following form:

$$\omega_1^2 = \frac{c^2 k^2}{\varepsilon_0}, \quad \omega_2^2 = \frac{\varepsilon_0 \omega_{TO}^2}{\varepsilon_\infty} + \frac{c^2 k^2}{\varepsilon_\infty} = \omega_{LO}^2 + \frac{c^2 k^2}{\varepsilon_\infty}.$$ (3.72)

For large values of k $\left(k \gg \omega_{TO} \sqrt{\varepsilon_0} / c \right)$, but still satisfying the macroscopic description condition $(ka < 1)$, we have:

$$\omega_1^2 = \omega_{TO}^2, \quad \omega_2^2 = \frac{c^2 k^2}{\varepsilon_\infty}.$$ (3.73)

Thus, there are two branches of elementary excitations (polaritons): the first branch with frequencies in the range $0 \le \omega_1(k) < \omega_{TO}$ and the second branch with frequencies in the range $\omega_{LO} \le \omega_2(k) < \infty$ (see Fig. 3.8). For large values of k, the excitations of the first branch coincide with the transverse phonons, and the excitations of the second branch – with photons in a medium with dielectric constant ε_∞. However, in the neighbourhood of the values $k = \omega_{TO} \sqrt{\varepsilon_0} / c$ the polaritons are a very complex 'mixture' of photons and phonons. Polariton excitations are stationary electromagnetic waves inside the crystal. The energy of these waves includes the energy of polarization fluctuations of the crystal.

In the area of transparency of the crystal $(\varepsilon(\omega) > 0)$ the polaritons with longer wavelengths coincide identically with the photons in the crystal, differing from the free photons only by the smaller (by a factor of n, where n is the refractive index) wavelength. The first branch of the polariton describes photons with frequencies below ω_{TO}, and the second – the photons with frequencies exceeding ω_{LO}.

Note that in this case we did not consider the variance of the transverse optical phonons. In addition, it should be borne in mind that equation (3.62) for $\varepsilon(\omega)$ is valid only outside the absorption region $\omega \ne \omega_{TO}$, therefore, the found polariton branches correspond to the electromagnetic waves only in

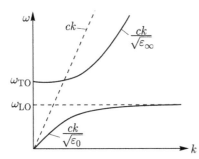

Fig. 3.8. The dispersion curves of the phonon–polaritons [136].

the transparency range of the crystal. Accounting the finite lifetime of the transverse phonons leads to the final lifetime of the polariton states. In addition, the effects of radiation of the electromagnetic field by the crystal leads to an additional reduction of the lifetime of the polaritons.

In the case of a finite crystal there is a possibility of the existence of electromagnetic waves of a different type – propagated along the surface and not removed from it in the perpendicular direction, i.e. a kind of 'tied' to the surface. Such waves are naturally called the *surface waves*, as opposed to *bulk waves*, which exist in the depth of the crystal and are not affected by its surface. The lack of the effect of distribution and, accordingly, spreading, diffraction in the direction normal to the surface of the crystal, automatically entails the attenuation of the field along the normal in the interior of the crystal. This also shows the surface character of the wave. It turns out that surface electromagnetic waves are excited in a certain vicinity of the polar transition frequency and are actually of a mixed electromagnetic-mechanical nature. Therefore, they are usually called *surface polaritons*. Thus, the surface polariton is an electromagnetic surface wave in the crystal, which, as in the bulk case, satisfies the macroscopic Maxwell equations. However, the electromagnetic field near the boundary of the crystal must also satisfy the boundary conditions that connect the components of the vector field on both sides of the border. This implies that the field outside the crystal, i.e. in an environment bordering on it, is also different from zero. The field of the surface polaritons is localized at both sides of the surface of the crystal and decreases with distance from it in any direction. At the same time, the flow of energy to the wave of the surface polaritons is tangential the crystal surface.

We obtain the dispersion relation for the surface polaritons propagating along a plane boundary of an isotropic non-magnetic ionic crystal ($z < 0$) – vacuum ($z > 0$). We describe the crystal in the continuum approximation by the frequency-dependent dielectric constant, given by (3.62). Maxwell's equations to be solved are the following:

$$\operatorname{div}(\varepsilon \mathbf{E}) = 0, \quad \operatorname{div}\mathbf{H} = 0, \quad \operatorname{rot}\mathbf{H} = \frac{1}{c}\varepsilon\dot{\mathbf{E}}, \quad \operatorname{rot}\mathbf{E} = -\frac{1}{c}\dot{\mathbf{H}}. \qquad (3.74)$$

We seek solutions in the form

$$\mathbf{E} = \mathbf{E}(z)\exp[i(k_x x - \omega t)], \quad \mathbf{H} = \mathbf{H}(z)\exp[i(k_x x - \omega t)]. \tag{3.75}$$

Substituting these expressions into equations (3.74), after transformations we get

$$\frac{\partial^2 E_x}{\partial z^2} = \alpha^2 E_x, \quad \frac{\partial E_z}{\partial z} = -ik_x E_x, \tag{3.76}$$

where $\alpha^2 = k_x^2 - \varepsilon(\omega / c)^2$.

From these equations we obtain the solution for the crystal

$$E_x(z) = \exp(\alpha z), \quad E_z(z) = -i\frac{k_x}{\alpha}\exp(\alpha z) \tag{3.77}$$

and for vacuum

$$E_x(z) = A\exp(-\alpha_0 z), \quad E_z(z) = i\frac{k_x}{\alpha_0}A\exp(-\alpha_0 z), \tag{3.78}$$

where $\alpha_0^2 = k_x^2 - (\omega / c)^2$.

The requirement for continuity of E_z and $D_x = \varepsilon(\omega)E_x$ on the surface $z = 0$ leads to the condition

$$\varepsilon(\omega) + \frac{\alpha}{\alpha_0} = 0, \tag{3.79}$$

from which, after substituting the expressions for α and α_0, we obtain

$$\omega = k_x c\sqrt{\frac{\varepsilon(\omega)+1}{\varepsilon(\omega)}}. \tag{3.80}$$

If the values of k_x are large, then from (3.80) it implies that $\varepsilon(\omega) \to -1$ and therefore, in this limit the relation (3.63) is satisfied.

Condition (3.79) of the existence of surface polaritons gives the negative value of the dielectric constant for frequencies of surface polaritons. According to (3.62), this can satisfied only when $\omega_{TO} < \omega < \omega_{LO}$, i.e. infrequency range in which there are no bulk polaritons.

Note that surface polaritons can have multiple branches. This is due to the fact that, in general, the elementary cells of the crystals contain several ions, so that the transverse optical branches, the phonons of which interact with the photons emitted with optical vibrations, can also be a few. An example of system in which this phenomenon takes place is an yttrium–iron garnet (see Fig. 3.9) [239].

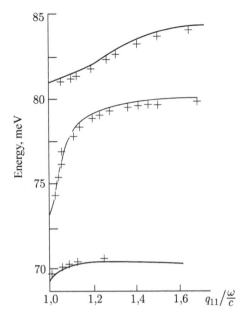

Fig. 3.9. Phonon dispersion of surface polaritons in Y_3FeSO_2, measured by the method of disrupted total internal reflection (DTIR) [239]

3.4. Surface magnons

The elementary excitations of the magnetization of a ferromagnet or antiferromagnet are known as spin waves or magnons. Consider a system in which each lattice node corresponds to a a definite spin moment S_n. Between neighbouring spins there is a short-range quantum-mechanical exchange interaction given by the Heisenberg Hamiltonian

$$\widehat{H} = -\sum_{n,n'} J_{nn'} \mathbf{S}_n \mathbf{S}_{n'}. \tag{3.81}$$

A necessary condition for the ferromagnetic ordering is the positivity of the integral of the exchange interaction between the neighbouring spins of the atoms, i.e. $J \equiv J_{n,n \pm 1} > 0$, and for the antiferromagnetic ordering, respectively, $J < 0$.

Long-wavelength spin waves in ferromagnets can be described as a coherent precession of the transverse component of the magnetization of the crystal around the direction of an external constant magnetic field. Quantization of spin waves leads to the introduction of the concept of elementary excitations – *the magnon*.

Consider, following [224], the simplest one-dimensional lattice of spins. Then the equation of motion of the spin $\mathbf{S}_n$ in the n-th node has the form

$$\frac{d\mathbf{S}_n}{dt} = J\left(\mathbf{S}_n \times \mathbf{S}_{n+1} - \mathbf{S}_{n-1} \times \mathbf{S}_n\right). \tag{3.82}$$

In the semiclassical approximation the spins S_n can be considered as normal vectors. Equation (3.82) is non-linear, but it can be linearized by assuming that the component S_{nx} is constant: $S_{nx} = S_x$. Then, in the linearized approximation, the equation (3.82) becomes the following system of equations:

$$\frac{dS_{ny}}{dt} = JS_x\left(2S_{nz} - S_{n+1,z} - S_{n-1,z}\right),$$

$$\frac{dS_{nz}}{dt} = JS_x\left(2S_{ny} - S_{n+1,y} - S_{n-1,y}\right). \tag{3.83}$$

Since the spin waves represent the spin rotation about the axis z, then the solution of equations (3.83) will be sought in the form

$$S_{ny} = A\sin(\omega_m t)\exp(inak),$$

$$S_{nz} = A\cos(\omega_m t)\exp(inak), \tag{3.84}$$

where a is the lattice constant, k is the wave vector.

Substituting (3.84) into (3.83), we obtain for the spin-wave frequencies the following expression:

$$\omega_m = 2JS_x\left[1 - \cos(ak)\right]. \tag{3.85}$$

In the long-wave approximation $(ka \ll 1)$ we have $\varepsilon(k) = \hbar\omega_m \simeq JS_x a^2 k^2$, i.e. the energy of spin waves (magnons) in ferromagnets is proportional to the square of the wave vector. This branch of the spin waves is called *acoustic*. In magnetic alloys when the composition of the unit cell can contain several magnetic ions, the spin waves, together with the acoustic branch can be characterized by optical branches of vibrations, the frequency of which, as in the case of optical phonon branches, tends to some constant with the wave vector of excitations tending to zero.

In the three-dimensional spin lattice the dispersion law of acoustic spin waves in ferromagnetics in the approximation of the interaction of the nearest spins is characterized by the expression

$$\omega_m(k) = JSv\left(1 - \frac{2}{v}\left[\cos(ak_x) + \cos(ak_y) + \cos(ak_z)\right]\right), \tag{3.86}$$

where v is the lattice coordination number equal to the number of the nearest neighbours of an atom in the crystal.

In a two-sublattice antiferromagnet, the dispersion law of acoustic spin waves is characterized by a more complex dependence

$$\omega_m(k) = 2JSv\sqrt{1 - \gamma^2(k)}, \quad \gamma(k) = \frac{1}{v}\sum_a e^{i(\mathbf{ka})}. \tag{3.87}$$

For a simple cubic lattice with $v = 6$ the expression for $\gamma(k)$ takes the form

$$\gamma(k) = \left[\cos(ak_x) + \cos(ak_y) + \cos(ak_z) \right]/3 \ .$$

In the long-wave approximation $(ka \ll 1)$ we obtain $\left[1 - \gamma^2(k) \right]^{1/2} \simeq ka/\sqrt{3}$ and, therefore, $\omega_m(k) \simeq 4\sqrt{3}JSka$, i.e. the acoustic magnons in the antiferromagnets are characterized by a linear dependence on the wave vector, in contrast to the quadratic dependence in ferromagnets.

Figure 3.10a shows schematically a spin wave in a one-dimensional case. Figure 3.10b shows the surface spin wave whose amplitude decreases with distance from the surface. Presented is a case of the acoustic spin wave in the long-wave approximation, when the orientation of the spins in lattice sites distant from each other varies only slightly. In the optical spin wave the projections of the precessing spins in neighbouring sites are oriented oppositely.

In [512] the authors presented a detailed description of the surface spin waves in polyatomic isotropic ferromagnets. Dispersion relations were obtained for acoustic $\omega_s^{(ac)}(k)$ and optical $\omega_s^{(op)}(k)$ branches of the surface spin waves, which are characterized by the following expressions:

$$\begin{aligned}
\omega_s^{(ac)}(\mathbf{K}) &= \omega_m(\mathbf{K}) - 2JS(ch\beta n - 1), \\
\omega_s^{(op)}(\mathbf{K}) &= \omega_m(\mathbf{K}) + 2JS(ch\beta n + 1),
\end{aligned} \tag{3.88}$$

where $\mathbf{K}$ is the wave vector tangential to the surface, $\omega_m(\mathbf{K})$ is the dispersion relation for bulk magnons propagating along the surface, which has the form

$$\omega_m(\mathbf{K}) = 2JS\left[2 - \cos(aK_x) - \cos(aK_y) \right]. \tag{3.89}$$

The parameter $\beta > 0$ in (3.88) characterizes the damping of the amplitude of the surface magnons at a distance from the surface described by the law $\exp(-\beta n)$, where n is the number of the site, measured from the normal to the surface.

In accordance with (3.88), the energy of acoustic magnons is lower than that of the corresponding bulk magnons, and the energy of optical

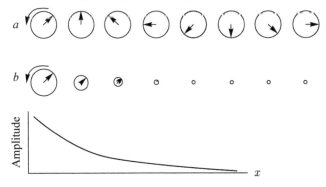

Fig. 3.10. Representation of a spin wave as a precession about the axis (a) and the surface spin wave (b) [203].

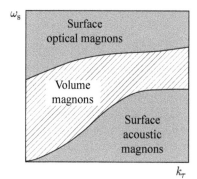

Fig. 3.11. Qualitative diagram of the dispersion of the magnon bands [203].

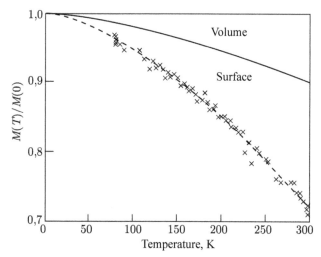

Fig. 3.12. Comparison of surface and bulk magnetizations in $Ni_{40}Fe_{40}B_{20}$ [438].

magnons is higher. Schematically this is shown in Fig. 3.11, which shows the qualitative dispersion diagram of magnons. The finer structure of the dispersion diagram is analyzed in [512], where the characteristics of spin waves in magnetic materials of different nature are described. Typical values of the wave vector for spin waves usually exceed 10^6 cm^{-1}.

However, the spin waves with significantly smaller wave vectors can also be excited. In [512] it is noted that microelectronic devices often use spin waves with a wave vector of 10^3 cm^{-1}. A typical frequency for the spin waves is $10^9 \div 10^{10}$ Hz, which indicates a relatively low energy of the magnons: they are easily excited even at low temperatures.

There are various methods for the detection of surface spin waves. They appear, for example, at Brillouin light scattering on the surface of the magnetic material [95]: in this case, the frequency of the waves scattered by the surface changes by $\pm\omega_s$ due to absorption or excitation of a magnon with energy $\hbar\omega_s$ by incident light. Another method is to measure the surface magnetization of the material. Since the excitation of the magnons leads to a decrease in the magnetization, the change in magnetization can be related to the number of excited magnons. Since the acoustic surface magnons have a lower energy than the bulk magnons, the rate of decrease of surface magnetization with increasing temperature will be greater than for bulk magnetization. This difference in the rate of decrease of the surface and bulk magnetization with temperature is confirmed by experimental studies (Fig. 3.12) [438].

4

SURFACE PROPERTIES OF SOLIDS AND METHODS FOR DESCRIBING THEM

Introduction

Over the past 50 years there has been significant progress in the study and understanding of the bulk electronic properties of solids. With respect to the surface properties of solids such tendency started only recently. The absence of three-dimensional symmetry complicates any calculations, and the formulation and conduct of experiments with clean reproducible surfaces have been made possible only recently. The development of ultra-high vacuum technology led to the preparation of samples with single-crystal surfaces, close to ideal, and precision quantitative methods have been developed for the analysis of thin surface properties. At the same time, the contribution of theoretical physics to the development of the theory of the surface properties of solids, based on which their electronic properties should be calculated, is still insignificant.

However, many practically important phenomena are determined by processes occurring at the interfaces between solids and determined by the very existence of the surface, the processes of its creation, changes of the surface size, etc. The effect of the energy state of the surface is important for the processes of ion-plasma deposition of coatings, in processing of metals with the removal of chips, in powder metallurgy and sintering processes during wetting, in the production of composite materials, in the formation and closing of cracks, in heterogeneous catalysis, corrosion processes, formation of radiation defects and of pores with a very large surface area, etc. Talking about the role of the surface, it should be noted that a large number of effects that are used in modern semiconductor microelectronics is based on the phenomena occurring in the surface layer and the interfacial area of the contacting solids.

Information on such surface characteristics as the distribution of electron density near the surface, the surface energy per unit area of material, the electron work function on the metal surface, interfacial energy and the intensity of interaction of solids along the boundary of the section, makes it possible to understand and explain the essence of the revealed surface phenomena and predict new properties.

However, even when considering the relatively simple (compared with the surfaces involved in the above processes) surface models, the researcher is faced with fundamental challenges that are absent or neglected in the study of the bulk properties of solids:

– Translational symmetry of the crystal is broken in a direction perpendicular to the surface;

– The electron gas in the surface layer becomes strongly non-uniform, changing its density at the wavelengths of the order of the lattice constant from the value characteristic of the volume of material to nearly zero;

– A process of surface lattice relaxation, i.e. displacement of the surface ionic planes characteristic of the bulk material, from their equilibrium position takes place.

This may be accompanied either by changes of the lattice parameter, or by the appearance of other structures on the surface. The change of the interplanar spacing in the surface layer of the metal are indicated by the experimental data [360], obtained by low-energy electron diffraction.

In response to these challenges, detailed research in the field of the theory of the electronic structure of solid surfaces was initiated relatively recently – in the late 1960s. Such studies are usually associated with a large volume of numerical calculations, since we cannot confine ourselves to relatively simple model systems. Currently, the theoretical description of surfaces of solids is carried out using different approaches and concepts (the formalism of the density functional [101, 181, 220, 299, 361], the pseudopotential method [101, 229, 240], the Green function method [479, 517], LCAO [237] etc.), previously developed to describe the bulk properties of solid bodies and the properties of atomic nuclei. However, to date, the changes of the properties of solids associated with the existence of surfaces, have been adequately described only for some model systems and simple metals [181, 451].

4.1. Main stages and directions of development of the theory of metal surfaces

In the early stages of development of the physics of the surface the theoretical description of the surface properties of solids was usually carried out within a phenomenological thermodynamic approach. Only in recent years, attention was given to the development of microscopic quantum statistical methods and of the physics and chemistry of surface phenomena in metals.

The thermodynamics of surface phenomena, whose foundations were laid in the work of J.W. Gibbs, has been further developed in numerous works of foreign [291] and Russian [206] researchers. The basic idea used in varying degrees in these studies is that the surface area is presented as a separate phase of finite or zero thickness, characterized by its thermodynamic parameters.

Another part of the studies, which appeared somewhat later, was based on the consideration of the surface layer as the region between two phases in which the density or some other value is characterized by a non-zero gradient, i.e. changes continuously throughout the surface layer, becoming permanent within the phases. In these studies, the electronic density was considered as the main variable determining the surface properties.

Thus, Ya.I. Frenkel' [312] calculated the surface tension of solids, based on the idea that electrons revolving around the nuclei situated on the surface, create a spatial charge distribution and thus the electric double layer. As shown later in [228], Frenkel' theory is applicable only to insulators, because in metals the main role in the surface phenomena is played by free (non-localized) conduction electrons.

The notion that the conduction electrons form an electron cloud close to the metal surface and in conjunction with the ionic charge create an electric double layer, proved fruitful. A number of modern electronic theories of surface phenomena in metals has been developed on this basis.

Further development of the theory associated with the Thomas–Fermi equation for the electron density is reflected in the works of L. Glauberman [53] and I.M. Spitkovskii [210]. They first investigated the dependence of surface energy on the orientation of the external faces. A.G. Samoilovich [204] in the calculation of the surface tension used a thermodynamic formula which follows from Fuchs' theory. The pressure of the electron gas in the surface area, included in this formula, was calculated in [204], taking into account the kinetic, exchange, electrostatic energies of the electron gas and the Weizsacker correction. The ionic component of surface tension was virtually ignored. A.H. Breger and A.A. Zhukhovitskii [23] examined the surface properties using a model of a two-dimensional electron gas and the infinite potential barrier at the boundary of the metal with the environment.

It should be noted that these initial statistical theories of surface tension of metals took into into account either only the excess energy of the electrons at the boundary with the vacuum, ignoring the contribution of the ionic component, or, conversely, determined the surface tension by the excess energy of surface ions, neglecting the contribution of the electrons. Reassessing the role of one component and ignorance of the other component led to the fact that these values of surface tension were very weakly consistent with experiment.

In early theoretical studies main attention was paid to the electron-ion interaction. In this case the volume potential of the lattice acted up to the surface, where it is was then connected directly with the constant potential characterizing the vacuum [85]. Many papers focus on the electronic states localized at the surface. In semiclassical consideration these states were compared with dangling bonds. Unfortunately, the results of all such calculations were greatly modified as a result of any potential deviation near the surface of a solid from its bulk behaviour.

Another common approach to the study of the electronic properties of solid surfaces takes int account mainly the electron–electron interaction [380]. The lattice is described by a homogeneous distribution of the positive charge, and leakage of electrons in the vacuum from the potential well inside the metal is calculated. This leakage, however, affects the shape and height of the potential barrier so that any subsequent calculations can be self-consistent. The calculation results show that the assumption of the conservation of the volume behaviour of the potential up to the surface does not match reality. Subsequently, these two approaches were actually combined. Now self-consistent calculations that take into account the electron–electron and electron–lattice interactions are carried out.

A significant step forward in the development of the microscopic electron theory of surface phenomena in metals has been made in the works of S.N. Zadumkin [88–90]. When calculating the surface characteristics metals, he took into account more extensively the energy of the electron and ion subsystems. This was accomplished by applying the method of the Wigner-Seitz cell [509] in his semi-empirical theory. The author concluded that a significant contribution to the surface tension is introduced by three components – the intrinsic energy of the electrons (Coulomb, kinetic, and correlation), the difference of the electrostatic energy of their interaction with ions in the surface layer from the energy of this interaction, which is calculated in the framework of the 'jelly' model, and the interaction energy of the Wigner–Seitz cell. It should be noted that the electron distribution near the surface was found from the solution of the Thomas–Fermi equation. However, the approach in which all interactions except the electrostatic were ignored, is not justified. As a result, the law of decreasing electron density near the surface, obtained in [88], turned out to be a power law in contrast to the exponential law which results from the more accurate theory based on the Hartree–Fock method [208].

A fundamentally new approach to study the surface properties of metals was proposed by N.D. Lang and W. Kohn [378, 380, 381]. Using the formalism of the theory of an inhomogeneous electron gas, developed in [345, 367], they solved the Schrödinger equation for the wave function of an electron near the surface in some effective potential. This potential was presented as a functional which depends on the electron density. Further, through iterations, the solution was made self-consistent. The numerical solution of the Schrödinger equation by Lang and Kohn yielded the values of the electron density as a function of the distance to the surface in the form of the curve, decaying outside the metal in accordance with the law close to exponential. Inside the metal near the surface the electron density undergoes Friedel oscillations, damped in depth. In addition, in [381] using the Ashcroft pseudopotential calculations were carried out of the surface energy of metals taking into account the corrections, associated with the discrete distribution of the ions. The method of Lang and Kohn was used to obtain reasonable values of the surface energy for some metals. The surface energy included three components. The first was the surface energy of the 'jelly' model, the second – the electron–ion interaction, calculated using the Ashcroft pseudopotential, the third – the ion–ion interaction, which was located by calculating the lattice sums [381]. In the first component the gradient energy is ignored, and the contribution from the second component does not depend on the orientation of the faces.

Further development of the theory of surface energy, which is generalization of the quantum-mechanical method of Lang and Kohn, was reflected in [409, 433, 434]. These authors showed that the disturbance caused by taking the crystallinity of the ion lattice, is small not for all metals, indicating that incorrect approximations were made in [381]. Based on the method in which the dependence on the variational parameter is included in the effective potential, they obtained the best agreement with the experimental values of surface energy and other surface characteristics for a number of metals. It was shown that the electron density distribution is significantly different from those obtained in the 'jelly' model, its dependence on the orientation of the crystal faces was revealed, and the influence of non-local effects of the exchange–correlation energy of the electron gas was also taken into account.

The disadvantages of this method are considerable mathematical difficulties associated with numerical methods for solving the Schrödinger equation. In this sense, the semi-classical theory of surface energy and work function, developed by J.R. Smith [462] proved to be more convenient for applications. He used the variational principle of minimizing the functional of the total energy of the electron gas (MFT). Test functions were represented by exponential functions proposed initially by R. Smolukhovskii [468]. However, in [462] no account was made of the discrete distribution of the ions of the metal lattice that led to a negative value of the surface energy of metals with high electron density.

The discreteness of the ionic core in the calculation of the surface energy of metals using the method proposed in [462], is taken into account in [110, 336, 367, 425, 426]. In these studies, an analytical expression was derived for the pseudopotential contribution to the surface energy associated with the influence of the crystallinity of the ionic lattice, and gradient corrections to the exchange–correlation energy were taken into account. In [336, 435, 488] the authors examined the effect of surface relaxation of ion planes on the surface energy and electrostatic potential. In [110, 367] it was concluded that the direct variational method [462] can be used to study the surface characteristics of liquid metals.

The dependence of the test functions of the electron distribution and the electrostatic potential, as well as of the parameter of their decay with distance from the strength of the external electric field is reflected in [185, 186, 223]. In [218, 223] the surface energy in a strong electric field was calculated. In [218] the additional 'field' contribution to the calculation of surface energy was taken into account, in contrast to [223], where the dependence on the field was taken into account only the variational parameter. In [183–185] it is noted that due to the smooth decay of the electron density the electric field near the surface itself should be determined because it depends both on the strength of the external field, and on the perturbation of the surface charge caused by the field.

The analysis of these studies leads to the conclusion that the quantum statistical method of the density functional is most suitable for the calculation of the surface properties of metals. It is distinguished by the possibility of obtaining analytical expressions for the surface characteristics, which facilitates their study.

4.2. The original equations of the density functional method for the study of the surface properties of metals

The density functional is one of the most popular concepts in many fields of physics. Atomic and nuclear physics, physics of crystals and liquids, the physics of surfaces and adsorption phenomena – this is not a complete list of those areas of modern physics in which is the density functional method is widely used.

The density functional method (DFM) thus introduces the particle density, defined in the usual three-dimensional space (not the many-particle wave function) as a basis for calculating the ground state energy and thermodynamic properties of the system. It is clear that dealing with the function of three variables is immeasurably easier than dealing with the total wave function of the system. The difficulty lies in the fact that the exact form of the density functional theory, which determines the ground-state energy of the system, is unknown and its construction is beyond the scope of the functional method density.

4.2.1. Density functional theory in the Hohenberg–Kohn formulation

The starting point of the DFM is the theorem (proven by P. Hohenberg and W. Kohn [345]) on the existence of a universal energy functional of the system of electrons in the field of the external potential $v(\mathbf{r})$. The density $n(\mathbf{r})$ of the ground state of the associated system of interacting electrons in an external potential $v(\mathbf{r})$ uniquely determines this potential.

Comments.

1. The term 'unique' here means 'up to the accuracy of the additive contants that is of no interest'.

2. In the case of a degenerate ground state the theorem is true for the density $n(\mathbf{r})$ of any of the ground state.

3. This statement is mathematically rigorous.

The proof is very simple. We give it for a non-degenerate ground state.

Let $n(\mathbf{r})$ be the density of the non-degenerate ground state of a system of N electrons in the potential $v_1(\mathbf{r})$, corresponding to the ground state Ψ_1 and energy E_1. Then

$$E_1 = (\Psi_1, H_1\Psi_1) = \int v_1(\mathbf{r})n(\mathbf{r})d\mathbf{r} + (\Psi_1, (T+U)\Psi_1), \qquad (4.1)$$

where H_1 is the full Hamiltonian corresponding to v_1, a T and U are the operators of kinetic energy and interaction energy. Now assume that there is a second potential $v_2(\mathbf{r})$, *not equal* to $v_1(\mathbf{r})$ + const, to which the ground state $\Psi_2 \neq \exp(i\theta)\Psi_1$ corresponds, and leads to the same density $n(\mathbf{r})$. Then

$$E_2 = \int v_2(\mathbf{r})n(\mathbf{r})d\mathbf{r} + (\Psi_2, (T+U)\Psi_2). \qquad (4.2)$$

Since the state Ψ_1 is assumed to be non-degenerate the Rayleigh–Ritz minimum principle for Ψ_1 leads to an inequality

$$E_1 < (\Psi_2, H_1\Psi_2) = \int v_1(\mathbf{r})n(\mathbf{r})d\mathbf{r} + (\Psi_2, (T+U)\Psi_2) =$$
$$= E_2 + \int [v_1(r) - v_2(\mathbf{r})]n(\mathbf{r})dr. \qquad (4.3)$$

Similarly,

$$E_2 \leq (\Psi_1, H\Psi_1) = E_1 + \int [v_2(\mathbf{r}) - v_1(\mathbf{r})]n(\mathbf{r})dr, \qquad (4.4)$$

where the sign $\leq$ is used because non-degeneracy of Ψ_2 is not assumed. Addition of the inequalities leads to a contradiction: $E_1 + E_2 < E_1 + E_2$. Consequently, according to the proof by contradiction, the assumption of existence of the second potential $v_2(\mathbf{r})$, which is not is equal to $v_1(\mathbf{r})$ + const, but gives the same density $n(\mathbf{r})$, can not be correct. Thus, for the theorem is proved for the non-degenerate ground state.

Since the density $n(\mathbf{r})$ determines both the number of particles N and the potential $v(\mathbf{r})$ (up to an unimportant additive constant), it gives a full Hamiltonian H and the particle number operator N for the electronic system. Consequently, $n(\mathbf{r})$ implicitly defines *all* the properties obtained from H by solving the time-independent or dependent Schrödinger equation (even in the presence of additional disturbances such as electromagnetic fields), such as many-body eigenstates $\psi^{(0)}(\mathbf{r}_1...\mathbf{r}_N)$, $\psi^{(1)}(\mathbf{r}_1...\mathbf{r}_N)$, the two-particle Green's function $G(\mathbf{r}_1, t_1; \mathbf{r}_2, t_2)$, the frequency-dependent electric polarizability $\alpha(\omega)$, etc. It should be recalled again that all this information is implicitly contained in $n(\mathbf{r})$ – the electronic density of the ground state.

Finally, we come to the next question: can any positive function $n(\mathbf{r})$ with the regular behaviour the integral of which is equal to the positive integer N, represent the ground state density corresponding to some potential $v(\mathbf{r})$? Such a density is called v-representable (VR). On the one hand, by expanding in powers of λ, it is easy to see that almost every real homogeneous density $n(\mathbf{r}) = n_0 + \lambda \sum n(\mathbf{q}) \exp\left[i(\mathbf{qr})\right]$ is v-representable or that for some particle any normalized density $n(\mathbf{r}) = |\psi(\mathbf{r})|^2$ is v-representable. On the other hand, M. Levi [388] pointed on an example of a hypothetical system with a degenerate ground state that there may exist densities with the regular behaviour that are not v-representable. The study of the topology of regions of v-representability in an abstract space of all $n(\mathbf{r})$ is being continued. But so far it seems to us that the fact that this problem has not been solved is not an obstacle for practical applications of DFM.

4.2.2. The Hohenberg-Kohn variational principle

The most important characteristic of the electronic ground state is its energy E. Using the methods of wave functions, E can be found directly by the approximate solution of the Schrödinger equation $H\psi = E\psi$, or the Rayleigh–Ritz minimum principle

$$E = \min_{\widetilde{\psi}} (\widetilde{\Psi}, H\widetilde{\Psi}), \tag{4.5}$$

where $\widetilde{\Psi}$ is the normalized trial function for a given number of electrons N. The formulation of the minimum principle not through trial function $\widetilde{\Psi}$ but through trial densities $\tilde{n}(\mathbf{r})$ was first given in [345].

Here we present a more concise conclusion by Levy [388], called the *conventional search method*. Any trial function $\widetilde{\Psi}$ corresponding to test density $\tilde{n}(\mathbf{r})$, obtained by integrating $\widetilde{\Psi}^*\widetilde{\Psi}$ with respect to all variables except the first, and by multiplying the result by N. (4.5) can be minimized in two stages. First, we define trial density $\tilde{n}(\mathbf{r})$ and denote by $\widetilde{\Psi}_{\tilde{n}}^{\alpha}$ the class of trial functions for a given density $\tilde{n}$. For a fixed $\tilde{n}(\mathbf{r})$ we define the conditional minimum energy as

$$E_v[\tilde{n}(\mathbf{r})] \equiv \min_{\alpha}(\widetilde{\Psi}_{\tilde{n}}^{\alpha}, H\widetilde{\Psi}_{\tilde{n}}^{\alpha}) = \int v(\mathbf{r})\tilde{n}(\mathbf{r})d\mathbf{r} + F[\tilde{n}(\mathbf{r})], \tag{4.6}$$

where

$$F[\tilde{n}(\mathbf{r})] \equiv \min_{\alpha}(\widetilde{\Psi}_{\tilde{n}(\mathbf{r})}^{\alpha}, (T+U)\widetilde{\Psi}_{\tilde{n}(\mathbf{r})}^{\alpha}). \tag{4.7}$$

The functional $F[\tilde{n}(\mathbf{r})]$ does not require precise knowledge of the potential $v(\mathbf{r})$. It is a universal density functional $\tilde{n}(\mathbf{r})$ (does not matter whether it is v-representable or not). In the second step we minimize the (4.6) over all $\tilde{n}$:

$$E = \min_{\tilde{n}(\mathbf{r})} E_v[\tilde{n}(\mathbf{r})] = \min_{\tilde{n}(\mathbf{r})}\left(\int v(\mathbf{r})\tilde{n}(\mathbf{r})d\mathbf{r} + F[\tilde{n}(\mathbf{r})]\right). \tag{4.8}$$

In the case of the non-degenerate ground state the minimum is reached when $\tilde{n}(\mathbf{r})$ is the density of the ground state, and for the degenerate ground state – when $\tilde{n}(\mathbf{r})$ is a (any) density of the ground state. Thus, the formidable problem of finding minimum $(\tilde{\psi}, H\tilde{\psi})$ for the $3N$-dimensional trial functions

$\tilde{\Psi}$ is transformed into a *seemingly* trivial (as regards the form) problem of finding the minimum $E_v[\tilde{n}(\mathbf{r})]$ for three-dimensional trial functions $\tilde{n}(\mathbf{r})$. In fact, the definition of (4.7) for $F[\tilde{n}(\mathbf{r})]$ leads us back to minimization for $3N$-dimensional trial wave functions. Nevertheless, considerable formal progress has been made: *rigorous formulations* completely in the language of the density distribution $\tilde{n}(\mathbf{r})$, for the problem of finding the densities and energies of the ground state, as well as a well-defined, though not known in an explicit form, density functional $F[\tilde{n}(\mathbf{r})]$, which is the sum of $T + U$ of the kinetic energy and the interaction energy and is related to the density $\tilde{n}$ (see formula (4.7)).

4.2.3. Self-consistent Kohn–Sham equations

In [330] D.R. Hartree introduced a system of self-consistent one-particle equations for an approximate description of the electronic structure of the atoms. It is believed that each electron moves in some effective one-particle potential

$$v_H(\mathbf{r}) = -\frac{Z}{r} + \int \frac{n(\mathbf{r}')}{|\mathbf{r} - \mathbf{r}'|} dV', \qquad (4.9)$$

where the first term represents the potential due to the nucleus with atomic number Z, and the second – the potential associated with the averaged density distribution $n(\mathbf{r})$ (taking the negative charge of the electron into account). Thus, the motion of each electron is governed by the one-particle Schrödinger equation

$$\left(-\frac{1}{2}\nabla^2 + v_H(\mathbf{r}) \right) \varphi_j(\mathbf{r}) = \varepsilon_j \varphi_j(\mathbf{r}), \qquad (4.10)$$

where j denotes both both spatial and spin quantum numbers. The average density is written as

$$n(\mathbf{r}) = \sum_{j=1}^{N} |\varphi_j(\mathbf{r})|^2. \qquad (4.11)$$

Here (in the ground state), the sum is over the N lowest eigenvalues to satisfy the Pauli exclusion principle. Equations (4.9)–(4.11) are called *self-consistent Hartree equations*. Their solution begins with choosing a first approximation for $n(\mathbf{r})$ (for example, according to the Thomas–Fermi theory), then $v_H(\mathbf{r})$ is plotted, equation (4.10) is solved for $\varphi_j(\mathbf{r})$, and equation (4.11) is used to calculate new density $n(\mathbf{r})$, which should be equal to the initial $n(\mathbf{r})$. If they are not equal, then the whole procedure is repeated again, and so long until they are equal.

The Hartree differential equation (4.10) *looks like* the Schrödinger equation for *non-interacting* electrons moving in an external potential v_{eff}. Is it possible to find something useful in the formulation of DFM for the non-interacting electrons moving in a given external potential $v(\mathbf{r})$? For such a system the Hohenberg–Kohn variational principle takes the form

$$E_{v(\mathbf{r})}[\tilde{n}] \equiv \int v(\mathbf{r})\tilde{n}(\mathbf{r})d\mathbf{r} + T_s[\tilde{n}(\mathbf{r})] \geq E, \qquad (4.12)$$

where (assuming that $\tilde{n}(\mathbf{r})$ for non-interacting electrons is v-representable) $T_s[\tilde{n}(r)]$ *is identically equal to the kinetic energy of the ground state of non-interacting electrons with the distribution of density* $\tilde{n}(\mathbf{r})$. We write the Euler–Lagrange equation, considering that the expression (4.12) is stationary with respect to such variations in $\tilde{n}(\mathbf{r})$ at which the total number of electrons remains unchanged:

$$\delta E_{v(\mathbf{r})}[\tilde{n}] \equiv \int \delta\tilde{n}(\mathbf{r})\left(v(\mathbf{r}) + \frac{\delta}{\delta\tilde{n}(\mathbf{r})}T_s[\tilde{n}(\mathbf{r})]\big|_{\tilde{n}=n} - \varepsilon \right)d\mathbf{r} = 0, \qquad (4.13)$$

where $\tilde{n}(\mathbf{r})$ is the exact ground state density for $v(\mathbf{r})$. Here ε is the Lagrange multiplier that ensures the preservation of the particles. In this case solved without interaction, we know that we can find the energy and density of the ground state by evaluating their eigenfunctions $\varphi_j(\mathbf{r})$ and eigenvalues ε_j of the following one-particle equations without interaction:

$$\left(-\frac{1}{2}\nabla^2 + v(\mathbf{r}) - \varepsilon_j \right)\varphi_j(\mathbf{r}) = 0. \qquad (4.14)$$

As a result, we obtain:

$$E = \sum_{j=1}^{N}\varepsilon_j; \quad n(\mathbf{r}) = \sum_{j=1}^{N}|\varphi_j(\mathbf{r})|^2. \qquad (4.15)$$

Here, j labels both the spatial quantum numbers and spin indices ± 1.

Returning to the problem of *interacting* electrons, which previously discussed by us with the approximate equations of the single-particle Hartree equations we now write specifically the functional $F[\tilde{n}(\mathbf{r})]$ (4.7) in the form

$$F[\tilde{n}(\mathbf{r})] \equiv T_s[\tilde{n}(\mathbf{r})] + \frac{1}{2}\int\frac{\tilde{n}(\mathbf{r})\tilde{n}(\mathbf{r}')}{|\mathbf{r}-\mathbf{r}'|}d\mathbf{r}d\mathbf{r}' + E_{\text{xc}}[\tilde{n}(\mathbf{r})], \qquad (4.16)$$

where $T_s[\tilde{n}(\mathbf{r})]$ is the kinetic energy functional of *non-interacting* electrons (4.12). The last term $E_{\text{xc}}[\tilde{n}(\mathbf{r})]$ is the so-called *functional of exchange-correlation energy*, and the formula (4.16) is its *definition*. Then the Hohenberg–Kohn variational principle for *interacting* electrons takes the form

$$E_v[\tilde{n}(\mathbf{r})] \equiv \int v(\mathbf{r})\tilde{n}(\mathbf{r})d\mathbf{r} + T_s[\tilde{n}(\mathbf{r})] +$$

$$+ \frac{1}{2}\int \frac{\tilde{n}(\mathbf{r})\tilde{n}(\mathbf{r}')}{|\mathbf{r}-\mathbf{r}'|}d\mathbf{r}d\mathbf{r}' + E_{xc}[\tilde{n}(\mathbf{r})] \geq E. \quad (4.17)$$

The corresponding Euler–Lagrange equation for a given total number of electrons has the form

$$\delta E_{v(\mathbf{r})}[\tilde{n}] \equiv \int \delta\tilde{n}(\mathbf{r})\left(v_{\text{eff}}(\mathbf{r}) + \frac{\delta}{\delta\tilde{n}(\mathbf{r})}T_s[\tilde{n}(\mathbf{r})]\big|_{\tilde{n}=n} - \varepsilon \right)d\mathbf{r} = 0, \quad (4.18)$$

where

$$v_{\text{eff}}(\mathbf{r}) \equiv v(\mathbf{r}) + \int \frac{n(\mathbf{r}')}{|\mathbf{r}-\mathbf{r}'|}d\mathbf{r}' + v_{xc}(\mathbf{r}) \quad (4.19)$$

and

$$v_{xc}(\mathbf{r}) \equiv \frac{\delta}{\delta\tilde{n}(\mathbf{r})}E_{xc}[\tilde{n}(\mathbf{r})]\big|_{\tilde{n}=n}. \quad (4.20)$$

Now the equation (4.18) is identical *in form* to equation (4.13) for non-interacting particles moving in an effective external potential v_{eff} instead of $v(\mathbf{r})$ and thus we find that the minimizing density $n(\mathbf{r})$ is determined by solving the one-particle equation

$$\left(-\frac{1}{2}\nabla^2 + v_{\text{eff}}(\mathbf{r}) - \varepsilon_j \right)\varphi_j(\mathbf{r}) = 0 \quad (4.21)$$

taking into account the relations

$$n(\mathbf{r}) = \sum_{j=1}^{N}|\varphi_j(\mathbf{r})|^2, \quad (4.22)$$

$$v_{\text{eff}}(\mathbf{r}) \equiv v(\mathbf{r}) + \int \frac{n(\mathbf{r}')}{|\mathbf{r}-\mathbf{r}'|}d\mathbf{r}' + v_{xc}(\mathbf{r}), \quad (4.23)$$

where $v_{xc}(\mathbf{r})$ is the *local* exchange-correlation potential that is functionally dependent on the full distribution of the density $\tilde{n}(\mathbf{r})$, according to the (4.20). Now, these self-consistent equations are called Kohn–Sham equations (KS).

The ground state energy is defined as

$$E = \sum_{j=1}^{N}\varepsilon_j + E_{xc}[n(\mathbf{r})] - \int v_{xc}(\mathbf{r})n(\mathbf{r})d\mathbf{r} - \frac{1}{2}\int \frac{n(\mathbf{r})n(\mathbf{r}')}{|\mathbf{r}-\mathbf{r}'|}d\mathbf{r}'. \quad (4.24)$$

If we generally neglect energy E_{xc} and potential v_{xc}, then the KS equations (4.21)–(4.24) are reduced to the self-consistent Hartree equations. The KS theory can be formally regarded as reducing the Hartree theory to the exact form. If E_{xc} and v_{xc} are known *precisely*, all many-body effects, in principle, are taken into account. Clearly, this fact draws focus to the functional $E_{xc}[\tilde{n}(\mathbf{r})]$. The practical usefulness of DFM for the calculation of systems in the ground state is determined entirely by whether it is possible to find such approximations for the functional $E_{xc}[\tilde{n}(\mathbf{r})]$, which are simple enough and at the same time sufficiently accurate.

Comments.

1. The exact effective one-particle potential $v_{eff}(\mathbf{r})$ (4.23) of the KS theory can be regarded as a fictitious external potential, which in the case of non-interacting particles gives the same electron density $n(\mathbf{r})$, as well as the density of interacting electrons in the real external potential $v(\mathbf{r})$. Thus, if the true density $n(\mathbf{r})$ is known from an independent source (e.g. experiment or, in the case of small systems, from the exact calculation by the many-body techniques with the use of wave functions), then $v_{eff}(\mathbf{r})$ and, hence, $v_{xc}(\mathbf{r})$ can be determined directly from the density $n(\mathbf{r})$.

2. Because of the close connection with the exact physical density $n(\mathbf{r})$ the one-particle wave KS functions $\varphi_j(\mathbf{r})$ can be considered 'optimal with respect to density', while the Hartree–Fock (HF) wave functions $\varphi_j^{HF}(\mathbf{r})$ are 'optimum as regards total energy' in the sense that the normalized HF determinant gives the lowest energy of the ground state which can be obtained with a single determinant. Since the DFM appeared, the term 'exchange energy' has been often used for the exchange energy, calculated using exact KS wave functions $\varphi_j(\mathbf{r})$, and not the HF wave functions $\varphi_j^{HF}(\mathbf{r})$. (For a homogeneous electron gas, both definitions are the same, the differences are usually very small.)

3. Strictly speaking, neither the exact KS wave functions φ_j nor energies ε_j have any real physical meaning, except that: a) there is a correlation (4.22) between φ_j and the true physical density $n(\mathbf{r})$; b) the energy of the topmost filled eigenvalue ε_j, measured from the vacuum zero, is the same as the ionization energy.

Finally, we discuss Friedel oscillations which is an important feature of the electron distribution near the surface. Their origin is due to the wave properties of electrons and is explained by the diffraction of Bloch waves at the metal. This effect was considered W. Kohn and L.G. Sham [367]. Choosing in (4.21) the wave function of the electron in the form of the plane wave as the initial solution, they have found the appropriate electron density. The solution of (4.21) with the effective potential gave the following order of approximation for the wave function. Thus, using the iterative procedure and the self-consistency potential, the solution for electron density oscillating in the vicinity of the surface was obtained. However, for metals with high electron density such oscillations are very small.

4.2.4. Method of test functions. Thomas–Fermi approximation

In the practical use of the DFM the energy of the ground state of the system in the Hohenberg–Kohn formulation can be written in the form of a functional

$$E[n(\mathbf{r})] = \int w[n(\mathbf{r})]d\mathbf{r}, \qquad (4.25)$$

introducing $w[n(\mathbf{r})]$ – the energy density, which takes into account the set of components. First, take into account the electrostatic interaction energy of the electrons. It includes the energy of the electrons in an external potential $v(\mathbf{r})$, which, *inter alia*, may include the pseudopotential of the ions:

$$w_1(\mathbf{r}) = v(\mathbf{r})n(\mathbf{r}). \qquad (4.26)$$

Another part of the electrostatic energy takes into account the Coulomb interaction of electrons with each other:

$$w_2(\mathbf{r}) = \frac{1}{2}\int\frac{n(\mathbf{r})n(\mathbf{r}')}{|\mathbf{r}-\mathbf{r}'|}d\mathbf{r}'. \qquad (4.27)$$

Another component is the kinetic (Fermi) energy of the degenerate electron gas with density $n(\mathbf{r})$, which will be calculated by the formulas connecting it with the density of the ideal homogeneous electron gas $n(\mathbf{r})$. It is assumed that the inhomogeneity of the electron gas is small, so that there remains a local relationship between energy and density (this is true in the case of slowly varying densities $|\nabla n(r)| < n_0/r_s$, where r_s is the radius of the Wigner–Seitz cell defined by $4/3\pi r_s^3 = n_0^{-1}$). In other words, in atomic units ($e = m_e = \hbar = 1$):

$$\varepsilon_F(r) = p_F^2(r)/2 = 1/2(3\pi^2)^{2/3}n^{2/3}(r). \qquad (4.28)$$

Thus, the kinetic energy density is given by

$$w_{kin}(r) = 0.3(3\pi^2)^{2/3}n^{5/3}(r). \qquad (4.29)$$

Introduction of amendments to the kinetic energy due to the inhomogeneity of the electron gas and originating from quantum effects in the decomposition of energy in respect of the quasi-classical parameter, leads to the Weizacker–Kirzhnits formula [104, 116, 504], taking into account the first term of the gradient expansion of the energy density of the inhomogeneous electron gas

$$w_g(\mathbf{r}) = \frac{|\nabla n(\mathbf{r})|^2}{72n(\mathbf{r})}. \qquad (4.30)$$

The decrease of the electrostatic energy of the electron gas, conditioned by the exchange interaction of electrons, will be described using the formula derived in the Hartree–Fock approximation [106]

$$w_x(\mathbf{r}) = -\frac{1}{4\pi^4} J(\mathbf{r}). \tag{4.31}$$

Here $J(\mathbf{r}) = 4\pi p_F^4(\mathbf{r})$ is the exchange integral, which depends on the density of electrons in a given point in space. Thus

$$w_x(\mathbf{r}) = -0.75(3/\pi)^{1/3} n^{4/3}(\mathbf{r}). \tag{4.32}$$

Consequential amendments to the heterogeneity of the exchange energy can be found in [335, 394].

Finally, for the correlation energy we can use the Wigner interpolation formula [508]

$$w_c(\mathbf{r}) = \frac{-0.056 n^{4/3}(\mathbf{r})}{0.079 + n^{1/3}(\mathbf{r})}, \tag{4.33}$$

sufficiently accurate for the electron densities that occur in real metals ($2 < r_s < 6$), whereas the Hellmann–Brueckner formula [318] is correct only at high densities, when r_s is small. Other approximations are presented in [317, 351, 423].

In summary, and denoting the Coulomb potential

$$\phi(\mathbf{r}) = v(\mathbf{r}) + \frac{1}{2} \int \frac{n(\mathbf{r}')}{|\mathbf{r} - \mathbf{r}'|} d\mathbf{r}', \tag{4.34}$$

we write the energy functional

$$E[n] = \int \phi(\mathbf{r}) n(\mathbf{r}) dV + G[n], \tag{4.35}$$

where $G[n]$ contains the kinetic, exchange, gradient and correlation energy.

In accordance with the DPM, to determine the 'correct' function of the density of the electron distribution $n(\mathbf{r})$ it is necessary to solve the following variational problem [462] for finding the minimum values of $E[n]$:

$$\frac{\partial}{\partial n}(E[n] - \mu N) = 0. \tag{4.36}$$

Here $\mu = \partial E[n]/\partial N$ is the Lagrange multiplier while maintaining the total number of particles $N = \int n(\mathbf{r}) dV = \text{const}$. The value of μ corresponds to the chemical potential of the system of electrons. To obtain a closed system of equations, it is necessary to find another equation that relates the value of $\phi(\mathbf{r})$ and $n(\mathbf{r})$. Under the assumption that $v(\mathbf{r})$ is created by the positively charge of the ions, distributed with density $n_0(\mathbf{r})$, we have

$$v(\mathbf{r}) = \int \frac{n_0(\mathbf{r}')}{|\mathbf{r} - \mathbf{r}'|} dV'. \tag{4.37}$$

Consequently, the potential $\phi(\mathbf{r})$ obeys the Poisson equation with the total charge density of electrons and ions:

$$\Delta\phi(\mathbf{r}) = 4\pi[n_0(\mathbf{r}) - n(\mathbf{r})]. \tag{4.38}$$

Consider a plane boundary of the metal, the outer normal to which is regarded as the axis z, so that the region of space occupied by the metal corresponds to $z < 0$. The effects associated with the three-dimensionality of the metal will not be considered. Thus, the electron density and electrostatic potential can be regarded as functions only of z.

To determine the electron distribution near the surface of the metal we apply the method of test functions: we enter some test functions $n = n(\alpha_i, z)$ and $\phi = \phi(\alpha_i, z)$, where α_i are some of the uncertain parameters whose values are, for example, determined from the condition of the minimum total energy of the system. The substitution of $n(z)$ and $\phi(z)$ in (4.35) by this procedure gives the total energy of the system. The selected test functions must satisfy the requirement of exponential decay outside the metal, tend to the bulk value of the electron density inside the metal and be continuous at the metal–environment interface, resulting in the condition of continuity of the normal component of the electrostatic induction. To clarify approximately the nature of the test functions it is necessary to solve the simplified problem, in which $G[n]$ (4.35) takes into account only the kinetic energy of the electrons (Thomas–Fermi approximation (TF) [490, 491]). The electron distribution functions found in the TF approximation should be used as test solutions for the full variational problem.

Using the 'jelly' model for the metal in which the discrete distribution of the ions is replaced by the background of the uniformly distributed positive charge density $n_0(z)$, we rewrite the Poisson equation (4.38) as

$$\frac{d^2}{dz^2}\phi(z) = 4\pi[n_0\Theta(-z) - n(z)], \tag{4.39}$$

where $\Theta(z)$ is the Heaviside step function. In the TF approximation

$$E[n(z)] = \int_{-\infty}^{\infty} \left[0.5\phi(z)n(z) + 0.3(3\pi^2)^{2/3} n^{5/3}(z) \right] dz. \tag{4.40}$$

Substituting (4.40) into (4.36) and varying, we find the ratio

$$\phi(z) + 1/2(3\pi^2)^{2/3} n^{2/3}(z) = \mu. \tag{4.41}$$

Since μ is constant, equation (4.41) is the condition of the constant electrochemical potential at any point of the system, in particular,

$$\phi(-\infty) + \alpha n_0^{2/3} = \phi(z) + \alpha n^{2/3}(z), \tag{4.42}$$

where $\alpha = 1/2(3\pi^2)^{2/3}$. Formula (4.42) shows that with decreasing electrostatic energy density the kinetic energy density must increase. From the condition (4.42) we find that

$$n(z) = n_0 \left(1 - \frac{\phi(z) - \phi(-\infty)}{\xi_0}\right)^{3/2}, \tag{4.43}$$

where $\xi_0 = \alpha n_0^{2/3}$. Substituting (4.43) into (4.38), we obtain the TF equation:

$$\frac{d^2}{dz^2}\phi(z) = 4\pi n_0 \left[1 - \left(1 - \frac{\phi(z) - \phi(-\infty)}{\xi_0}\right)^{3/2}\right], \quad z < 0,$$

$$\frac{d^2}{dz^2}\phi(z) = 4\pi n_0 \left(1 - \frac{\phi(z) - \phi(-\infty)}{\xi_0}\right)^{3/2}, \quad z > 0. \tag{4.44}$$

It is known [380] that the solution of the TF equation gives the power law of the decay of electron density with distance from the surface of the metal. Therefore, the solution of the exact TF equation is not suitable as test functions.

We carry out a formal linearization of the TF equation, i.e. we decompose its right side in powers of $\phi(z) - \phi(-\infty)$ and discard terms of higher orders of magnitude, starting from quadratic. The expansion parameter is the ratio of change in the electrostatic potential to the Fermi energy $[\phi(z) - \phi(-\infty)]/\xi_0$. Formula (4.39) shows that this parameter is less than one, and at $z \to -\infty$ it tends to zero, and in the tail of the electron distribution at $z \to +\infty$ is close to unity. In other words, the linearization procedure is strictly correct only in the depth of the metal. Introducing the notation

$$\beta = \frac{6\pi n_0}{\xi_0}^{1/2}, \tag{4.45}$$

we write the linearized TF equation in the form

$$\frac{d^2}{dz^2}\phi(z) = \beta^2[\phi(z) - \phi(-\infty)], \quad z < 0,$$

$$\frac{d^2}{dz^2}\phi(z) = \beta^2[\phi(z) - \phi(-\infty)] - 4\pi n_0, \quad z > 0. \tag{4.46}$$

The expression (4.46) is a linear inhomogeneous differential second-order equation. Its solution is as follows:

$$\phi(z) = C_1 e^{\beta z} + C_2 e^{-\beta z} + \phi(-\infty), \quad z < 0,$$

$$\phi(z) = C_3 e^{\beta z} + C_4 e^{-\beta z} + \phi(-\infty) + 4\pi n_0 / \beta^2, \quad z > 0. \tag{4.47}$$

The condition of finiteness of the potential at $z = +\infty$ and $z = -\infty$ gives $C_2 = C_3 = 0$. Constants C_1 and C_4 are found from the continuity conditions of the potential and its derivative at the metal boundary $z = 0$:

$$C_1 = 2\pi n_0 / \beta^2, \quad C_4 = -2\pi n_0 / \beta^2.$$

Finally, choosing a point of reference of the potential $\phi(\infty) = 0$, we obtain $\phi(-\infty) = -4\pi n_0/\beta^2$. As a result, the solution of (4.47) has the form

$$\phi(z) = 2\pi n_0 / \beta^2 e^{\beta z} - 4\pi n_0 / \beta^2, \quad z < 0,$$
$$\phi(z) = -2\pi n_0 / \beta^2 e^{-\beta z}, \quad z > 0. \tag{4.48}$$

Substituting (4.48) into Poisson's equation for the function density, we have the following expression:

$$n(z) = n_0 \left[\left(1 - \frac{1}{2} e^{\beta z} \right) \Theta(-z) + \frac{1}{2} e^{-\beta z} \Theta(z) \right]. \tag{4.49}$$

The functions (4.48) and (4.49) will be used as test functions in minimizing the energy functional (4.25). Parameter β will be regarded as a variational parameter. From a physical point of view, the value of β represents the characteristic thickness of the surface layer in which the electron density drastically changes. By order of magnitude β^{-1} is the Debye screening length of the impurities in the metal [106]. It should be noted that the trial functions (4.48) and (4.49) were used by J.R. Smith in [462] without discussion of their derivation and validation.

We will postulate the applicability of the test functions, obtained as solutions of the linearized TF equation, in the entire region near the metal surface. These functions, although they are solutions of the exact TF equation only in the depth of the metal, satisfy the requirements on the test functions listed above. In particular, being exponentially decaying outside the metal, they no correspond more accurately to the solution of the quantum-mechanical problem for the electron density than the power-law solution of the exact TF equation. The disadvantage of these test functions is the absence of Friedel oscillations, decaying from the surface into the metal.

4.3. Application of the density functional method (DFM) for calculation of the work function of the electron from the metal surface

The *work function* is the minimum energy which must be supplied to the electron at $T = 0$ so that it can be removed from the metal. The highest occupied levels in the conduction band correspond to the Fermi energy $\varepsilon_F = \mu$ (μ – chemical potential), therefore, in the approximation of non-interacting particles it can be expected that the work function is defined as

$$W = v_{eff}(\infty) - \mu. \tag{4.50}$$

In this expression, $v_{eff}(\infty)$ is the crystal potential at a large distance from the metal. It is not immediately clear whether it is justified to use this expression to take into account many-body effects. If we remove one electron the metal is singly positively charged and therefore attracts an electron. It is also questionable whether the contribution essential for overcoming such forces is taken into account accurately in determining the work function. This issue was resolved by N.D. Lang and W. Kohn [380] in the framework of the density functional formalism.

Thus, taking into account the electric field existing outside the crystal, the work required to move an electron from point $\mathbf{r}_i$ to point $\mathbf{r}_j$, located at a small distance from the surface compared to the size of the surface of the crystal face, but large compared to the lattice constant, is $W_j - W_i$. In this regard, the work function is given by

$$W_i = [\phi(\mathbf{r}_i) + E_{N-1}] - E_N, \tag{4.51}$$

where E_N is the ground-state energy of the neutral N-electron crystal; E_{N-1} is the ground state energy of the crystal from which one electron is removed; $\phi(\mathbf{r}_i)$ is the electrostatic potential at the point $\mathbf{r}_i$.

Since the chemical potential $\mu = E_N - E_{N-1}$, we have

$$W_i = \phi(\mathbf{r}_i) - \mu. \tag{4.52}$$

This formula was obtained F.K. Schulte [456]. Both terms in (4.52) can be measured from an arbitrarily chosen zero potential; as the origin it is convenient to take the average electrostatic potential in the metal

$$\bar{\phi} \equiv \Omega^{-1} \int_\Omega \phi(r) dr, \tag{4.53}$$

where Ω is the volume of the metal, and the integration is performed over this volume. Thus, we can write:

$$\overline{D}_i \equiv \phi(r_i) - \bar{\phi}, \qquad \bar{\mu} \equiv \mu - \bar{\phi}, \tag{4.54}$$

so

$$W_i = \overline{D}_i - \bar{\mu}. \tag{4.55}$$

Often for the two terms in equation (4.55) it is convenient to use another origin of the potential that is different from the average potential $\bar{\phi}$. In Wigner–Seitz computations [509], for example, a natural reference point is the electrostatic potential on the Wigner–Seitz sphere.

If the DFM is used to solve the variational problem for finding the minimum of the energy functional, written in the form (4.35), then at the constant total number of electrons after averaging over the volume of the metal we find

$$\bar{\mu} = \mu - \bar{\phi} = \Omega^{-1} \int_\Omega \frac{\delta G[n]}{\delta n(r)} dr. \tag{4.56}$$

Hence we see that $\bar{\mu}$ is a characteristic of the bulk system. This is the bulk chemical potential, measured from the average electrostatic potential of the metal; it does not depend on the latter, which follows from the definition of $G[n]$ as total non-electrostatic energy. Thus, formula (4.55) divided the work function into two components: bulk and surface [380]. It is essential that these values due to the universality of the functional $G[n]$ are implicitly dependent on the potential of the crystal lattice. The result (4.55) indicates that the assumption (4.50) for the work function is valid. Many-body effects are reflected in the volume and correlation contributions to $G[n]$ and, as a consequence of (4.54), (4.55) – also in the value of D_i.

Let us turn to the consideration introduced in section 4.2 of the homogeneous background model for the metal surface ('jelly' model). The electric double layer which is formed due to the fact that the electron density different from zero behind the formal (sharp) boundary of the background, leads to the fact that, in general, the electrostatic potential in vacuum $\phi(\infty)$ will be higher than inside the metal $\phi(-\infty)$. Thus, the electrons are trying to fly out of metal and are subjected to the effect of an electrostatic barrier having a height of

$$D = \phi(\infty) - \phi(-\infty) = 4\pi \int_{-\infty}^{\infty} z[n(z) - n_0(z)]dz. \qquad (4.57)$$

Since in this case $\bar{\phi} = \phi(-\infty)$, then the height D is the quantity defined by (4.54). From (4.54)–(4.55) we find the work function:

$$W = D - \bar{\mu}, \qquad (4.58)$$

$$\bar{\mu} = \frac{d\varepsilon(\bar{n})}{d\bar{n}}, \qquad (4.59)$$

where $\varepsilon(\bar{n})$ is the average energy per particle of a homogeneous electron gas having a density n.

Quite often, the potential barrier at the metal surface is determined using the effective potential, not its electrostatic component. This is due to the fact that the formation of the exchange–correlation hole around an electron inside the metal lowers its energy relative to the vacuum level. However, only the electrostatic component of the full height of the surface barrier depends directly on the surface properties.

In [380] self-consistent calculations were conducted of the work function for polycrystalline simple metals (the model of a homogeneous background is not suitable for transition or noble metals) in the 'jelly' model, using the local density approximation to describe the exchange and correlation interactions and dependences of the quantities W, $\bar{\mu}$ and D on the density parameter r_s. It is shown that with a decrease in r_s the work function reaches a maximum and then begins to decrease. Although according to V. Peuckert [436] the magnitude of W tends to a finite limit, equal to approximately 1.3 eV, the individual components of W diverge at $r_s \to 0$. This means that at high values of electron density the 'jelly' model is inadequate. This is confirmed by the calculations of the surface energy of metals, leading to negative values of the surface energy at high values of the electron density. In addition, calculations show that the dipole barrier for small r_s is very large compared with the values obtained with the semi-empirical approach. Both of these discrepancies show that to develop a good model of the metal surface it is essential to introduce in the appropriate manner the potential of the ions.

The above-described 'jelly' model completely ignores the effects of the discrete ionic lattice, though they are quite substantial. The experimental values of the work function of a single crystal often strongly depend on the selected face of the crystal. Close-packed surfaces turned out to have higher work functions compared with less filled ones. Even in cases where the average work function is well described by the 'jelly' model, this model is not able to satisfactorily predict the value of the surface energy. It was shown [380] that the situation can be corrected by adding to the pseudopotential the perturbation caused by the discrete lattice potential.

The hypothesis of the origin of the anisotropy of the work function was proposed by the first time by R. Smolukhovskii [468]. He used the model of a corrugated open surface of the jelly. In order to minimize their kinetic energy, the electrons tend to follow a smooth boundary. This leads to the dipole potential the sign of which is opposite to the electrostatic potential, calculated for a flat surface. The appearance of such potential, of course, lowers the work function. J.R. Smith showed [463] that semi-quantitative results can be obtained for the (110) face of a tungsten crystal using this general idea of the density functional formalism. N.D. Lang W. Kohn [380] also studied the anisotropy of the work function in the 'jelly' model using the potential of the ion in the first approximation of the perturbation theory. However, until now no satisfactory agreement has been obtained between calculated and experimental results of the work function [39].

4.4. The phenomenon of adhesion and ways of describing it

The phenomenon of formation of a bond between the surface layers of dissimilar condensed systems, brought into contact, is called adhesion. Adhesion depends on the nature of the contacting bodies, the properties of their surfaces and the contact area. From a physical viewpoint, adhesion is determined by intermolecular interaction and the presence of ionic, covalent, metallic and other types of bond. There is a need to define the characteristics of the adhesive interaction of various materials in terms of both applied and fundamental science of surface phenomena. Films and coatings are widely used in production and are usually deposited on the substrate. The resulting bond, i.e. the adhesion between the film (adhesive) and the base (substrate) can make the surface to have lubricating, anti-friction and other properties, to prevent corrosion.

There are various theories that describe the adhesion of solids. In terms of the formation of bonds between the adhesive and the substrate we should consider the adsorption theory of adhesion [15]. Among the theories that determine the magnitude of adhesion, depending on the nature and number of bonds per unit contact, we must include the diffusion theory [42]. For the implementation of the diffusion processes it is necessary to observe two conditions: thermodynamic which is reduced to the mutual solubility of the adhesive and the substrate and their compatibility, and kinetic, which is achieved by the mobility of the molecules. In diffusion into the depth the interface eroded. Diffusion takes place in adhesion of limited number of systems. The adsorption and diffusion are the consequences of exposure two bodies and cannot occur outside the contact [121].

The microrheology theory of adhesion has also been proposed [73]; according to this theory, the process of film formation from the melt is accompanied by filling of pits in the rough surface, increasing the real contact area and hence the number of bonds between adhesive and substrate. This leads to increased adhesion and adhesion strength. Thus, the diffusion and microrheological theories consider the mechanism of formation of the area of the actual contact surfaces. But the methods discussed above represent only one aspect of the adhesion process and does not allow us to describe the causes of the phenomenon.

Currently, the analytical methods for describing the processes of adhesion, based on the determination of the energy state of the bodies in contact with the involvement of methods of quantum mechanics, are being intensively developed. This approach is general and allows us to functionally link the work of adhesion to the internal parameters of the contacting materials and, if developed sufficiently, to predict the value of adhesion strength. The theories explaining the causes of adhesion strength include the electron theory. The presence of the electric double layer and its effect on adhesion strength have been proved experimentally [80].

Theoretical treatment of the metal–insulator interface has been the focus of several studies [56, 113, 184, 219, 339], where test functions depending on the dielectric constant of the medium were proposed and then used to calculate the electronic distribution. The effect of extending the electron cloud with increasing ε was theoretically detected and investigated in [113]. In [219] the authors calculated the dependence of the surface energy and the electrostatic potential of ε and showed that the first of these values vary insignificantly, while the second increases at the interface with increasing ε.

In [56], the dependence of the variational parameter of the electron distribution on ε was calculated using the sum rule [272], which has been proposed as a criterion for the correctness of the DFM.

However, the agreement of the absolute value of the parameter of density decay in [56] with the values obtained self-consistently was achieved through an artificial process.

In studying the interaction between two metal surfaces, various authors used the variational principle [110, 308, 406]. In [309, 310] a similar problem is solved on the basis of the quantum mechanical equations. In

[485, 496] the exact results following from the sum rule were used. In [110, 366], the dependence of the variational parameter of the electron distribution on the distance between the metals was studied and the adhesion energy and force of interaction between the metals were calculated. In [308], such dependence is not taken into account. In [115] the 'jelly' model was used to calculate the forces of interaction between two identical metals when the gap between them filled with a dielectric medium. It turned out that the strength of interaction decreases with increasing distances more smoothly. Comparison with the results of the calculation of forces in direct contact of metals, when the distance between them is zero [115, 332, 465], shows that the dependence of the force on the electron density of metals is of the same nature and has the same order of values. In [310, 366] it was pointed out that taking into account the discreteness of the ionic lattice increases the maximum interaction force and shifts the equilibrium position in the system towards smaller gaps.

4.5. Adsorption on metal surfaces

The formalism of the density functional method [213] is very convenient for the study of adsorption on metal surfaces, as well as the chemical bonds between the atom or molecule and the metal surface. A classic case was considered in some studies where the surface contains an atom or molecule of the adsorbate, which violates the periodicity of the substrate. In other studies, we consider the case when a layer of atoms is adsorbed on the surface, usually with the same lattice constant as that of the substrate; in this case the symmetry of the 'bare' substrate is preserved.

Usually the calculations determine the difference between the local densities of states for the metal–adsorbate system (MA) and 'bare' metal (M):

$$\delta n(\varepsilon, \mathbf{r}) = n^{MA}(\varepsilon, \mathbf{r}) - n^{M}(\varepsilon, \mathbf{r}). \tag{4.60}$$

Expressed in terms of $\delta n(\varepsilon, \mathbf{r})$, the density of states associated with the presence of an adsorbed atom is equal to

$$\delta n(\varepsilon) = \int \delta n(\varepsilon, \mathbf{r}) d\mathbf{r}, \tag{4.61}$$

and the extra electron density is given by

$$\delta n(\mathbf{r}) = \int_{-\infty}^{\varepsilon_F} \delta n(\varepsilon, \mathbf{r}) d\varepsilon. \tag{4.62}$$

In this section we consider the studies of the metal–adsorbate system. We begin with the case in which both the substrate and the adsorbate are described by the model of a homogeneous background, and conclude with a case in which the full nuclear potential (or the ion pseudopotential) of the atoms of both the substrate and the adsorbate was taken into account in calculations.

4.5.1. The model of a homogeneous background for the substrate and the adsorbate

One of the most popular tasks in previous studies of adsorption was the study of systems consisting of a layer of alkali metal atoms adsorbed on a substrate with a high work function. Typically, in such a system we measured the change in the work function when the surface is coated with a layer of adsorbed atoms. As adsorption of atoms takes place the work function decreases rapidly, reaching a minimum, and then increases, approaching the bulk work function of the alkali metal, when the first layer is completely filled with adsorbate atoms.

Such a system was considered in [379] using the model of a homogeneous background for both the substrate and the adsorbate. The positive density of a homogeneous background for this problem is taken as a

$$n_+(z) = \begin{cases} n^M, & z \le 0; \\ n^A, & 0 < z \le d; \\ 0, & z > d. \end{cases} \qquad (4.63)$$

Since the alkali metals are monovalent, $n^A d = N$, where N is the number of adsorbed atoms per unit area. The thickness d of the adsorbate is assumed to be fixed and equal to the interplanar distance between the most close-packed planes in an alkaline metal, and n^A changes. Thus, the changes in the filling N are simulated. The thickness of the adsorbate layer d is fixed because it related to the distance between the nuclei of the substrate and the adsorbate in the real systems, which should not change significantly during filling.

In [379], the electronic density was determined from the Kohn–Sham equation using the local density approximation for the exchange and correlation. The differential equation (4.21) was integrated with respect to the direction from vacuum to the metal. The density of the substrate n^M corresponding to the value of $r_s = 2$ ($4\pi r_s^3 / 3 = n^{-1}$) was selected, i.e. fairly typical density for the substrates with a high work function (a discussion based on this model was also published by K.F. Wojciechowski [510, 511]). S. Warner [502] discussed a similar model (but with $d \sim N$) in the framework of the generalized Thomas–Fermi model.

The work function curves $\Phi(N)$, calculated in the framework of this model, clearly demonstrate the minimum observed in the experiment. The values of the degree of filling, corresponding to the minimum, are smaller than the corresponding values obtained in experiments with adsorption on transition metal substrates in which the work function is much higher than in the model. However, the actual value of the work function at the minimum is almost independent of the work function of the substrate so the results can be compared with experiments carried out on transition metals [379].

4.5.2. The model of a homogeneous background for the substrate

In this section we consider the classical problem of adsorption – an atom, bonded with the surface of a semi-infinite metal. The use of the homogeneous background model for the metal leads to a problem with cylindrical symmetry, which makes a significant simplification.

Let d be the distance between the nucleus of the adsorbed atom (adatom) and the edge of the positive background. This distance is calculated by minimizing the total energy. This model is completely characterized by two numbers – the density parameter r_s of the homogeneous substrate and the charge z of the adsorbed atom nucleus.

We are interested in two quantities: the dipole moment μ and the binding energy of the atom ΔE_a. If we choose the origin at the centre of the nucleus of the adsorbed atom and let z tend to infinity in the direction of vacuum, the dipole moment will be determined by the formula

$$\mu = \int z\delta n(\mathbf{r})d\mathbf{r}. \tag{4.64}$$

The binding energy of an atom ΔE_a is the difference between the energy of the metal (M) plus a single atom (A) and the energy of the metal–adatom (MA) bonded system, i.e.

$$\Delta E_a = E^{\mathrm{M}} + E^{\mathrm{A}} - E^{\mathrm{MA}}. \tag{4.65}$$

The generalized Thomas–Fermi model. In fact, the Thomas–Fermi model (4.40) can not provide good results for the electronic structure of the surface (negative surface energy and zero work function for all values of density r_s [213]). J.R. Smith attempted in [463] to address this defect, using for the functional $G[n]$ the gradient expansion in respect of the density parameter $n(\mathbf{r})$. He left the first gradient term of the series and used for it the coefficient of the simplest approximation which takes into account only the contribution of kinetic energy. Thus, he used the following expression:

$$G[n(\mathbf{r})] = \int \left[w(n(\mathbf{r})) + \frac{1}{72} \frac{|\nabla n(\mathbf{r})|^2}{n(\mathbf{r})} \right] d\mathbf{r}, \tag{4.66}$$

where $w(n)$ is the energy density of the homogeneous electron gas with density n, taking into account exchange and correlation. For the electron density Smith used the expression (4.49) and minimized the total energy functional $E[n]$ with respect to β. The functional $G[n]$, written in the form (4.66), corresponds to the generalized Thomas–Fermi approximation.

J.R. Smith, S.G. Ying and W. Kohn [466, 515] studied the adsorption of hydrogen on a substrate regarded as a homogeneous background. They found a linear function of the density–potential response of the substrate by using the generalized Thomas–Fermi model for the energy functional $G[n]$ (4.66), and used it to determine the induced density and other characteristics of the ground state by considering a point charge of unit magnitude ($z = 1$) as a perturbation. A comparison of the results S.C. Ying and others with the results of fully quantum-mechanical non-linear calculations [213], the use of the linear response, coupled with the generalized Thomas–Fermi model, is a very rough approximation.

Calculations were performed for a high density substrate. The resultant equilibrium position is fully supported by quantum-mechanical analysis [382]. The energy of desorption of the ion $\Delta E_i = 9$ eV was also determined. The experimental data for hydrogen in tungsten give $\Delta E_i \approx 11.3$ eV [466] (despite the fact that the model of a homogeneous background does not describe the transition metals with sufficient accuracy). J.R. Schrieffer [455] showed, however, that the binding energy of an atom $\Delta E_a = \Delta E_i + \Phi - I$ is more important, where Φ – the work function of the substrate, I – the ionization potential of the free atom (for H/W measurements give $\Delta E_a = 3.0$ eV). S.C. Ying et al [515], using the measured work function for the most close-packed faces of tungsten (5.3 eV) and the exact value of I (13.6 eV), found that $\Delta E_a = 0.7$ eV. They concluded that the method which gives a good value of ΔE_i is not accurate enough to determine ΔE_a. It was also found that these approximations cannot provide the desired accuracy of the values of the dipole moment or electron density at the nucleus [382].

W. Kohn and S.C. Ying [362] studied the adsorption of alkali metals, using the response function and describing the skeleton of an adsorbed atom using the pseudopotential. They assumed that the approximation of the linear response should be more suitable for larger atoms that can transmit their valence electron of the substrate than for the hydrogen atom. In [362] the authors also used a different energy functional in which the coefficient of 1/72 at the gradient term in the expression (4.66) was replaced by another, depending on the density. The calculated energy of ion desorption and the dipole moment are in reasonable agreement with the values obtained in the study of adsorption on transition metals. It was shown that the use of the coefficient, which depends on the density, at the gradient term in the energy functional significantly affects the magnitude of the dipole moment.

In [354] the nonlinear response approximation was used to study the adsorption of a Na atom using the generalized Thomas–Fermi model. The variational form was used for the electron density and the total energy was minimized. The skeleton of the adsorbed atom was described using the pseudopotential. The density of the substrate was assumed equal to the density of metallic sodium.

Quantum-mechanical treatment. The quantum-mechanical consideration of the problem of adsorption of atoms on the substrate, described by a model of a homogeneous background, was published by N.D. Lang and

A.R. Williams [382, 383] and also by O. Gunnarsson, H. Hjelmberg and B.I. Lundquist [325, 343]. In studies [382, 383], the Kohn–Sham equations for continuous states was written in the form of the Lipmann–Schwinger equation:

$$\Psi^{MA}(\mathbf{r}) = \Psi^{M}(\mathbf{r}) + \int G^{M}(\mathbf{r},\mathbf{r}';\varepsilon)\delta v_{eff}(\mathbf{r}')\Psi^{MA}(\mathbf{r}')d\mathbf{r}'. \qquad (4.67)$$

Here G^{M} is Green's function of the 'bare' metal, and δv_{eff} is the difference of v_{eff} potential values for the metal–adatom system and the 'bare' metal. Now the differential Kohn–Sham equation can be solved by direct numerical integration of the area outside the nucleus of the adsorbed atom. Solutions $\Psi_i(\mathbf{r})$ thus obtained are characterized by a certain angular dependence near the nucleus of the adsorbed atom and in the general case do not satisfy the boundary conditions corresponding to the Lipmann–Schwinger equation. The required solution, however, is a linear combination of these fundamental solutions:

$$\Psi^{MA}(\mathbf{r}) = \sum_i C_i \Psi_i(\mathbf{r}), \qquad (4.68)$$

where the coefficients C_i are found by substituting (4.68) into (4.67). Using Lipmann–Schwinger equation in a natural way allows one to take into account the short-range effect of the quantities $\delta n(\mathbf{r})$ and $\delta v_{eff}(\mathbf{r})$ determined by screening near the adsorbed atom.

For the nucleus states of the adsorbed atom the intrinsic functions of equation (4.21) satisfy the boundary conditions which have a simpler form than in the case of continuous states and, therefore, the Lipmann–Schwinger equation is not used. Just as in the case of continuous states, the direct numerical integration of the Kohn–Sham equation outside the nucleus of the adatom comprises a solution of the radial equations for different values of l (thus, the state of the nucleus can be polarized in a totally non-spherical potential of the metal–adatom system).

In [325, 343], the eigenfunctions of the metal–adatom system were decomposed over the basis containing the eigenfunctions of the 'bare' metal Ψ^{M} and a number of localized functions. The problem of overflow of the basis was solved by introducing an additional condition on the coefficients of the functions of the 'bare' metal. The problem of solving the Kohn–Sham equation leads to the problem of finding Green's function which takes into account just the matrix elements of localized functions, the value of $\delta n(\mathbf{r})$ was determined from this Green's function. The presence of the short-range order enables us to express $\delta n(\mathbf{r})$ only through localized functions.

Calculations for Li, Si, Cl, adsorbed on a substrate with a high density ($r_s = 2$), were carried out by N.D. Lang and A.R. Williams [382]. The states, which form $3p$-resonance in Cl, are situated below the Fermi level and are therefore occupied; the states also forming the same resonances ($2s$) in Li are completely free. This means that Cl attracts the charge of the substrate, and Li gives up its charge to the substrate, as should be expected on the basis of the electronegativity of atoms. A large amount of energy is required to fill or release the $3p$-level in Si, so its resonance is close to the Fermi level. The result is a covalent rather than ionic bond. Of great importance is also the change of the dipole moment μ (4.64) as a function of the distance from the adsorbed atom to the metal surface d.

In particular, the dynamic charge of the adatom can be defined as the slope of the curve $\mu(d)$, i.e. $\mu'(d)$. For distances d of the order of 2.0 atomic units (a.u.) $\mu'(d)$ is ~0.4 for Na, ~ −0.5 for the Cl and ~0 for Si (in units of electron charge). It should not, however, be assume that these numbers are a good approximation for the value of the statistical charge transfer, because there are other important contributions to the value of $\mu'(d)$, in particular, the contribution due to the dependence of the polarization effects on the distance [402].

These studies [325, 343, 382, 383] investigated both the equilibrium state of the metal–adatom system and the dependence of the properties of the metal–adatom system on the distance between the atom and the metallic surface. This has important implications for understanding the dynamics of such systems. It is especially interesting to consider the change of the difference in the density of states $\delta n(\varepsilon)$. In fact, at the longest distance the metal–adatom interaction is small, and the resonance $\delta n(\varepsilon)$, which is located below the Fermi level, is very narrow. When the atom approaches the surface, the interaction intensity increases and the resonance condition significantly broadens. It also moves away from the Fermi level, i.e. it corresponds to the change in the of the 'bare' metallic surface. When an atom is located closer to the surface, the broadening due to the increasing intensity of the metal–adatom interaction is replaced by contraction, which is due to a decrease of the density of the metallic states. It appears that in this case the resonant state is moving towards the bottom of the metallic zone and the resonance again becomes smaller. A more detailed analysis of changes of the behaviour of the resonant state is given in [326, 413].

4.5.3. Lattice model of the substrate

The model of a homogeneous background at high electron density (transition metals) does not give satisfactory results for the electronic surface structure [213], and we must take into account the discreteness of the crystal lattice. The presence of a discrete lattice allows one to consider the anisotropy of quantities such as work function and surface energy, and, moreover, leads to a positive surface energy for metals with a high electron density in contrast the opposite to the model of the homogeneous background.

Examination on the basis of the perturbation theory. The influence of the discrete lattice of the substrate on adsorption can be studied by introducing the lattice into the model of the homogeneous background using the perturbation theory. For a simple metal, the effect of the lattice of ions on the electrons can be described by the quantity $v_{pseudo}(\mathbf{r})$ – a superposition of ionic pseudopotentials. In relation to the potential of the homogeneous background $v_{+}(\mathbf{r})$ it causes a perturbation

$$\delta v(\mathbf{r}) = v_{pseudo}(\mathbf{r}) - v_{+}(\mathbf{r}). \tag{4.69}$$

Such an analysis uses the first order of the perturbation theory and gives the value of the total energy of the system depending on the location of the adatom and its distance from the metal. The minimum total energy determines the energetically favourable sites of the bonds (nodes) and the equilibrium adsorbate–substrate bond length and also the adsorption energy of the adsorbate (heat of adsorption).

This analysis was first conducted by O. Gunnarsson, H. Hjelmberg B.I. Lundquist [325, 343]. They calculated the binding energy of the hydrogen atom on Al (100). Similar calculations for the Si/Al (111) system were carried out N.D. Lang and A.R. Williams [383]. The distance between the metal and the adatom in the energetically most favourable locations of adsorption is reduced in comparison with the values obtained for the substrate of the model of the homogeneous background; the binding energy at these locations naturally increases.

Review outside the framework of perturbation theory. Here we consider the study of adsorption on the metal surface by the DFM, in the electron distribution (and hence other properties) is determined taking into account the full three-dimensional lattice potential. These studies are mainly conducted by the self-consistent solution of the Kohn–Sham equations in the local density approximation. The potential was taken either or as the toal nuclear potential or as a superposition of the ionic pseudopotentials (in this case the core states of the metal are not taken into account).

The adsorption of nitrogen on Cu (100) was studied in [464]. The system consisted of a copper plate with three layers and monolayers of nitrogen made on both sides. Calculations were fully self-consistent with the total lattice potential. As for Cu, the basis of atomic orbitals, which are approximated by Gaussian orbitals, was also used for N. The location of adsorption sites and the distance between the adsorbate and the substrate were taken from the experiment. The quantum-mechanical Kohn–Sham equation was solved in the local density approximation for exchange-correlation components of energy. It was shown that the charge density in the central layer, obtained from these calculations, and in a clean copper plate are almost identical. This means that the central (second) layer of the copper plate is already showing bulk properties, i.e. there is a rapid screening by the metal of the influence of adsorption on the substrate. The experimentally discovered general features of the density of states (for example, its increase or decrease compared to the clean surface) were reproduced by calculations. It was shown that the self-consistency of such calculations is important – the total density of the states of the system with a finite self-consistent potential dramatically differs from the density of states for the initial potential, constructed by the superposition of atomic charge densities.

A number of studies by M. Scheffler et al [419, 475, 476, 506] were concerned with the adsorption of alkali metals on Al (111) and Al (100). Calculations within the density functional formalism were performed using fully self-consistent methods based on surface Green's functions (for single adatoms) and pseudopotential plane waves (for monolayer coatings). The quantum-mechanical Kohn–Sham equation (4.21) was solved in the approximation of the local density for exchange and correlation. The system under consideration consisted of aluminum plates (4, 7 and 10 layers) with monolayers of Na and K atoms on both sides. It was shown that the calculation of the density of distribution of the charge, the adsorption energy and work function of the surface of the substrate for the plate with 10 Al layers gives almost the same results as for the plate with 4 layers. Therefore, in order to simplify computing resources, most calculations were done for the Al plate with 4 layers. In these works, the formation of stable and metastable adsorption structures, depending on the coverage parameter Θ, was predicted for the first time on the basis of comparison of the calculated values of adsorption energy for different adsorption structures characteristic of unactivated, activated, substitutional and islet-like adsorption.

In accordance with the calculations, the adsorption structures formed in non-activated adsorption for all values of Θ ($\Theta < 1$) are characterized by the lowest values of adsorption and are metastable. Experimental studies of the structural distribution of adatoms recorded the formation of such adsorption structures only at very low temperatures ($T \leq 100$ K) [474], whereas at higher temperatures, close to room temperature, the formation is only recorded with mixing structures of adatoms and surface atoms of the substrate (substitution and activated adsorption). Such structures, known as *stable*, are characterized by higher values of adsorption energy. The authors obtained a fairly good agreement between the calculated and experimental values of adsorption energy and the electron work function of the substrate surface.

In the 2000–2003 the group of M. Scheffler described in [432, 452, 453] similar calculations for the Co/Cu(111) and Co/Cu(100) systems, as well as for the alkali metals on Al(111). The copper substrate was made of a plate with 3 layers, the Al substrate – 4 layers. For the first time in solving the Kohn–Sham quantum mechanical equations (4.21) calculations were carried out outside the local density approximation and the gradient amendments to the inhomogeneous uniformity of the electronic system for exchange and correlation components of energy were taken into account. It was shown that to determine the geometry of the adsorption structures of transition metal atoms it is important to take these corrections into account and the use of only local density approximations to describe the adsorption of transition metals leads to incorrect results. At the same time, to describe the adsorption of alkali metals, it was shown that in this case moving outside the local density approximation does not qualitatively alter the received results concerning the prediction of the stability or metastability of adsorption structures it is reflected only in a slight decrease in the calculated values of the adsorption energy adatoms.

4.6. Conclusions

The analysis of the theory of surface phenomena in solids leads to several conclusions.

1. The most consistent and developed method for describing the surface properties of metals is the quantum-statistical method of the electron density functional method (DFM). However, DFM, as applied to the description of the surface properties of metals, is not a fully completed theory. Thus, there is no universal theory which could be used for various metals to adequately describe and obtain values of the surface characteristics that are consistent with experiment, as well as to reflect their dependence on the orientation of the surface facets.

2. The strong heterogeneity of the electron gas in the subsurface area makes it necessary to go beyond the approximation of the local electron density and take into account the amendments in the gradient expansion for the kinetic and exchange–correlation energies. But so far the nature of the convergence of the gradient decomposition is still unclear. The question of the order to which the terms of the gradient decomposition can be restricted in the description of a particular group of metals has not been answered. It is also unclear which of the numerous amendments to the inhomogeneity of the exchange–correlation energy obtained by using different approximations should be used when calculating the surface energy for each type of metal.

3. The metal surface is characterized by the violation of the translational symmetry of the crystal in a direction perpendicular to the surface, – a process of lattice relaxation of the surface takes place. Change of the interplanar spacing leads to a change in both surface energy and electron work function from the surface of metals. Influence of the displacement of surface ion planes on the surface characteristics must be found self-consistently, taking into account the mutual influence of the profile of the electron density near the surface and the relaxing ion plane.

4. Only the further development of methods for determining the surface properties will make it possible to use them to calculate the adhesive characteristics of various metals, to identify the main parameters determining both the structural and energy states of the surfaces of different materials and the characteristics of their adhesive interactions depending on the condition of the surface and intermediate layer having the dielectric properties.

5. It should be noted that the description of the adhesive properties of the complex compounds and alloys in the framework of the electronic density functional would require a large number of empirical parameters,

and the presence of a complex band structure casts doubt on the possibility of describing the adhesion of semiconductors by this method. Although most of the effects that are used in modern semiconductor microelectronics are based on the phenomena occurring in the surface layer and the interfacial contact area of the contacting semiconductor compounds, they can not be described within the framework of existing theories, even at a qualitative level, which requires the development of new approaches based on the use of various collective excitations localized on the surface of solids.

6. Only the further development of methods for describing the surface properties can ensure that they can be used to calculate the energy characteristics of adsorption of atoms and molecules on various metals, identify key parameters that determine both the structural and energy state of the surface of various materials and the energy adsorption characteristics depending on the surface.

7. The ability of many adsorbates at room temperature to replace the near-surface atoms of the substrate, and not simply adsorb on the surface, and the experimentally observed effects of mixing adatoms and substrate atoms to form surface binary solutions make it necessary to develop theoretical methods for describing these adsorption structures and their properties.

In subsequent chapters of this book we described the relatively simple and intuitive approaches and methods developed by us in the framework of the electronic density functional theory and dielectric formalism. These methods make it possible, in good agreement with experiment, to calculate the surface and adhesion characteristics of a wide range of materials – metals, alloys, semiconductors, and complex compounds. In particular, the following methods are described:

– Methods of self-consistent calculation and determination of the values the surface energy of metals, adjusted for heterogeneity of the electronic system and the discreteness of the crystal structure in the framework of electronic density functional theory;

– Methods for calculating the electron work function of metal surfaces as one of the most important surface characteristics, which determines the energy state of the surface, and the results obtained by comparing the values of the electron work function for various metals with the results of experimental studies;

– Methods for calculating the interfacial energy of interaction, the adhesion energy and force of adhesion interaction of different metals in the framework of the electronic density functional theory taking into account the gradient corrections for the heterogeneity of the system and the displacement of ionic planes in the interfacial region between the media;

– Methods for calculating the adhesive interaction of metals and semiconductors, as well as complex compounds based on the dielectric formalism and the use of the concept of surface plasmons;

– The results of calculating the energy of adhesion and adhesive strength of interaction for a number of metals, semiconductors, and complex compounds;

– Methods for describing and calculating the adhesion component of the friction force of metals under dry friction conditions;

– Methods for calculating the characteristics of friction materials with solid lubricants, oxide and diamond-like wear-resisting coatings, and recommendations for their use to select the optimum friction pairs for the dry friction sections;

– Models and methods for describing the non-activated and activated adsorption of metal atoms on metal surfaces;

– The results of calculation of the effect of adsorption on the work function of an electron from the surface of the adsorption system.

5

CALCULATION OF SURFACE CHARACTERISTICS OF METALS USING THE METHOD OF THE ELECTRONIC DENSITY FUNCTIONAL

Introduction

In this chapter, the method of self-consistent calculation of the surface energy and work function of metals by the density functional method (DFM) taking into account gradient amendments for the inhomogeneity of the electron gas in the surface region for the kinetic and exchange–correlation energies is gradually described. The effect of various approximations for the exchange–correlation corrections of the energy density of the inhomogeneous electron system to the values of surface energy and work function of both simple and transition metals is investigated.

The method and results of the self-consistent calculation of the displacement of the surface ion plane in different metals are presented, and the influence of these relaxation effects on the values of surface characteristics for a number of metals is analyzed.

The electron–ion interaction in metals is widely described using the pseudopotential method. One of the simplest types of pseudopotentials is the Ashcroft pseudopotential characterized by only one adjustable parameter. The self-consistent variational procedure of the density functional method allows us to uniquely identify the parameter of the Ashcroft pseudopotential for various metals, and thus to calculate the influence of the electron–ion interaction on the magnitude of the surface energy and the work function of metals. This chapter presents a comparative analysis of the results obtained using the DFM, with the results obtained with the assistance of other models, in particular, the more complex generalized Heine–Abarenkov pseudopotential characterized by two parameters. The introduction of an additional parameter in the pseudopotential determined by comparing the results of calculation of the surface energy with the experimental values, makes it possible to vary the calculated values of the work function while comparing the values of both surface characteristics of metals.

The material of the chapter draws heavily on the results obtained directly by the authors of this book and on published original research articles [25, 140–145].

5.1. Methods of calculating the surface energy

5.1.1. The calculation of the surface energy of metals in the framework of the 'jelly' model

The density functional method (DFM) is one of the most popular at present, and is widely used in atomic and nuclear physics, physics of crystals and liquids, physics of surface and adsorption phenomena, etc. In this method, the ground-state energy of interacting particles in an external field and the thermodynamic potential of the system are represented as valued functionals, which depend only on the particle number density $n(\mathbf{r})$. The minimum of such a functional gives the true distribution of density in the system. Thus, as a basis for calculating the ground state energy and thermodynamic properties of the system, instead of introducing the particle density, defined in the usual three-dimensional space, we introduce the many-body wave function which depends on a large number of variables in the configuration space. It is obvious that it is immeasurably easier to deal with the function of three variables than work with the complete wave function of the system. The complexity of the same approach lies in the fact that the exact form of the density functional, which determines the ground-state energy of the system is unknown and its construction is beyond the scope of the density functional method. The density functional components are determined, as a rule, in the framework of the quantum theory of many-electron systems. In the mathematical treatment of the DFM, for the known dependence of the energy on the particle number density $E[n]$ we solve the following variational problem is solved

$$\frac{\partial}{\partial n}(E[n] - \mu N) = 0,$$

which allows to obtain the true distribution of the particles of the system and determine its energy characteristics. Here $\mu = \dfrac{\partial E[n]}{\partial N}$ is the Lagrange multiplier. While maintaining the full particle number $\left(N = \int n(\mathbf{r})dV = const\right)$ factor μ corresponds to the chemical potential of the system.

To study the surface characteristics, we use the model which is a semi-infinite metal, bordering with vacuum. Consider a plane boundary of the metal, the outer normal to this boundary is regarded as the z axis so that the region of space occupied by metal corresponds to $z < 0$. In this model, electronic density $n(\mathbf{r})$ and electrostatic potential $\phi(\mathbf{r})$ are considered as a function only of z.

We represent the total energy functional in the form of the gradient expansion

$$E[n(z)] = \int_{-\infty}^{\infty} \left\{ w_0[n(z)] + w_2[n(z), |\nabla n|^2] + O(|\nabla|^4) \right\} dz, \qquad (5.1)$$

where $w_0[n(z)]$ is the energy density of the homogeneous electron gas, which takes into account the following components (all expressions are given in atomic units, where $e = m_e = \hbar = 1$):

1) The electrostatic energy of the electrons, which includes the potential created by the distribution of the electrons and the background of the positive charge (it is also possible to take into account the potential for the external field)

$$w_{cul}(z) = 0.5\phi(z)n(z); \qquad (5.2)$$

2) kinetic (Fermi) energy of the electron gas density with density n, neglecting its heterogeneity

$$w_{kin}(z) = 0.3(3\pi^2)^{2/3} n^{5/3}(z); \qquad (5.3)$$

3) decrease in the electrostatic energy of the electron gas due to the exchange interaction of electrons (in the Hartree–Fock approximation)

$$w_x(z) = -3/4(3/\pi)^{1/3} n^{4/3}(z); \qquad (5.4)$$

4) the correlation energy (using the Wigner interpolation formula)

$$w_c(z) = \frac{-0.056 n^{4/3}(z)}{0.079 + n^{1/3}(z)}, \qquad (5.5)$$

and the expression

$$w_2[n(z), |\nabla n|^2] = \frac{|\nabla n(z)|^2}{72n(z)} + w_{2,xc}(z)[n(z), |\nabla n|^2] \qquad (5.6)$$

is the first member of the gradient expansion, which takes into account inhomogeneity of the electron gas. Here the first term is the amendment for the inhomogeneity of the electron gas to kinetic energy (Weizsacker-Kirzhnits formula) and the second – an amendment to the heterogeneity of for the exchange and correlation energies. For $w_{2,xc}$ we used the following approximation:

1) random-phase approximation (RPA)

$$w_{2,xc}(z) = 0; \qquad (5.7)$$

2) Hubbard' approach [352]

$$w_{2,xc}(z) = \frac{|\nabla n|^2}{29.16n^{4/3} + 12n + 1/235n^{2/3}};\qquad(5.8)$$

3) the Singwi–Scholander–Tozi–Landa (SSTL) approximation [461]

$$w_{2,xc}(z) = \frac{A(n)B^2(n)\,|\nabla n|^2}{3^{4/3}\,\pi^{5/3}n^{4/3}},\qquad(5.9)$$

where $A(n) = 1.063 - 0.11537k_F(n)$, $B(n) = -0.2736 + 0.618k_F^{1/5}(n)$, $k_F = (3\pi^2 n)^{1/3}$ is the Fermi wave vector;

4) the Langref–Mehl (LM) approximation [385]

$$w_{2,xc}(z) = \frac{0.00214\,|\nabla n|^2}{n^{4/3}}\left[2\exp\left(-\frac{0.262\,|\nabla n|}{n^{7/6}}\right) - 7/9\right];\qquad(5.10)$$

5) the Vashishta–Singwi approximation (VS) [497]

$$w_{2,xc}(z) = \frac{A(n)B^2(n)\,|\nabla n|^2}{3^{4/3}\,\pi^{5/3}n^{4/3}},\qquad(5.11)$$

where $A(n) = 0.4666 + 0.3735k_F^{-2/3}(n)$, $B(n) = -0.0085 + 0.3318k_F^{1/5}(n)$.

These expressions for $A(n)$, $B(n)$ are obtained by interpolating the values given in the relevant papers.

The distribution function of electron density $n(z)$ near the surface must satisfy the requirement of exponential decay outside the metal and to tend to the bulk value of the electron density inside the metal. We apply the method of trial functions, i.e. we introduce some function $n = n(\alpha_i, z)$ and $\phi = \phi(\alpha_i, z)$, where α_i are some of the parameters determined from the condition of the minimum total energy of the system. Substituting thus determined $n(z)$ and $\phi(z)$ into (5.1), we find the total energy of our system.

As trial functions for the electron density and the electrostatic potential we choose the solutions of the linearized Thomas–Fermi equation [462]:

$$n(z) - n_0\left[\left(1 - \frac{1}{2}e^{\beta z}\right)\Theta(-z) + \frac{1}{2}e^{-\beta z}\Theta(z)\right],$$

$$\phi(z) = -\frac{4\pi n_0}{\beta^2}\left[\left(1 - \frac{1}{2}e^{\beta z}\right)\Theta(-z) + \frac{1}{2}e^{-\beta z}\Theta(z)\right],\qquad(5.12)$$

where n_0 is the bulk electron density, $\Theta(z)$ is the stepped Heaviside function, β is a variational parameter. From the physical viewpoint the value of β is the characteristic thickness of the surface layer at which the electron density changes dramatically.

It should be noted that, formally, the integral (5.1) diverges, or in other words, the total energy of the electron gas of the semi-bounded system is infinite. However, we are only interested in the effects associated with the presence of the surface. Therefore, given the dimensionality of the problem, it is convenient to introduce the surface energy, which in the 'jelly' model is the difference between the total energy of the system, when the electrons are distributed in accordance with the function $n(z)$, and when their distribution is given by a 'step', i.e., $n(z) = n_0(z) = n_0\Theta(-z)$:

$$\sigma_0(\beta_{min}) = \int_{-\infty}^{\infty} \left\{ w[n(z)] - w[n_0\Theta(-z)] \right\} dz, \qquad (5.13)$$

where the values of the parameter β_{min} can be obtained from the condition of minimum surface energy

$$\left. \frac{\partial \sigma_0(\beta)}{\partial \beta} \right|_{\beta_{min}} = 0. \qquad (5.14)$$

The program of numerical integration in (5.13) in the implementation of the simultaneous minimization of the surface energy by the Hooke–Jeeves method (see Appendix) allows us to calculate the surface energy σ_0 of metals in the 'jelly' model as a function of the parameter $r_s = (4\pi n_0/3)^{-1/3}$ for the different exchange–correlation corrections to the heterogeneity of (5.7)–(5.11). The results of calculations in [35] are presented in Fig. 5.1, which shows that the approach can be more or less applied only to describe the surface energy of metals with a sufficiently low concentration of free electrons ($r_s > 2.5$), that is, in fact, only for a series of alkali metals. At low values of the electron density the surface energy is positive and tends to zero at $r_s \to \infty$. At the same time, the main contribution to the surface energy is associated with the exchange–correlation interaction. For large values of electron density corresponding to most other metals, the surface energy begins to fall rapidly and moves into the region of non-physical negative values. Thus, we can conclude that it is essential to make a significant adjustment to the 'jelly' model going from a continuous uniform distribution of charge ions (positive 'background') of the 'jelly' model to take into account the discrete distribution of ions in the crystal lattice.

5.1.2. Influence of the discreteness of the lattice on the value of surface energy

In accordance with [381] taking into account the discreteness in the distribution of ions over the sites of the crystal lattice of metals leads to amendments to the electrostatic interaction energy due to both ion–ion and ion–electron interactions. As a result, the surface energy of the metal can be written as follows:

$$\sigma = \sigma_0 + \sigma_{ii} + \sigma_{ei}, \qquad (5.15)$$

where σ_0 is the contribution of the electronic system in the framework of the 'jelly' model, σ_{ii} is the contribution arising from the electrostatic

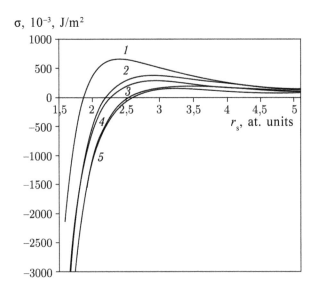

Fig. 5.1. The dependence of surface energy σ_0 of metals in the framework of the 'jelly' model on the density parameter r_s for different exchange–correlation amendments to the heterogeneity: *1* – Hubbard approximation, *2* – VS; *3* – SSTL; *4* – LM; *5* – RPA.

interaction of ions with each other; σ_{ei} is the contribution due to the difference in the electrostatic interaction between electrons and discrete ions and with a homogeneous 'jelly' background. For σ_{ii} we can use the analytical expression from [308]:

$$\sigma_{ii} = \frac{\sqrt{3}Z^2}{c^3} \exp\left(-\frac{4\pi d}{\sqrt{3}c}\right), \tag{5.16}$$

where Z is the charge of the ions; c is the distance between the nearest neighbours in a plane parallel to the surface; d is the interplanar distance.

To calculate the corrections for the electron–ion interaction, we must specify the potential of the individual ions in the metal. Since currently we cannot define the explicit form of this potential, such calculations in solid state theory are carried out using various methods and approximations, in particular, the so-called method of pseudopotentials [101]. We use the Ashcroft pseudopotential (potential of empty spheres) widely used in metal physics; according to the definition of this pseudopotential, in the vicinity of each i-th node with the valence ion Z the conduction electrons 'feel' the electrostatic potential given by the following expression:

$$V_{ei}(\mathbf{r}) = \begin{cases} 0, & |\mathbf{r}-\mathbf{r}_i| < r_c; \\ -\dfrac{Z}{|\mathbf{r}-\mathbf{r}_i|}, & |\mathbf{r}-\mathbf{r}_i| > r_c, \end{cases} \tag{5.17}$$

where the r_c is the so-called *cutoff radius*.

As seen from (5.17), an ion in the Ashcroft model is an empty sphere with radius r_c which attracts electrons to itself under the law of the usual Coulomb interaction.

The question of the value of the parameter r_c for the Ashcroft pseudopotential of a metal can be solved as follows. For this we consider the sphere with the atomic radius R_{at}, inside which there is an ion which creates an electrostatic field with potential (5.17). We write the three-dimensional electron gas energy per atom as follows: $E_V = E_{cul} + E_{kin} + E_x + E_c$, where E_{cul}, E_{kin}, E_x, E_c are, respectively, Coulomb, kinetic, exchange and correlation components. The expressions for E_{kin}, E_x, E_c are easily obtained if in (2.3)–(2.5) we move from the density of the particle number n to the parameter $r_s : n = 3/4\pi r_s^3$ and multiply each component by the volume of one atom $V_{at} = 4Z\pi r_s^3/3$.

The Coulomb component of energy E_{cul} can be written as

$$E_{cul} = 1/2 \int_0^{R_a} nV_{el}(\mathbf{r})d\mathbf{r} + \int_0^{R_a} nV_i(\mathbf{r})d\mathbf{r}, \qquad (5.18)$$

where the first integral describes the energy of the interaction of the electron gas with density n with the potential created by the electron gas, and the second integral is directly responsible for the electron–ion interaction, $R_a = Z^{1/3}r_s$ is the radius of the atom in the metal. Factor $V_{el}(\mathbf{r})$ can be defined as the potential created by a sphere of radius R_a, filled with free electrons:

$$V_{el}(\mathbf{r}) = \begin{cases} \dfrac{3Z}{2R_a}\left(1 - \dfrac{r^2}{3R_a^2}\right), & r \le R_a; \\ Z/r, & r > R_a, \end{cases} \qquad (5.19)$$

$V_i(\mathbf{r})$ in (5.18) is defined in the form of the Ashcroft pseudopotential (5.17). The procedure for elementary integration in (5.18) yields the following expression for E_{cul}:

$$E_{cul} = -\frac{9Z^2}{10R_a} + \frac{3Z^2 r_c^2}{2R_a^3}. \qquad (5.20)$$

Carrying out the summation of all components of the energy we can finally obtain the following expression for the total energy of the electron of gas per atom as a function of density parameter r_s:

$$E_V = w(n_0)Z n_0 =$$
$$= Z\left(\frac{1.105}{r_s^2} - \frac{0.458}{r_s} - \frac{0.056}{1+0.127r_s} + \frac{1.5r_c^2}{r_s^3} - \frac{0.9Z^{2/3}}{r_s}\right). \qquad (5.21)$$

Using the condition of the minimum total energy of the electron gas in relation to r_s, we obtain the relation for an unambiguous determination of r_c for different metals:

$$r_c = (\sqrt{2}/3)r_s\left(0.458 - \frac{2.21}{r_s} + 0.9Z^{2/3} + \frac{0.0071r_s^2}{(1+0.127r_s)^2}\right)^{1/2}. \qquad (5.22)$$

To describe the electron–ion component of the surface energy σ_{ei} we use the technique developed in [111]. According to [111], potential (5.17) generated by the ions is homogeneously 'smeared' over the crystal planes parallel to the surface, with coordinates $z = -1/2d, -3/2d, -5/2d$, etc. (interplanar distance $d > r_c$). Then the correction σ_{ei} will be calculated as follows:

$$\sigma_{ei} = \int_{-\infty}^{\infty} \delta V(z)\left[n(z) - n_0\Theta(-z)\right]dz, \qquad (5.23)$$

where $\delta V(z)$ is the sum, averaged over the crystal planes, of ionic pseudopotentials less the potential of the semi-infinite homogeneous background of the positive charge, namely,

$$\delta V(z) = \left\langle \sum_i V_{Ash}^i(\vec{r}) \right\rangle - \phi_+(z). \qquad (5.24)$$

The magnitude of $\delta V(z)$ is conveniently represented as the sum of two components:

$$\delta V(z) = \delta V_1(z) + \delta V_2(z),$$

where

$$\delta V_1(z) = \sum_i \langle V_{Ash}^i(\mathbf{r}) + Z/|\mathbf{r}-\mathbf{r}_i| \rangle,$$

$$\delta V_2(z) = \left\langle \sum_i -Z/|\mathbf{r}-\mathbf{r}_i| \right\rangle - \phi_i(z). \qquad (5.25)$$

The value of $\delta V_1(z)$ represents the average value of the set of point Coulomb potentials, which are different from zero only in the spheres of radius r_c, described around each i-th node. Averaging means smearing the ion charge on the crystal planes in accordance with condition $\sum Z = \int \rho dz$, where $\rho = n_0 d$ is the surface charge density. As a result, for example, for the area $-d/2 - r_c < z < -d/2 + r_c$ the electrostatic field can be defined by the potential of the uniformly charged plane $z = -d/2$, and then 'matching' at the boundaries of the region we find that for $-d < z < 0$:

$$\delta V_1(z) = 2\pi n_0\{(d(r_c - |z+d/2|)\Theta(r_c - |z+d/2|). \qquad (5.26)$$

Calculation of $\delta V_2(z)$ is also a simple electrostatic problem, and as a result for the magnitude of $\delta V(z)$ in the first subsurface ion plane $(-d < z < 0)$ we obtain the following expression:

$$\delta V(z) = 2\pi n_0 \{d(r_c - |z + d/2|)\Theta(r_c - |z + d/2|) - [z + d\Theta(-z - d/2)]^2\}.$$
(5.27)

Using distribution (5.12) for $n(z)$, and in the summation over the ionic planes with $z = -(i + d/2)$, $i = 1, 2, ...$, the periodicity of the potential $\delta V(z - d) = \delta V(z)$, as a result from (5.23) we obtain the following expression for the electron–ion component of the surface free energy:

$$\sigma_{ei} = \frac{2\pi n_0^2}{\beta^3}\left(1 - \frac{\beta d \exp(-\beta d/2)}{1 - \exp(-\beta d)}\operatorname{ch}(\beta r_c)\right).$$
(5.28)

Equations (5.15), (5.16), (5.28) form the basis for calculating the surface energy of the metal, taking into account the discreteness of the distribution ions in the lattice sites. The value of β as the variational parameter is determined by minimizing the full expression (5.15) for the surface energy σ, i.e. the solution of the equation

$$\left.\frac{\partial\sigma(\beta)}{\partial\beta}\right|_{\beta_{min}} = \left.\frac{\partial(\sigma_0 + \sigma_{ii} + \sigma_{ei})}{\partial\beta}\right|_{\beta_{min}} = 0.$$
(5.29)

Components σ_{ii} and σ_{ei} in the total surface energy can be used to analyze its dependence on the parameters and the type of symmetry of the crystal lattice, as well as to explain the observed, in a number of experiments [230], dependence of the surface energy of metals on the crystallographic orientation of the surface facet.

As a result, taking into account the discreteness of the crystal structure, the value σ and β values become explicit functional dependences on parameters n_0, Z, d, c, r_c, which characterize the electronic properties of the metal, its symmetry and orientation of surface facets. The values of these parameters are used in the calculations are given in Table 5.1.

It is known in melting of most metals the properties such as electrical conductivity, thermal conductivity, specific heat, the surface energy, etc., vary slightly (~10%). This means that the electronic state remains practically unchanged and, consequently, the electronic component of the surface surface energy of metallic melts can still be calculated in the framework of the above variant of the electron density functional. Moreover, for liquid metals at $T = T_m$ the bulk electron density n_0 due to expansion effects of the expansion is lower values of n_0 for solid metals. Because of this, in accordance with (5.22), the Ashcroft pseudopotential parameter r_c for liquid metals is characterized by higher values than for solid metals. Taking into account the isotropic nature of the liquid metals, the parameters c and d, defining, respectively, the distance between the nearest neighbours in a

Table 5.1. The values of input parameters used to calculate the surface characteristics, and the experimental values of the surface energy σ_{exp} and of the parameter of relative displacement δ/d_{exp} of the surface ion plane for a number of metals

Metal	Face	Z	n_0, a.u.	r_c, a.u.	d, a.u.	c, a.u.	δ/d_{exp}, %	σ_{exp}, mJ/m²
Na	110				5.71	6.99		
BCC	100	1	0.0038	1.74	4.03	8.06		220
	111				2.33	11.41		
	liquid		0.0036	1.77	5.71	5.71		191
K	110				6.99	8.55		
BCC	100	1	0.00196	2.34	4.94	9.88		149
	111				2.85	13.98		
	liquid		0.00186	2.38	6.99	6.99		114
Al	111				4.92		+2.5	
FCC	100	3	0.0269	1.11	3.71	5.25		1140
	110				2.63		−5.0	
	liquid		0.0237	1.173	4.65	4.65		856
Cu	111				3.92		−4.1	
FCC	100	2	0.0252	0.938	3.39	4.80		1750
	110				2.40		−10.0	
	liquid		0.0219	1.017	4.72	4.72		1350
Fe	110				3.84	4.70		
BCC	100	4	0.0504	0.945	2.71	5.43	−1.4	2.170
	111				1.57	7.67		
Cr	110				3.85	4.72		
BCC	100	4	0.0492	0.956	2.72	5.44		2200
	111				1.57	7.69		
Pb	111				5.38			
FCC	100	4	0.0194	1.46	4.66	6.589		560
	110				3.29			
	liquid.		0.0184	1.49	6.13	6.13		453

plane parallel to the surface, and the interplanar distance is taken for melts equal to the parameter of the short range order determined on the basis of the position of the first peak of the radial distribution function.

Calculation of the parameter β, characterizing the change in the electron density near the surface, and of the surface energy σ was carried out by us on a PC using the above procedure for a number of metals (Na, Pb, Cu, Al) using the data in Table 5.1 and the results are presented in Table 5.2. Calculations of these quantities were performed for different orientations of the surface face of the metals in the crystalline phase and also for metals in the molten state.

Table 5.2. The results of calculation of values of the parameter of the decrease the electron density β (a.u.) and the surface energy σ (mJ/m²) for a number of metals, taking into account different exchange–correlation corrections for the heterogeneity

Metal		RPA	Hubbard	SSTL	VS	LM
Na 110	β	1.24	0.85	0.98	0.88	1.12
	σ	139	284	209	265	160
Na 100	β	0.85	0.71	0.77	0.73	0.89
	σ	167	280	219	264	187
Na 111	β	0.77	0.66	0.71	0.67	0.78
	σ	327	431	374	417	349
Na (liq.)	β	1.21	0.84	0.96	0.86	1.11
	σ	122	257	188	242	142
Pb 111	β	1.48	1.12	1.29	1.26	1.44
	σ	558	1620	966	1064	635
Pb 100	β	1.08	0.94	1.01	1.00	1.09
	σ	537	1381	847	925	620
Pb 110	β	0.82	0.78	0.81	0.80	0.82
	σ	1838	2512	2081	2142	1933
Pb (liq.)	β	1.85	1.31	1.55	1.50	1.64
	σ	455	1660	931	1050	566
Cu 111	β	1.45	1.08	1.24	1.19	1.38
	σ	275	931	542	639	331
Cu 100	β	1.25	1.00	1.12	1.08	1.26
	σ	244	836	480	567	295
Cu 110	β	1.05	0.90	0.98	0.96	1.07
	σ	451	971	654	731	507
Cu* 111	β	1.53	1.2	1.36	1.34	1.57
	σ	230	1693	772	872	346
Cu* 100	β	1.28	1.08	1.19	1.17	1.21
	σ	132	1413	595	682	315
Cu* 110	β	1.04	0.95	1.00	0.99	0.97
	σ	968	2059	1353	1427	1902
Al* 111	β	1.71	1.30	1.50	1.48	1.63
	σ	530	2246	1167	1269	643
Al* 100	β	1.39	1.14	1.27	1.26	1.40
	σ	696	2161	1227	1313	793
Al* 110	β	1.06	0.97	1.02	1.02	1.07
	σ	2745	3933	3162	3231	2858

Al 111	β	1.47	1.18	1.32	1.31	1.46
	σ	292	1812	847	937	389
Al 100	β	1.18	1.04	1.12	1.11	1.20
	σ	191	1458	650	726	296
Al 110	β	0.96	0.89	0.93	0.92	0.96
	σ	1761	2848	2139	2202	1881
Fe 110	β	1.6	1.33	1.48	1.47	1.61
	σ	−473	2683	598	635	−304
Fe 100	β	1.11	1.04	1.09	1.08	1.12
	σ	1259	3616	2028	2055	1822
Fe 111	β	0.98	0.94	0.96	0.96	0.96
	σ	10169	12270	10849	10873	10356
Cr 110	β	1.58	1.31	1.46	1.46	1.59
	σ	−433	2615	604	644	−260
Cr 100	β	1.1	1.03	1.08	1.08	1.09
	σ	1322	3606	2069	2099	1846
Cr 111	β	0.97	0.93	0.96	0.95	0.96
	σ	10361	12399	11022	11055	10796

Cu – monovalent copper, Cu* – bivalent copper
Al – r_c = 1.11 a.u., Al* – r_c = 0.96 a.u.

From the experiments it is known that the values of σ decrease with increasing density of packing of the surface edge. For liquid metals, the surface energy is even lower than for close packed faces of the same metals. Table 5.2 shows that for sodium with a body-centered cubic (bcc) lattice, the calculated values are $\sigma_{110} < \sigma_{100} < \sigma_{111}$, while for metals with the face-centered cubic (fcc) lattice it was found that $\sigma_{100} < \sigma_{111} < \sigma_{110}$, although the {111} faces are more closely packed than the {100}. Investigation of the dependence of the surface energy on the value of r_c indicates that with decreasing r_c the surface energy increases and the increment of σ is larger for looser faces. As an example, Table 5.2 summarizes the values of σ for Al obtained as a result of the correction with r_c = 0.96 a.u. For the liquid metals all calculated values of σ with r_c, given in [111], were higher than the experimental ones. In this connection, calculations were carried out using used lower values of the parameter r_c (see Table 5.1). The corresponding values of σ are given in Table 5.2, which also lists results of a study of the effect of different exchange–correlation corrections arising in an inhomogeneous electron system, the value of the parameter β and the surface energy σ. It is seen that the exchange–correlation corrections lead to a significant increase in the values of the surface energy of metals. The approximations used in their calculation form the following series in order of increasing σ: LM, SSTL, VS, Hubbard approximation. The character of

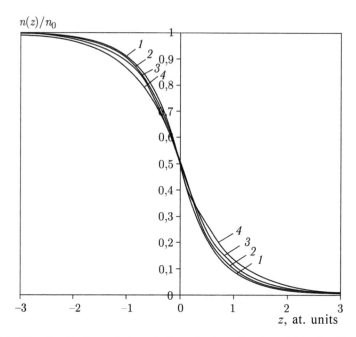

Fig. 5.2. The distribution of the relative electron density $n(z)/n_0$ in Al near the surface for different exchange–correlation corrections to the heterogeneity: *1* – RPA; *2* – LM; *3* – VS, SSTL; *4* – Hubbard approximation.

the influence of these approximations on the value of β is illustrated in in Fig. 5.2, which shows graphs of the electron density distribution in Al (r_c = 0.96 a.u.) near the surface for different exchange–correlation corrections to the heterogeneity.

Comparison of the calculated values of the surface energy σ with the available experimental data (Table 5.1) shows that none of the approximations used is universal for all the metals. Thus, for the simple metals the best approximation is VS (5.11). In the case of transition metals the best agreement with experiment (40–50%) is achieved using the correction for Hubbard's heterogeneity (5.8). Copper in its compounds and electronic properties can act as a monovalent and divalent metal. We calculated the surface energy for both cases. However, for monovalent copper all the approximations gave low results, whereas for the bivalent copper good agreement was obtained between the calculated and experimental values of the surface energy using the correction for Hubbard's heterogeneity.

5.1.3. Accounting for the effects of lattice relaxation of metallic surfaces

The above expressions for the components of surface energy do not take into account the effects of the displacement of surface ionic planes characteristic of the bulk material, from their equilibrium position. The metal surface is characterized by violation of the translational symmetry of the crystal in a direction perpendicular to the surface – a process of lattice relaxation of the surface takes place. The change in the interplanar spacing in the surface layer of the metal is indicated by the experimental data obtained by low-energy electron diffraction [360]. Changing the distance between planes causes changes in the surface energy of metals. The influence of the displacement of surface ion planes on the surface energy must be found self-consistently, taking into account the mutual influence of the profile of electron density near the surface and the relaxing ion plane.

Consider what changes are caused by taking into account the displacement of a single ion the plane in the value of δ ($\delta > 0$ corresponds to the extension) relative to the position of this plane in the undisturbed material, characterized by the coordinate $z = -d/2$. We assume that in this case the trial function for the electron density $n(z)$ in (5.12) continues to be characterized only by the parameter β and does not acquire an explicit dependence on δ. In this case, the component of the surface energy $\sigma_0(\beta)$, reflecting the contribution from the electronic system within the framework of the 'jelly' model, also becomes explicitly dependent on δ. As for the contribution σ_{ii}, arising from the electrostatic interaction of ions with each other, using the method described in [308], in [35] we obtained the following expression:

$$\sigma_{ii}(\delta) = \frac{\sqrt{3}Z^2}{c^3}\exp\left(-\frac{4\pi(d-2\delta)}{\sqrt{3}c}\right) = \sigma_{ii}(0)\exp\left(\frac{8\pi\delta}{\sqrt{3}c}\right), \qquad (5.30)$$

where $\sigma_{ii}(0)$ is given by (5.16). In the component $\Delta\sigma_{ei}(\beta, \delta)$, reflecting the effects of electrostatic interaction of electrons with discrete ions, one can distinguish two contributions: $\Delta\sigma_{ei}^{(1)}(\delta)$ and $\Delta\sigma_{ei}^{(2)}(\beta,\delta)$. The first relates to the difference in interaction energies of the displaced ionic surface plane and the non-displaced ionic plane with a uniform electron background in the layer with the coordinates $-d < z < 0$. In accordance with [336]

$$\Delta\sigma_{ei}^{(1)}(\beta,\delta) = 2\pi n_0^2 d\delta^2. \qquad (5.31)$$

The contribution of $\sigma_{ei}^{(2)}(\delta)$ can be obtained using the ratio (5.23) with

$$\delta V(z) = 2\pi n_0\left[(d(r_c-|z+d/2-\delta|)\Theta(r_c-|z+d/2-\delta|)-\right.$$
$$\left.-(z+d)^2\Theta(-z-d/2+\delta)-(z^2+2d\delta)\Theta(z+d/2-\delta)\right] \qquad (5.32)$$

for $z > -d$. For $z < -d$ we used relation (5.32) with $\delta = 0$ and carry out summation over planes $z = -(i + d/2)$, $i = 1, 2, ...$, taking into account the periodicity of the potential $\delta V(z - d) = \delta V(z)$. As a result,

$$\sigma_{ei}(\beta,\delta) = \sigma_{ei}(\beta,0) + \Delta\sigma_{ei}^{(1)}(\delta) + \Delta\sigma_{ei}^{(2)}(\beta,\delta), \qquad (5.33)$$

where

$$\Delta\sigma_{ei}^{(2)}(\beta,\delta) = \frac{2\pi n_0^2}{\beta^2} d[1 - \exp(\beta\delta)]\exp(-\beta d/2)\mathrm{ch}(\beta r_c),$$

and $\sigma_{ei}(\beta, 0)$ from (5.28).

As a result, the total surface energy $\sigma(\beta, \delta)$ as a function of variational parameters is determined by the sum of the contributions given by expressions (5.13), (5.30), (5.33) and (5.34), with the energy density of the inhomogeneous electron gas $w[n(z)]$ from (5.2)–(5.6). The values of variational parameters β, δ, used in determining the surface energy of the metal, are found from the condition of minimality. The results of calculation of these quantities are presented in Table 5.3.

Analysis of calculation results shows that taking into account the effects of lattice relaxation near the surface leads to a slight decrease in surface energy values for the close-packed faces and to a maximum decrease σ of $3\div7\%$ for the loosest edges. The displacement δ of the surface plane of ions

Table 5.3. The results of calculation of the parameter of decrease of the electron density β(a.u.), the surface relaxation parameter δ(a.u.) and the surface energy σ (mJ/m^2) for a number of metals, taking into account different exchange–correlation corrections for heterogeneity and surface relaxation effects

Metal		RPZ	Hubbard	SSTL	VS	LM
Na 110	β	1.20	0.79	0.93	0.82	1.11
	δ	0.0278	0.1183	0.075	0.108	0.04
	σ	138	275	205	258	158
Na 100	β	0.83	0.68	0.74	0.69	0.88
	δ	0.027	0.097	0.063	0.09	0.011
	σ	166	275	217	260	187
Na 111	β	0.85	0.70	0.79	0.71	0.89
	δ	−0.200	−0.146	−0.198	−0.1519	−0.212
	σ	310	422	384	406	331
Na (liq.)	β	1.15	0.76	0.89	0.78	1.07
	δ	0.0464	0.1456	0.0984	0.1356	0.0587
	σ	121	246	182	232	140
Pb 111	β	1.47	1.08	1.26	1.22	1.43
	δ	−0.0061	0.0428	0.0213	0.0248	0.0083
	σ	558	1590	958	1054	634
Pb 100	β	1.00	0.86	0.93	0.92	1.01
	δ	0.053	0.092	0.0689	0.0725	0.0493
	σ	503	1276	789	861	589
Pb 110	β	0.87	0.80	0.84	0.83	0.88
	δ	−0.068	−0.048	−0.060	−0.056	−0.069
	σ	1773	2478	2028	2093	1866

Pb	β	1.85	1.29	1.54	1.49	1.64
(liq.)	δ	0.0005	0.0186	0.0068	0.0086	0.004
	σ	455	1884	1055	1190	565
Cu	β	1.4	1.00	1.18	1.12	1.35
111	δ	0.0123	0.0926	0.0565	0.066	0.0343
	σ	271	892	526	618	325
Cu	β	1.17	0.90	1.03	0.99	1.19
100	δ	0.0548	0.1255	0.0862	0.0968	0.052
	σ	232	774	450	530	284
Cu	β	1.09	0.90	1.00	0.97	1.11
110	δ	−0.0045	−0.002	−0.026	−0.020	−0.049
	σ	442	971	650	729	496
Cu*	β	1.48	1.13	1.3	1.28	1.01
111	δ	0.023	0.065	0.04	0.04	0.09
	σ	219	1614	740	836	313
Cu*	β	1.18	0.9	1.09	1.07	0.913
100	δ	0.052	0.055	0.07	0.075	0.012
	σ	89	1820	516	596	274
Cu*	β	1.09	0.96	1.04	1.03	0.91
110	δ	−0.045	−0.018	−0.034	−0.032	−0.005
	σ	932	2054	1332	1408	1887
Al	β	1.42	1.10	1.26	1.25	1.43
111	δ	0.022	0.058	0.036	0.039	0.021
	σ	280	1728	814	899	377
Al	β	1.09	0.94	1.02	1.01	1.10
100	δ	0.056	0.095	0.070	0.074	0.0540
	σ	131	1313	553	622	242
Al	β	1.03	0.93	0.99	0.98	1.04
110	δ	−0.069	−0.047	−0.061	−0.060	−0.071
	σ	1648	2792	2050	2117	1764
Fe	β	1.55	1.26	1.42	1.41	1.18
110	δ	0.019	0.047	0.03	0.031	0.06
	σ	−504	2515	530	565	−354
Fe	β	1.09	1.00	1.05	1.05	0.98
100	δ	0.018	0.041	0.026	0.027	0.041
	σ	1239	3520	1988	2014	1783
Fe	β	1.09	1.01	1.06	1.06	1.02
111	δ	−0.138	−0.121	−0.133	−0.132	−0.118
	σ	9217	11513	9967	9993	9136

Cr 110	β	1.53	1.24	1.40	1.39	1.16
	δ	0.0217	0.05	0.031	0.03	0.06
	σ	−466	2443	532	571	−342
Cr 100	β	1.08	0.99	1.05	1.05	0.98
	δ	0.017	0.038	0.025	0.025	0.041
	σ	1307	3523	2036	2065	1816
Cr 111	β	1.09	1.01	1.06	1.06	0.99
	δ	−0.148	−0.132	−0.142	−0.142	−0.132
	σ	9300	11540	10034	10063	9978
Cu − monovalent copper, Cu* − bivalent copper						

relative to the volume position depends on the orientation of surface facets and takes smaller values than the more closely packed face. The value of δ is maximum in magnitude and has the character of compression ($\delta < 0$) for the loosest face.

Comparison of the results of calculation of the relative magnitude of surface relaxation δ/d (%) with its experimental values [360] shows that the values of δ/d, calculated for Al in the framework of VS approximation, are in qualitative agreement with experimental values (Table 5.1), but differ several times quantitatively.

5.1.4. Influence of gradient corrections of the fourth order in the calculation of the surface energy of metals

In order to improve the quantitative agreement of the values of surface energy for the transition and noble metals we can consider the effect on the surface characteristics of a member of the fourth order in powers of the gradient of $n(z)$ to the kinetic energy density [395]:

$$w_{4,kin} = \frac{1.336n}{540(3\pi^2 n)^{2/3}}\left[\left(\frac{\nabla^2 n}{n}\right)^2 - \frac{9}{8}\left(\frac{\nabla^2 n}{n}\right)\left(\frac{|\nabla n|}{n}\right)^2 + \frac{1}{3}\left(\frac{|\nabla n|}{n}\right)^4\right],$$

(5.34)

and the density of the exchange–correlation energy [61]:

$$w_{4,xc} = 2.94 \cdot 10^{-4}\exp\left(-0.2986n^{-0.26}\right)\left(\frac{\nabla^2 n}{n}\right)^2.$$

(5.35)

Table 5.4 reflects the results of our [35] study of the effect on the values of surface energy σ of the gradient fourth-order corrections in the different approximations for the exchange–correlation energy in an inhomogeneous electron system. Since the best approximation for simple metals is the VS approximation, and in the case of transition metals taking into account the terms of the gradient expansion of higher order leads to a significant increase (in comparison with simple metals) in the values of surface energy, in what follows we have used the VS approximation for the calculation of surface characteristics of metals. This approximation is regarded as most universal.

The results of the calculations are given in Table 5.5.

The results of calculations show that for simple metals it is best to use gradient corrections of only the second order. For Al it is necessary to take into account the gradient corrections of the fourth order to the kinetic energy, due to greater heterogeneity of the electron gas in the *sp*-hybridization of quantum states of the electrons. For noble and transition metals, the best agreement with the experimental values of surface energy was obtained by using gradient corrections to the kinetic and exchange-correlation energy up to the fourth order due to the effects of *spd*-hybridization.

Copper in its compounds and the electronic properties can act as both monovalent and divalent metal. The surface energy for both cases was calculated. However, the values obtained by all the approximation for monovalent copper were too low compared with the experimental values, while the values obtained for the bivalent copper were in good

Table 5.4. The results of calculating the values of surface energy σ (mJ/m^2) for a number of metals in the Ashcroft pseudopotential model taking into account gradient corrections up to fourth order for the most closely packed facets at different exchange–correlation corrections for heterogeneity

Metal	Gradient correction of 4-th order	RPA	Hubbard	SSTL	VS
Na	to kinetic energy	196	316	250	300
	to exchange-correlation energy	287	388	332	374
K	to kinetic energy	109	158	134	166
	to exchange-correlation energy	147	189	168	196
Al	to kinetic energy	579	1985	1078	1165
	to exchange-correlation energy	1361	2584	1785	1860
Cu	to kinetic energy	534	1871	1013	1104
	to exchange-correlation energy	1319	2469	1721	1799
Fe	to kinetic energy	176	2995	1008	1043
	to exchange-correlation energy	1506	4142	2365	2395
Cr	to kinetic energy	376	2915	997	1035
	to exchange-correlation energy	1469	4019	2301	2335

Table 5.5. The results of calculating the values of surface energy σ (mJ/m²) for a number of metals in the Ashcroft pseudopotential model taking into account the gradient corrections up to fourth order in the VS approximation for the most close-packed faces

Metal	Gradient correction of 2nd order	Gradient correction of 4-th order to kinetic energy	Gradient correction of 4-th order to exchange-correlation energy
Na	<u>267</u>	300	374
K	<u>151</u>	166	196
Al	941	<u>1165</u>	1860
Cu	872	1104	<u>1798</u>
Fe	635	1043	<u>2394</u>
Cr	641	1034	<u>2334</u>
Mo	876	1304	<u>2759</u>
Underlined values are in the best agreement with experimental results [225].			

Table 5.6. The results of calculation of the parameter of decrease of electron density β, the relaxation parameter δ, as well as the corresponding values surface energy, taking into account the effects of surface relaxation

Metal	Face	β, a.u.	σ, mJ/m²	β_δ, a.u.	δ, a.u.	σ_δ, mJ/m²
Na	110	0.876	266	0.814	0.109	259
(BCC)	100	0.726	265	0.689	0.09	261
	111	0.674	417	0.711	−0.152	407
K	110	0.728	151	0.666	0.153	146
(BCC)	100	0.605	150	0.573	0.114	147
	111	0.57	233	0.602	−0.206	226
Al	111	1.106	1165	1.146	0.053	1096
(HCC)	100	1.065	876	0.885	0.112	1136
	110	0.914	2295	0.955	−0.054	2224
Cu	111	1.049	1799	0.986	0.096	1621
(HCC)	100	0.986	1400	0.903	0.124	1148
	110	0.902	1934	0.904	−0.003	1935
Fe	110	1.213	2394	1.15	0.063	2088
(BCC)	100	1.016	2950	0.976	0.046	2818
	111	0.928	11531	0.99	−0.117	10822
Cr	110	1.202	2335	1.136	0.065	2025
(BCC)	100	1.008	2966	0.971	0.044	2848
	111	0.921	11687	0.988	−0.127	10879
Mo	110	1.23	2759	1.172	0.052	2468
(BCC)	100	0.981	4247	0.943	0.042	4098
	111	0.882	19153	0.956	−0.138	17780

agreement when using the fourth-order gradient corrections to the kinetic and exchange–correlation energies. The introduction of different types of gradient corrections in the description of the surface properties of simple and transition metals has made it possible to use a single exchange-correlation correction in the VS approximation (5.11). It should be noted that this fact greatly simplifies and clarifies calculations of the adhesion characteristics of metal contacts.

The estimated totals of surface energy and variational parameters for the different orientations of surface faces are shown in Table. 5.6.

5.2. Methods of calculating the electron work function of metal surfaces

Another important characteristic is the surface work function of an electron from the metal surface. The complexity of the theoretical description and comparison of the results of the calculation of the work function with the experimental values in comparison with a similar task for the surface energy of metals is expressed by the small experimental relative change in the work function for a number of materials with metallic properties. Thus, in the metals molybdenum possesses the largest value of surface energy, which is nearly twenty times greater than the surface energy for potassium, while the value of the work function for these metals differs by only a factor of 2. So far, no satisfactory agreement has been obtained between the calculated and experimental results of the work function [39].

In addressing the issue of research into the work function, it must be borne in mind that we have considered the pure surface of a metal placed in a vacuum.

Contamination of this surface, adsorption of foreign substances, and distortion of the lattice by foreign atoms will change the properties of the metal surface and thus affect the value of the work function. The surface of the metal may be adsorb for different reasons both positive and negative ions, neutral atoms, molecules representing hard or induced dipoles, as well as molecules affected by the van der Waals forces from the metal surface. The adsorption of positive ions on the the metal surface causes the formation of an electrical double layer and greatly reduces the work function W. The adsorption of the negative ions leads to an increase in W. The adsorption of molecules having a constant or induced electric dipole moment is also accompanied by the formation of the electrical double layer, which considerably lowers the work function. In addition, many atoms can be adsorbed on the metal surface by chemical means to form single and polyatomic layers, such as oxygen in the oxide layers. These layers on the surface can be polarized. In many cases, atoms or molecules of these layers are arranged in some preferred areas, determined by the lattice structure. This causes a significant change in W. Since metals tend to adsorb atoms and molecules on their surface, measurement of the work function requires the use of high-vacuum techniques and intensive surface treatment of metals with a view to removal of adsorbed particles. Knowing the value of the work function for a given metal makes it possible to control the purity of its surface by direct measurement of the work function.

The work function is defined as the minimum energy necessary to remove an electron from the solid. Its nature is connected with the existence of a potential barrier near the metal surface. The work function is determined by the difference in the potential barrier height and the chemical potential:

$$W = e\varphi = D - \mu. \tag{5.36}$$

The value of the potential of the electric dipole layer V_D, acting on an electron near the surface in the 'jelly' model, can be obtained from the Poisson equation

$$\Delta V_D(z) = -4\pi\rho(z) \tag{5.37}$$

given that

$$\rho(z) = e[n(z) - n_0\Theta(-z)] = en_D(z). \tag{5.38}$$

As a result,

$$V_D(z) = 4\pi e\left(z\int_z^\infty n_D(z')dz' - \int_z^\infty z'n_D(z')dz'\right). \tag{5.39}$$

Then the height of the dipole potential barrier D_0 in the 'jelly' model is given by

$$D_0 = e[V_D(\infty) - V_D(-\infty)] = 4\pi e^2\int_{-\infty}^\infty dz'z'n_D(z'). \tag{5.40}$$

We apply in these ratios the distribution of the electron density $n(z)$ in the form of the solution of the linearized Thomas–Fermi equation (5.12) which has been already been used to calculate the surface energy. Then, after integration in (5.40) we obtain that

$$D_0 = \frac{4\pi n_0 e^2}{\beta^2}. \tag{5.41}$$

This expression for the dipole potential barrier within the 'jelly' model must be supplemented by the amendments to the electronic interaction with allowance for the discrete charge distribution of the ions in the crystal lattice. The effect of the electron–ion interaction on the work function is due to the difference of electrostatic interactions of ions with the electron density in the ground state and in a state with one electron removed. Following [381], we write an additional contribution to the potential barrier as

$$D_{ei} = \frac{-\int\limits_{-d}^{\infty} \delta V(z) n_e(z) dz}{\int\limits_{-\infty}^{\infty} n_e(z) dz}, \tag{5.42}$$

where d is the interplanar spacing, and $\delta V(z)$ has the meaning of the average sum of the ion potential averaged over the crystal planes of potentials of minus the potential of the semi-infinite uniform positive background.

Since the expression (5.43) is uniform with respect n_e, n_e can be regarded as equal to the surface charge density of a system situated in a weak electric field with strength E_z in a semi-infinite model of a homogeneous background:

$$n_e(z) = \begin{cases} \left[n_0 - \left(\dfrac{\beta E_z}{8\pi} + \dfrac{n_0}{2} \right) \right] \exp(\beta z), & z < 0; \\[3mm] -\left(\dfrac{\beta E_z}{8\pi} - \dfrac{n_0}{2} \right) \exp(-\beta z), & z > 0. \end{cases} \tag{5.43}$$

To calculate the corrections to the electron–ion interaction, as in the calculation of surface energy, we use the Ashcroft pseudopotential (5.17). For $\delta V(z)$ with $-d < z < 0$ we use the previously obtained expression (5.27). Carrying out summation over ion planes with $z = -(i + d/2)$, $i = 1, 2, ...,$ and using the periodicity of the potential $\delta V(z - d) = \delta V(z)$, from (5.43) we obtain

$$D_{ei} = -\frac{4\pi n_0 \exp(-\beta d/2)}{\beta^2[2 - \exp(-\beta d)]}[\beta d ch(\beta r_c) - 2sh(-\beta d/2)]. \tag{5.44}$$

Considering the effects of displacement of the surface ion plane with respect its volume position on the value of δ leads to an additional contribution to the dipole barrier

$$D_\delta = -\frac{4\pi n_0 e^{\beta(\delta - d/2)}}{\beta^2(2 - \exp(-\beta d))} \left(2(1 - e^{\beta\delta})sh\frac{\beta d}{2} - \beta^2 \delta d \right) $$
$$+ D_{ei}(e^{\beta\delta} - 1) - 4\pi n_0 d\delta. \tag{5.45}$$

As a result, the dipole potential barrier is determined by the sum of the contributions: $D = D_0 + D_{ei} + D_\delta$.

The chemical potential μ of the electron gas is determined by its bulk value

$$\mu = \frac{\partial w}{\partial n},$$

where w is the bulk energy density of the crystal. In the approximation of local density the bulk energy of the metal per one atom and expressed in terms of the density parameter r_s, is defined by the expression

$$E_V = w(n_0)Zn_0 =$$

$$= Z \left(\frac{1.105}{r_s^2} - \frac{0.458}{r_s} - \frac{0.056}{1+0.127r_s} + \frac{1.5r_c^2}{r_s^3} - \frac{0.9Z^{2/3}}{r_s} \right). \quad (5.46)$$

As a result, the expression for the chemical potential μ with the exchange–correlation and pseudopotential corrections taken into account has the form

$$\mu = 0.5(3\pi^2 n_0)^{2/3} - \left(\frac{3n_0}{\pi} \right)^{1/3} - \frac{0.056n_0^{2/3} + 0.0059n_0^{1/3}}{(0.079 + n_0^{1/3})^2} -$$

$$-0.4Z^{2/3} \left(\frac{4\pi n_0}{3} \right)^{1/3} + 4\pi n_0 r_c^2. \quad (5.47)$$

Due to the need to take into account the terms of the gradient expansion of the fourth-order identified in the calculation of the surface energy of metals, the influence of gradient corrections of the fourth order on the work function was studied [146] (Table 5.7). The values of variational parameters β and δ, required to calculate potential barrier height D, were determined by minimizing surface energy (Table 5.8). It is seen that, excluding the effects of displacement of the surface ion plane, we obtain a good agreement of the calculation results with the experimental values of the work function of the simple metals and a significant overestimation for the transition metals. Accounting for the effects of surface relaxation leads to the corresponding values for the transition metals but to lower values for the simple metals.

In connection the much stronger dependence of the work function on the displacement of the ion plane (from 20 to 50%) than for surface energy (see section 5.1.3), found in [146], there is a need for more accurate determination of the relaxation parameter. To do this, apparently, it is

Table 5.7. The results of calculating the values of work function W(eV) for several metals in the model of the Ashcroft pseudopotential, taking into account gradient corrections up to fourth order for the most closely packed faces

Metal	Gradient correction of second order	Gradient correction of fourth order to kinetic energy	Gradient correction of fourth order to exchange-correction energy
Na	2.63	3.1	3.8
K	2.07	2.38	2.87
Al	3.25	4.2	6.3
Cu	3.98	4.95	7.07
Fe	3.44	4.56	7.31
Cr	3.5	4.57	7.27

Table 5.8. The results of calculation of the parameter of the Ashcroft pseudopotential r_c, the parameter of reduction of electron density β, the relaxation parameter δ, taking into account the work function (W_δ) and without (W) considering the effects of surface relaxation, and the experimental values of work function W_{exp} [217]

Metal	Face	r_c. a.u.	W_{exp}. eV	β. a.u	W. eV	β_δ. a.u.	δ. a.u.	W_δ. eV
Na	110			0.876	2.65	0.814	0.109	2.11
	100	1.74	2.35	0.726	2.98	0.689	0.09	2.77
	111			0.674	3.05	0.711	−0.152	3.22
	Liquid	1.77		0.858	2.65			
K	110			0.728	2.07	0.666	0.153	1.58
	100	2.34	2.137	0.605	2.24	0.573	0.114	2.07
	111			0.57	2.2	0.602	−0.206	2.37
Al	111		4.24	1.106	4.19	1.146	0.053	3.44
	100	1.11	4.41	1.065	5.35	0.885	0.112	4.76
	110		4.06	0.914	6.98	0.955	−0.054	5.60
Cu	111		4.98	1.049	7.07	0.986	0.096	4.88
	100	0.938	4.59	0.986	7.6	0.903	0.124	5.68
	110		4.48	0.902	8.41	0.904	−0.003	8.40
Fe	110		4.31	1.213	7.3	1.15	0.063	4.48
	100	0.945	4.24	1.016	10.19	0.976	0.046	9.42
	111			0.928	11.8	0.990	−0.117	12.5
Cr	110			1.202	7.27	1.136	0.065	4.45
	100	0.956	4.58	1.008	10.1	0.971	0.044	9.36
	111			0.921	11.76	0.988	0.127	12.4

necessary to take into account the effects of multiple ion displacement of the surface layers, as well as the effect of temperature on the value of δ.

5.3. Application of the Heine–Abarenkov potential in the calculation of the surface characteristics of metals

In order to achieve the universality of the theory of surface phenomena, in [145] we used for the first time the more general Heine–Abarenkov pseudopotential. The explicit form of the pseudopotential can be written as follows:

$$V_{ei}(\mathbf{r}) = \begin{cases} -V_0 Z, & |\mathbf{r}-\mathbf{r}_i| < R_m; \\ -\dfrac{Z}{|\mathbf{r}-\mathbf{r}_i|}, & |\mathbf{r}-\mathbf{r}_i| > R_m, \end{cases} \quad (5.48)$$

where Z is the valence of the ions.

Under the condition $V_0 = 0$ and $R_m = r_c$, the Heine–Abarenkov pseudopotential changes to the expression for the Ashcroft pseudopotential.

Corrections to the electron–ion interaction to the dipole barrier in this model is obtained using a technique identical with that used in the Ashcroft model. As a result, for $\delta V(z)$ at $-d < z < 0$ we can obtain the following expression:

$$\delta V(z) = 2\pi n_0 \{d(R_m - |z + d/2|) - \frac{V_0 d}{2}[R_m^2 - (z + d/2)^2] \times$$
$$\times \Theta(R_m - |z + d/2|) - [z + d\Theta(-z - d/2)]^2\}. \tag{5.49}$$

Carrying out in the formula (5.23) summation over the ionic planes with $z = -(i + d/2)$, $i = 1, 2, \ldots$, and using the periodicity of the potential $\delta V(z - d) = \delta V(z)$, we obtain

$$D_{ei}^{XA} = -\frac{4\pi n_0 \exp(-\beta d/2)}{\beta^2 [2 - \exp(-\beta d/2)]} \{\beta dch(\beta R_m) +$$
$$+ V_0 d[sh(\beta R_m) - \beta R_m ch(\beta R_m)] - 2sh(-\beta d/2)\}. \tag{5.50}$$

Consideration of pseudopotential amendments in the Heine–Abarenkov model leads to another form of the expression for the bulk energy of the metal per atom:

$$E_V = w(n_0)Zn_0 = Z\left(\frac{1.105}{r_s^2} - \frac{0.458}{r_s} - \frac{0.056}{1 + 0.127 r_s} + \right.$$
$$\left. + \frac{1.5R_m^2}{r_s^3} - \frac{0.9Z^{2/3}}{r_s} - V_0 \frac{R_m^3}{r_s^3}\right). \tag{5.51}$$

Since the chemical potential μ of the electron gas is determined by its bulk value

$$\mu = \frac{\partial w}{\partial n},$$

where w is the density of the bulk energy of the crystal, the resulting expression for the chemical potential μ in the framework of the Heine-Abarenkov pseudopotential takes the form

$$\mu = 0.5(3\pi^2 n_0)^{2/3} - \left(\frac{3n_0}{\pi}\right)^{1/3} - \frac{0.056n_0^{2/3} + 0.0059n_0^{1/3}}{(0.079 + n_0^{1/3})^2} +$$
$$+ 0.5Z^{2/3}\left(\frac{4\pi n_0}{3}\right)^{1/3} - 4\pi n_0 R_m^2 + \frac{8\pi V_0 R_m^3 n_0}{3Z}. \tag{5.52}$$

To determine the parameters of the Heine–Abarenkov pseudopotential, we can use the condition of the minimum bulk energy of the metal.

Minimization of (5.52) with respect to r_s leads to an expression linking V_0 and R_m:

$$V_0 = \frac{r_s^2}{3R_m^3}\left(\frac{2.21}{r_s} - 0.9Z^{2/3} + 4.5\frac{R_m^2}{r_s^2} - 0.458 - \frac{0.00711r_s^2}{(1+0.127r_s)^2}\right). \quad (5.53)$$

As a result, the problem of determining the second parameter of the potential (5.49) must be solved. It is usually determined by comparing the results of calculations performed using this pseudopotential with any empirical parameters. As an experimental characteristic we used in [145] the value of the surface energy. The following restrictions apply to the values of V_0 and R_m:

a) $0 < V_0 < Z/R_m$, which follows from general physical considerations (a positive ion in the metal should attract electrons at a rate commensurate with the size of its Coulomb charge);

b) $r_c < R_m$ – to this limit we arrive by comparing (5.21) and (5.52);

c) $R_m < d/2$ – a necessary condition for obtaining an amendment to the electron–ion interaction to the dipole barrier, as it is this range of variation of R_m in which we can enter an idea of 'smearing' of the charge of ions on the crystal planes.

The parameters β and δ which, in accordance with the above expressions, also determine the work function, are determined by minimizing the surface energy and are given in Table 5.9.

The application of the Heine–Abarenkov potential [229] leads to an additional term to the expression (5.33) for the electron–ion component of surface energy:

$$\Delta\sigma_{ei}^{XA}(\beta,0) = \frac{2\pi dV_0 n_0^2}{\beta^3}\frac{\exp(-\beta d/2)}{1-\exp(-\beta d)}[\beta R_m \mathrm{ch}(\beta R_m) - \mathrm{sh}(\beta R_m)],$$

$$\Delta\sigma_{ei}^{XA}(\beta,\delta) = \frac{2\pi n_0^2}{\beta^3}\left[\exp(\beta\delta)-1\right]dV_0 \times$$
$$\times\exp(-\beta d/2)[\beta R_m ch(\beta R_m) - sh(\beta R_m)]. \quad (5.54)$$

The values of the parameters V_0 and R_m, and the corresponding values of the electron work function of the metal surface, obtained in the process of comparing the calculated values of surface energy with the experimental ones are shown in Table 5.9. The values for V_0 and R_m are in accordance with the above restrictions.

The introduction of an additional parameter V_0 in the pseudopotential determined by comparing the results of the calculation of surface energy with the experimental values, allows the variation of the calculated values of the work function, while comparing the values of the surface characteristics of both metals.

Table 5.9. The results of calculation of parameters of the Heine–Abarenkov pseudopotential V_0 and R_m, and the corresponding values of the parameter of reduction of electron density β and work function without (W) and taking into account (W_δ) the effects of surface relaxation

Metal	Face	R_m, a.u.	V_0, a.u.	β, a.u.	W, eV	β_δ, a.u.	δ, a.u.	W_δ, eV
Na (BCC)	110			0.728	2.59	0.813	0.119	1.99
	100	2.7	0.326	0.743	2.81	0.706	0.092	2.58
	111			0.7	2.78	0.735	−0.146	2.95
	Liquid	3.1	0.326	0.858	2.56			
K (BCC)	110			0.737	2.13	0.679	0.145	1.66
	100	2.5	0.087	0.614	2.28	0.585	0.104	2.2
	111			0.58	2.22	0.614	−0.214	2.4
Al (FCC)	111			1.209	4.248	1.151	0.052	2.85
	100	1.2	0.19	1.069	5.38	0.982	0.082	4.05
	110			0.918	6.98	0.96	−0.055	7.38
Cu (FCC)	111			1.054	7.16	0.993	0.095	4.93
	100	1.4	0.603	0.993	7.654	0.912	0.122	5.69
	110			0.92	8.39	0.913	−0.005	8.4
Fe (BCC)	110			1.213	7.24	1.149	0.064	4.43
	100	1.2	0.475	1.02	9.98	0.977	0.046	9.38
	111			0.93	11.57	0.991	−0.117	12.5
Cr (BCC)	110			1.201	7.18	1.137	0.065	4.44
	100	1.3	0.53	1.013	9.82	0.97	0.043	9.25
	111			0.93	11.35	0.988	−0.126	12.4
Mo (BCC)	110			1.229	5.46	1.172	0.052	2.55
	100	1.2	0.212	0.983	9.8	1.023	0.036	8.5
	111			0.885	12.12	0.987	−0.138	13.13

CALCULATION OF ADHESION CHARACTERISTICS OF METALS AND THEIR MELTS BY THE DENSITY FUNCTIONAL METHOD

Introduction

In this chapter, the method of the electron density functional is used to examine the technique of calculating the adhesion characteristics of the metal contact, taking into account gradient corrections for the heterogeneity of the system in the interfacial region between the media. The results of calculating the interfacial energy of interaction, the adhesion energy and the force of adhesive interaction of various metals and their melts are used to analyze the adhesion characteristics depending on the distance between the surfaces of metals and dielectric constant of the intermediate layer. The calculation of the parameters of the displacement of surface ion planes of the contacting metals with respect to their bulk position in the interphase region of the interface is carried out, and the dependence of these displacement parameters on the distance between the surfaces of metals and dielectric permittivity of the interlayer is analyzed. The influence of these effects of surface relaxation in the interfacial region on the adhesion characteristics of the contact of different metals is investigated. Based on the analysis of components of the interfacial energy the physical features of interaction between the metals are studied.

The material of the chapter is based on the results obtained by the authors of the book in [26, 30, 31, 141, 142, 193, 194].

6.1. The basic equations. Calculation procedure

The application of the functional density method to describe the adhesion of various metals and metallic coatings based on the solution of the variational problem of finding the energy minimum of the generalized electronic system of two contacting metals studied on the background of the given distribution of the positive ion charge. As the trial functions to describe the electron distribution in the interacting subsystems of metals we choose solutions of the formally linearized Thomas–Fermi equation, and the variational parameter is assumed to be the inverse screening length β.

Consider the following geometric model of the contact of the two semi-infinite metals, which occupy the areas $z < -D$ and $z > D$, with the intermediate vacuum layer of thickness $2D$. Let the positive charge of a uniform background of ions ('jelly' model) be distributed in accordance with the formula

$$n_+(z) = n_1\Theta(-z-D) + n_2\Theta(z-D),\qquad(6.1)$$

where n_1 and n_2 are the charge densities of the background for each of the metals; $\Theta(z)$ is the step function.

The solution of the linearized Thomas–Fermi equation with the boundary conditions reflecting the continuity of the electrostatic potential $\phi(z)$ and electric induction $d\phi/dz$ at $z = \pm D$, and the finiteness of the potential at infinity, allows at the relationship $\phi(z) = -4\pi n(z)/\beta^2$ to obtain the following expression for the density of the electron distribution $n(z)$ in the system:

$$n(z) = \begin{cases} n_1 - 0.5n_1a_1 \exp[\beta(z+D)], & z<-D; \\ 0.5\{n_1(1-a_1)\exp[\beta(z+D)]+ \\ \quad + n_2(1-a_2)\exp[-\beta(z-D)]\}, & |z| < D; \\ n_2 - 0.5n_2a_2 \exp[-\beta(z-D)], & z > D; \end{cases}\qquad(6.2)$$

where $a_1 = 1 - (n_2/n_1)\exp(-2\beta D)$, $a_2 = 1 - (n_1/n_2)\exp(-2\beta D)$.

We define the interfacial energy of interaction of metals per unit area of contact as an integral over z from the bulk density of the ground state energy of the electron gas:

$$\sigma(\beta,D) = \int_{-\infty}^{\infty}\{w[n(z)] - w[n_+(z)]\}dz.\qquad(6.3)$$

In the 'jelly' model, the volumetric energy density of the inhomogeneous electron gas can be represented as a gradient expansion

$$w[n(z)] = w_0[n(z)] + w_2\left[n(z),|\nabla n(z)|^2\right] + O(\nabla^4),\qquad(6.4)$$

where

$$w_0[n(z)] = 0.5\phi(z)n(z) + 0.3(3\pi^2)^{2/3}n^{5/3}(z) - $$
$$-0.75(3/\pi)^{1/3}n^{4/3}(z) - 0.056\frac{n^{4/3}(z)}{0.079+n^{1/3}(z)}\qquad(6.5)$$

is the energy density of the homogeneous electron gas in atomic units, including consecutively the electrostatic, kinetic, exchange and correlation energy, and

$$w_2\left[n(z),|\nabla n(z)|^2\right] = \frac{1}{72}\frac{|\nabla n(z)|^2}{n(z)} + w_{2,xc}\left[n(z),|\nabla n(z)|^2\right] \tag{6.6}$$

is the first term of the gradient expansion, which takes into account the inhomogeneity of the electron gas. The first term in (6.6) is the correction for the inhomogeneity of the kinetic energy, and $w_{2,xc}$ denotes the correction for inhomogeneity for the exchange and correlation energies. Taking into account only the corrections for the kinetic energy without consideration of the effects $w_{2,xc}$ corresponds to the random-phase approximation (RPA). Expressions (5.7)–(5.11) for different exchange-correlation corrections $w_{2,xc}$ are given in chapter 5.

Amendments to the interfacial energy associated with the discrete distribution of the positive charge, as well as in calculating the surface energy, are calculated in the framework of the Ashcroft pseudopotential model, averaged over the crystal planes parallel to the metal surface, and are characterized by the cutoff radii r_1, r_2 and interplanar distances d_1, d_2. As a result, the correction, associated with the electron–ion interaction, takes the form

$$\sigma_{ei} = \frac{2\pi}{\beta^3}\sum_{j=1,2}n_j^2 a_j\left(1 - \frac{\beta d_j\exp(-\beta d_j/2)}{1-\exp(-\beta d_j)}\mathrm{ch}\,(\beta r_j)\right). \tag{6.7}$$

In this chapter, as in Sec. 5, the value of the cutoff radius r_c Ashcroft pseudopotential for each of the metals (r_1 and r_2, respectively) can be determined by minimizing the bulk energy of the metal at the observed equilibrium atomic volume $\Omega = Z/n_0 = 4\pi r_s^3 Z/3$ with the parameter $r_s = (4\pi n_0/3)^{-1/3}$. In accordance with the local density approximation of the bulk energy of the metal in terms of the density parameter r_s, takes the form

$$E = Z\left(\frac{1.105}{r_s^2} - \frac{0.458}{r_s} - \frac{0.056}{1+0.127r_s} + \frac{1.5r_c^2}{r_s^3} - \frac{0.9Z^{2/3}}{r_s}\right). \tag{6.8}$$

Then the expression (6.8) is differentiated by r_s and is equal to zero. The solution of the resultant equation with respect to r_c gives the values of the unknown parameters r_1 and r_2.

For corrections to the interfacial energy associated with the interaction of metal ions, we use the interpolation formula introduced in [112]. Then

$$\sigma_{ii} = \sqrt{3}\sum_{j=1,2}\frac{Z_j^2}{c_j^3}\exp\left(-\frac{4\pi d_j}{\sqrt{3}c_j}\right) - $$
$$-2\sqrt{3}\frac{Z_1 Z_2}{(c_1 c_2)^{3/2}}\exp\left[-\frac{2\pi}{\sqrt{3}}\left(\frac{d_1+D}{c_1} + \frac{d_2+D}{c_2}\right)\right], \tag{6.9}$$

where Z_1, Z_2 are the valences of the metals; c_1, c_2 are the distances between the nearest ions in planes parallel to the surfaces of metals.

In accordance with the method of density functional significance of the variational parameter β is found from the requirement of minimum total interfacial energy:

$$\frac{d\sigma(\beta, D)}{d\beta} = 0, \tag{6.10}$$

where $\sigma(\beta, D) = \sigma_0(\beta, D) + \sigma_{ei}(\beta, D) + \sigma_{ii}(D)$, and $\sigma_0(\beta, D)$ is the interfacial energy, calculated in the framework of the 'jelly' model. The solution of equation (6.10) defines the parameter β_{min} as a function of the gap $2D$. The result of solving the variational problem is the total interfacial energy of the system $\sigma(\beta(D), D)$. Knowing this energy, it is easy to find the energy of the adhesion system as the work necessary to remove the metals from each other to infinity, i.e.

$$E_a(D) = \sigma(\infty) - \sigma(D). \tag{6.11}$$

Then the strength of adhesive interaction of the system is defined as derivative D of the interfacial energy $\sigma(\beta, D)$ with $\beta = \beta_{min}$:

$$F_a(D) = \frac{d\sigma(\beta_{min}, D)}{dD}. \tag{6.12}$$

In contrast to other studies, in the solution of the variational problem and calculating the interfacial energy $\sigma(D)$, we gave up trying to find analytically the components of the interfacial energy $\sigma_0(\beta, D)$ in (6.5) and the numerical solution of the transcendental equation corresponding to (6.10), to determine $\beta_{min} = \beta(D)$. A program was developed for numerical integration in (6.5) and finding the total interfacial energy for given values of the parameters characterizing the metal contact, with β varying in some interval. Simultaneously, a program of numerical determination of parameter β_{min}, localizing the minimum total interfacial energy, and calculating the energy and all its components at β_{min} is applied. Minimization of the total interfacial energy $\sigma(\beta, D)$ was carried out by direct Hooke–Jeeves method. The subsequent program of calculation of the adhesion energy and the adhesion interaction force, based on the data on the interfacial energy, can close the cycle of calculations of the adhesive characteristics of the system.

6.2. The calculation results and discussion

Calculations of the adhesive properties using the above method were carried out for a number of metals and their melts under the assumption that the metals in the solid state are directed to the interfacial boundary between close-packed faces. The values of the initial parameters, used to calculate the adhesion characteristics of the metals, are given in Table 5.1.

The results of studies and calculations of the surface energy for different metals, presented in chapter 5, showed that in the group of the approximations used for exchange-correlation corrections to the heterogeneity, there is no universal approximation which can be used to calculate the surface energy and, hence, the interfacial energy of interaction of metals. Calculations of the surface energy (see Table 5.2) have shown that for Pb such approximation is the random-phase approximation (RPA) which does not take into account the exchange–correlation amendments to the heterogeneity; for Al it is the VS approximation. However, the above method of calculating the interfacial energy is based on the fact that the bulk energy density in (6.5) must be shared by the two metals in contact, and therefore the approximation used for the exchange–correlation corrections to the heterogeneity should also be shared. We can therefore expect that this calculation of the adhesive characteristics of contact of solid lead with its melt in the random phase approximation and of solid aluminum with its melt in the VS approximation gives the most reliable dependence of the energy and adhesion force on the gap $2D$ (curves *1, 6* in Fig. 6.1 and Fig. 6.2). At the same time, if we consider the contact of Pb and Al in the crystalline state by using the RPA or VS, in the first case the calculation gives too low (curves *2* in both figures) and in the second case excessive (curves *5* in both figures) values of the adhesion characteristics.

Besides, the investigations described in chapter 5, have shown that in the calculation of the surface energy of simple metals in a gradient expansion of the bulk energy density it is sufficient to restrict considerations to the second order corrections. At the same time, for

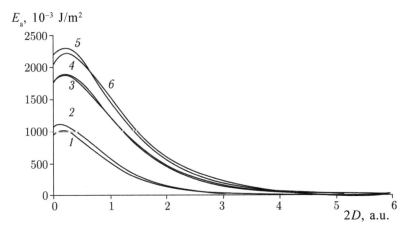

Fig. 6.1. Adhesion energy $E_a(D)$ of metals Pb(111), Al(111) and their melts in dependence on the size of the gap for the various approximations used for the exchange–correlation corrections for heterogeneity: *1.* Pb (solid) – Pb (liquid), RPA; *2.* Pb (solid) – Al (solid), RPA; *3.* Pb (solid) – Al (solid), mixed approximation; *4.* Pb (solid) – Al (solid), mixed approximation with adjustment of the boundaries; *5.* Pb (solid) – Al (solid), VS; *6.* Al (solid) – Al (liquid), VS.

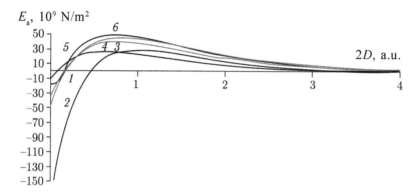

Fig. 6.2. Adhesion force $E_a(D)$ of metals Pb(111), Al(111) and their melts depending on the size of the gap for various approximations used to obtain the exchange-correlation corrections for heterogeneity: *1*. Pb (solid) – Pb (liquid), RPA; *2*. Pb (solid) – Al (solid), RPA; *3*. Pb (solid) – Al (solid), mixed approximation; *4*. Pb (solid) – Al (solid), mixed approximation with adjustment of the boundaries; *5*. Pb (solid) – Al (solid), VS; *6*. Al (solid) – Al (liquid), VS.

transition and noble metals we must consider a correction of the fourth order of the powers of the electron density gradient. Thus, for Al (see Table 5.3) it is necessary to take into account the gradient corrections of fourth order with respect to the powers of the gradient $n(z)$ to the kinetic energy density

$$w_{4,\text{kin}} = \frac{1.336n}{540(3\pi^2 n)^{2/3}} \left[\left(\frac{\nabla^2 n}{n} \right)^2 - \frac{9}{8} \left(\frac{\nabla^2 n}{n} \right) \left(\frac{|\nabla n|}{n} \right)^2 + \frac{1}{3} \left(\frac{|\nabla n|}{n} \right)^4 \right],$$

(6.13)

and for the noble (Cu, Ag, Au) and transition (Fe, Cr, Mo, W, etc.) metals the fourth-order gradient corrections also to the density of the exchange-correlation energy

$$w_{4,xc} = 2.94 \cdot 10^{-4} \exp\left(-0.2986 n^{-0.26}\right) \left(\frac{\nabla^2 n}{n} \right)^2.$$

(6.14)

Thus, inclusion of the fourth-order gradient corrections helps to avoid the problem of different approximations for the case of contact of the transition metals with each other. In the case of contact with simple metals, to produce more realistic values of the adhesion characteristics, it is proposed to divide the region of integration in (6.5) into three regions and in each of them to use the bulk density of the energy with the exchange–correlation corrections to the heterogeneity, calculated in various approximations. In particular, for the region $z < -D$, occupied by, for example, lead it is proposed to used in the bulk energy density the

exchange–correlation corrections in the RPA and to limit consideration to the
the gradient corrections of the second order, in the region $z > D$, occupied
by aluminum, use the correction in the VS approximation and gradient
corrections of the fourth order for the kinetic energy density, and in the gap
$|z| < D$ – the rms value of these corrections. The results of calculation by
this method are presented (curves *3*) in Fig. 6.1 and Fig. 6.2.

If we analyze the distribution of electron density $n(z)$ in the Pb–Al
metal system, obtained on the basis of these representations (see Fig.
6.3), we see that near the boundary of each metal in the interfacial section
there is a region of length $1/\beta$, in which the electron density is undergoing
significant changes and affects the interaction effects of metals. Therefore,
in [32] calculations were carried out with an additional correction of the
boundaries split into regions:

I – $z < -D - 1/\beta$, where RPA is used for lead;

III – $z > D + 1/\beta$, where the VS approximation is used for aluminum
and the fourth-order gradient correction for the density of kinetic energy;

II – $|z| < D + 1/\beta$, which uses the average of these approximations
(curves *4* in Fig. 6.1 and Fig. 6.2).

However, the figures show that for the Pb–Al contact this correction
of the boundaries leads to minor changes in the values of adhesion
characteristics of the contact of these metals, at the gap $2D \sim 1$ a.u.

The above method with mixed approximation and corrections of the
boundary was used to calculate the adhesion characteristics in the case of
exposure of other metals. Thus, Fig. 6.4 shows the values of interfacial
energy of both simple and transition metals, depending on the size of
the gap between the contacting surfaces of the metals. At direct contact
($D = 0$) the value of the interfacial energy of two different metals is
negative. It should be noted that in the case of cohesion (contact between
two similar metals) the plot of the dependence of the interfacial interaction
energy on the gap between the surfaces of metals starts from the origin.

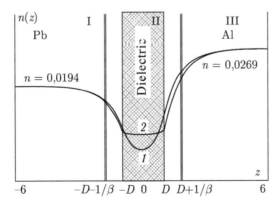

Fig. 6.3. The electron density distribution of the Pb(111)–Al(111) system at the
dielectric gap od $2D = 2$ a.u. with permeability: *1* – $\varepsilon = 1$; *2* – $\varepsilon = 15$.

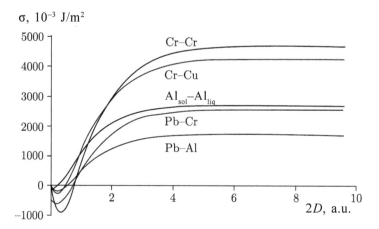

Fig. 6.4. Interfacial energy of the system of two metals, depending on the size of the gap.

For all graphs interfacial energy minimum is observed at magnitude of the gap $2D \sim 0{,}25$ a.u., then by increasing the gap monotonic increase of interfacial energy is characterized by a plateau. The largest asymptotic values of interfacial energy and bright pronounced minima are observed in the case of contact of transition metals with high electron density, for example, for Cr–Cr and Cr–Cu. Tables 6.1 and 6.2 show the results of the calculation of the adhesive characteristics for the gaps $D = 0$ and $D \to \infty$ for the cases of contact both simple and transition metals.

In order to achieve the universality of the theory of surface phenomena in calculating the corrections to the interfacial energy associated with the discrete distribution of the positive charge of the ions, we use the more general Heine–Abarenkov pseudopotential (5.49), averaged over the crystal planes parallel to the surface of metals. Just as in chapter 5, to determine the parameters of the Heine–Abarenkov pseudopotential V_1, V_2, R_1, R_2, we use the condition of the minimum bulk energy of the metal (5.54).

The values of these parameters for different metals, obtained by comparing the calculated values of the surface energy with the experimental

Table 6.1. The values of adhesion energy E_a(MJ/m^2) and adhesion force F_a (10^9 N/m^2) of the metals Pb, Al and their melts at a gap of $D = 0$, with the use of different approximations

Value	Pb(solid)– Pb(liquid) RPA	Pb(solid)– Al(solid) RPA	Pb(solid)– Al(solid) mixed approx.	Pb(solid)– Al(solid) VS approx.	Al(solid)– Al(liquid) VS approx.
E_a	966	1077	1783	2058	2202
F_a	−10.9	−175	−34	−47	−17.4

Table 6.2. The values of interfacial energy σ, parameter β, and the adhesion force F_a of various metals at the gap of $D = 0$ and $D \rightarrow \infty$

Value	Pb–Al	Pb–Cr	Al(solid)–Al(liquid)	Cr–Cu	Cr–Cr
$\sigma(D = 0)$, MJ/m²	−60	−476	−156	−30	0
$\sigma(D \rightarrow \infty)$, MJ/m²	1706	2545	2703	4261	4687
$\beta(D \rightarrow \infty)$, a.u.	1.26	1.24	1.27	1.15	1.2
$F_a(D = 0)$, 10^9 N/m²	−175	−41	−17.4	−66	−210

values, are presented in Table 5.9. As a result, the correction, associated with the electron–ion interaction, takes the form:

$$\sigma_{ei}(\beta, D) = \frac{2\pi}{\beta^3} \sum_{j=1,2} n_j^2 a_j \left\{ 1 - \frac{\beta d_j \exp(-\beta d_j / 2)}{1 - \exp(-\beta d_j)} \mathrm{ch}(\beta R_j) \right\} +$$

$$+ \frac{2\pi}{\beta^3} \sum_{j=1,2} n_j a_j^2 d_j V_j \frac{\exp(-\beta d_j / 2)}{1 - \exp(-\beta d_j)} \left\{ \beta R_j \mathrm{ch}(\beta R_j) - \mathrm{sh}(\beta R_j) \right\}. \quad (6.15)$$

Figures 6.5 and 6.6 show the comparative adhesion characteristics for a couple of the Al–Pb simple metals, calculated using the Heine-Abarenkov pseudopotential with the exchange–correlation correction in the VS approximation, and the Ashcroft pseudopotential (at $V_0 = 0$ the Heine–Abarenkov pseudopotential changes to the expression for the Ashcroft pseudopotential) with the exchange–correlation correction in the VS approximation and the RPA. The latter is due to the fact that in the Ashcroft pseudopotential model the surface energy of aluminium is most appropriately described by the exchange–correlation VS correction, and the surface energy of lead by RPA. But the method of calculating the adhesion characteristics requires the use of a single exchange–correlation

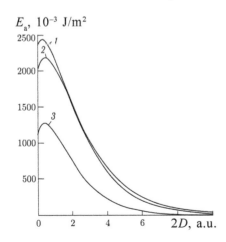

E_a, 10^{-3} J/m²

Fig. 6.5. Adhesion energy $E_a(D)$ for a Pb–Al pair as a function of the gap: *1* – Ashcroft model, VS approximation; *2* – Heine–Abarenkov model, VS approximation; *3* – Ashcroft model, RPA.

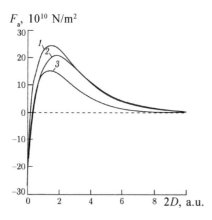

F_a, 10^{10} N/m²

Fig. 6.6. Adhesion force $F_a(D)$ for a Pb–Al pair as a function of the gap: *1* – Ashcroft model, VS approximation; *2* – Heine–Abarenkov model, VS approximation VS; *3* – Ashcroft model, RPA. Adhesion force $F_a(D)$, 10^{10} N/m² at $D = 0$: –26.8 (*1*); –25.7 (*2*) –18.0 (*3*).

correction for both contact metals. This requirement is most fully realized, as was shown in [145], using for various metals the Heine–Abarenkov pseudopotential with the exchange–correlation correction in the VS approximation. The calculations, performed in the Heine–Abarenkov pseudopotential approximation, significantly clarify the relevant calculations performed in the framework of the Ashcroft model, and can be extended to describe the adhesion of both simple and transition metals.

We analyze the changes with the distance between the surfaces of the contacting metals of all components of the interfacial interaction energy, namely: part of the interaction energy in the approximation of uniform electron density distribution in the two metals σ_{uni}, gradient corrections for the inhomogeneity to the kinetic and exchange energies of both second and fourth orders of magnitude, the electron–ion σ_{ei} and ion–ion σ_{ii} constituents. σ_{uni} includes exchange, kinetic, Coulomb and correlation constituents of the interfacial interaction energy of the homogeneous electron system.

Graphs of the contributions to the interfacial energy for the Cu–Cr pair are shown Figs. 6.7 and 6.8. The largest absolute values are the exchange and kinetic components, and their contributions are opposite in sign. The smallest absolute value is the correlation component. The contributions σ_{ei} and σ_{ii}, associated with taking into account the discreteness of the distribution of the ions, are positive; the first of them is comparable with the electrostatic contribution to the interfacial energy, and the second is an order of magnitude smaller and is comparable to the correlation contribution. It should be noted that for the case of contact of metals with a large electron density, such as, in particular, Cr, the contribution to the interfacial energy due to the gradient corrections to the heterogeneity of 4-th order, is quite significant and comparable in magnitude with the ion–ion component. With the distance between the metal surfaces increasing from zero to infinity, all the components of the interfacial energy vary from their minimum value at $D = 0$ to a constant at $D \to \infty$ corresponding to the system of the insulated metals.

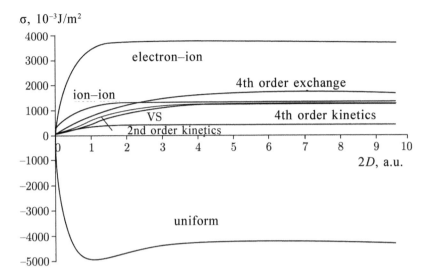

Fig. 6.7. The dependence of the components of the interfacial energy of the Cu–Cr system on the gap size.

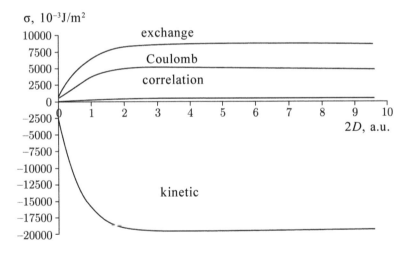

Fig. 6.8. The components of the interfacial energy of the homogeneous Cu–Cr system as a function of the distance between the metals.

6.3. Influence of the intermediate dielectric layer on the adhesion of metals

If in the previous section we studied the questions of the description of the adhesion of a system of two semi-infinite metals separated by a vacuum gap, here we consider the case where metals are separated by a dielectric layer with permittivity ε. Consideration of the dielectric interlayer simulates the possibility of the formation on the surface of the substrate during spraying on the films of a layer of oxide or a compound having the dielectric properties that have a noticeable effect on the adhesion of the coating.

It should be emphasized that the size of the metal layer near the surface where the electron density $1/\beta$ decreases sharply, is usually of the order of 1 atomic unit or the lattice constant. Therefore, the concept of dielectric permittivity as a certain averaged characteristics of the material in this area is not clearly defined. The difficulty of taking into account the atomic structure of the dielectric layers adjacent to the metal surface, requires that the first step in description of the system is the consideration of the continuum model, characterized by the dielectric constant.

Consider two semi-infinite metals occupying the regions $z < -D$ and $z > D$ with a dielectric interlayer with thickness of $2D$ and permeability ε. Using the 'jelly' model and the solution of the linearized Thomas–Fermi equation, with the boundary conditions reflecting the continuity of the electrostatic potential $\varphi(z)$ and electric induction $\varepsilon d\phi/dz$ at $z = \pm D$ taken into account, and also considering the finiteness of the potential at infinity allows at the relationship $\phi(z) = -4\pi n(z)/\beta^2$ to obtain the following expression for the density of the electron distribution $n(z)$ in the system:

$$n(z) = \begin{cases} \left| n_1 - \left(\dfrac{\Delta}{1+\Delta}\right) n_1 a_1 \exp[\beta(z+D)], \right. & z < -D; \\[2mm] 0.5\{n_1(1-a_1)\exp[\beta(z+D)/\Delta] + \\[1mm] \quad + n_2(1-a_2)\exp\left[-\beta(z-D[/\Delta]\right\}, & |z| < D; \\[2mm] n_2 - \left(\dfrac{\Delta}{1+\Delta}\right) n_2 a_2 \exp(-\beta(z-D)), & z > D, \end{cases} \tag{6.16}$$

where

$$\Delta = \sqrt{\varepsilon}, \quad a_1 = 1 - \frac{(\Delta-1)\exp(-2\beta D/\Delta)+(\Delta+1)(n_2/n_1)}{(1+\Delta^2)\mathrm{sh}(2\beta D/\Delta)+2\Delta\mathrm{ch}(2\beta D/\Delta)},$$

$$a_2 = 1 - \frac{(\Delta-1)\exp(-2\beta D/\Delta)+(\Delta+1)(n_1/n_2)}{(1+\Delta^2)\mathrm{sh}(2\beta D/\Delta)+2\Delta\mathrm{ch}(2\beta D/\Delta)}.$$

Just as before the introduction of the dielectric interlayer, we determine the interfacial energy of interaction of the contacting metals per unit area as the integral over z from the bulk density of the ground state energy of the electron gas:

$$\sigma(\beta, D, \varepsilon) = \int_{-\infty}^{\infty} \{w[n(z,\varepsilon)] - w[n_+(z)]\}dz. \tag{6.17}$$

In the 'jelly' model, all components of the gradient expansion of the bulk energy density of the inhomogeneous electron gas do not depend explicitly on ε. Changes occur in the corrections relating to taking into account the discreteness of the distribution of the positive charge. Thus, the correction for the electron–ion interaction takes the form

$$\sigma_{ei}(\beta,D,\varepsilon) = \frac{\Delta}{(1+\Delta)} \frac{4\pi}{\beta^3} \sum_{j=1,2} n_j^2 a_j \left(1 - \frac{\beta d_j \exp(-\beta d_j/2)}{1-\exp(-\beta d_j)} \mathrm{ch}(\beta r_j) \right). \quad (6.18)$$

The correction to the interfacial energy, associated with the ion–ion interaction in metals, separated by a dielectric layer, can be written as

$$\sigma_{ii} = \sqrt{3} \sum_{j=1,2} \frac{Z_j^2}{c_j^3} \exp\left(-\frac{4\pi d_j}{\sqrt{3}c_j} \right) - 2\sqrt{3} \frac{Z_1 Z_2}{(c_1 c_2)^{3/2}} \times$$
$$\times \exp\left[-\frac{2\pi}{\sqrt{3}} \left(\frac{d_1 + D/\Delta}{c_1} + \frac{d_2 + D/\Delta}{c_2} \right) \right]. \quad (6.19)$$

In accordance with the basic principles of the density functional method the value of the variational parameter β is determined from the requirement of the minimum total interfacial energy:

$$\frac{d\sigma(\beta,D,\varepsilon)}{d\beta} = 0, \quad (6.20)$$

where $\sigma(\beta, D, \varepsilon) = \sigma_0(\beta, D) + \sigma_{ei}(\beta, D, \varepsilon) + \sigma_{ii}(D, \varepsilon)$. The solution of equation (6.20) defines the parameter β_{min} as a function of the gap $2D$ and dielectric permittivity. The result of solving the variational problem is the total interfacial energy of the system $\sigma(\beta(D), D, \varepsilon)$. Knowing this value, it is easy to find the energy of adhesion of the system as the work required for removal of metals from each other at infinity, i.e.

$$E_a(D,\varepsilon) = \sigma(\infty,\varepsilon = 1) - \sigma(D,\varepsilon).$$

Then the strength of adhesive interaction of the system is defined as derivative D of the interfacial energy $\sigma(\beta, D, \varepsilon)$ with $\beta = \beta_{min}$:

$$F_a(D,\varepsilon) = \frac{d\sigma(\beta_{min}, D, \varepsilon)}{dD}. \quad (6.21)$$

On the basis of the above methods, in [32, 33] we carried out calculations and plotted graphs (Figs. 6.9 and 6.10) of the dependence of energy and force of adhesion for a Pb–Al pair in the solid state, on the value of the dielectric gap $2D$ at = 1, 4, 10, 15. From the figures it is clear that at short distances $\left(\beta D / \sqrt{\varepsilon} \ll 1 \right)$ there is a slight increase in adhesion energy, which corresponds to the attraction of the metal surfaces. The subsequent rapid

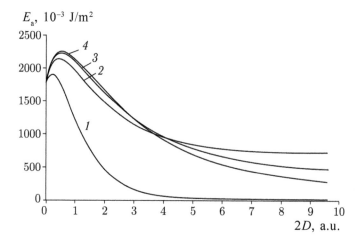

Fig. 6.9. Adhesion energy $E_a(D)$ of Pb(111)–Al(111), depending on the value of the dielectric permittivity of the gap: $1 - \varepsilon = 1$; $2 - \varepsilon = 4$; $3 - \varepsilon = 10$; $4 - \varepsilon = 15$

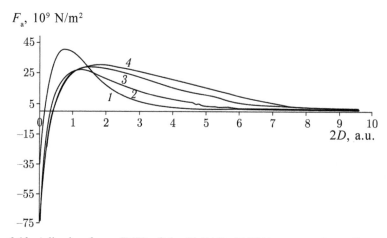

Fig. 6.10. Adhesion force $F_a(D)$ of the Pb(111)–Al(111) system, depending on the value of the dielectric permittivity of the gap: $1 - \varepsilon = 1$; $2 - \varepsilon = 4$; $3 - \varepsilon = 10$; $4 - \varepsilon = 15$.

decrease in adhesion energy with increasing size of the gap is accompanied by the repulsion of the metal surfaces. The force of electrostatic repulsion is characterized by maximum at $\beta D / \sqrt{\varepsilon} \approx 1$ and a strong decline at $\beta D / \sqrt{\varepsilon} \geq 1$. If $\beta D / \sqrt{\varepsilon} \gg 1$ the values of adhesion energy tend asymptotically to the difference of the total surface energies of metals at the interface with vacuum and at the boundary with the dielectric permittivity ε, i.e. $\Delta E_a(D, \varepsilon) = \sigma(\infty, \varepsilon = 1) - \sigma(\infty, \varepsilon)$. At the same time ΔE_a decreases with increasing ε.

The force of attraction due to the contacting surfaces is determined first of all by the kinetic component of the interfacial energy. But the contribution to the kinetic component decreases faster with distance than, for example, the contribution from the inhomogeneity of the electronic distribution and therefore with increasing distance attraction is replaced by repulsion. The physical nature of the change of the electrostatic force of adhesion, as noted in [115], is connected with the fact that at short distances the electronic 'tail' of one metal penetrates the ionic core of another metal and is attracted by the. With increasing gap the electronic 'tail' leaves the zone of interaction with the ionic core and interacts with the electronic 'tail' of the opposite metal. This causes the repulsion of metal surfaces. Analysis of the results shows that at $\beta D / \sqrt{\varepsilon} \geq 1$ the repulsive force increases with increasing ε, i.e. the increase in the polarity of the intermediate dielectric layer increases the drag force impeding the rapprochement of the metal surfaces.

6.4. Influence of lattice relaxation effects of the surface on adhesion of metals

The expressions for the components of the interfacial energy, given in section 6.3, do not take into account the effects of surface displacement of ion planes in the interfacial region between the media from their equilibrium positions characteristic of the bulk material. Consider what changes will be caused by taking into account the displacement of one ion plane in each of the metals by the value δ_j ($\delta_j > 0$ corresponds to extension) relative to the position of this plane in the undisturbed material, characterized by the coordinate $z_j = \pm D \pm d/2$. We assume that in this case the density function of the electron distribution $n(z)$ in (6.2) is still characterized only by parameter β and does not acquire an explicit dependence on δ. At the same time, the component of the interfacial energy $\sigma(\beta)$, reflecting the contribution from the electronic system within the 'jelly' model, also becomes apparently dependent on δ. As for the contribution σ_{ii}, arising from the electrostatic interaction of ions with each other, the following expression was obtained using the method described in [308]

$$\sigma_{ii} = \sqrt{3} \sum_{j=1,2} \frac{Z_j^2}{c_j^3} \exp\left(-\frac{4\pi(d_j - \delta)}{\sqrt{3}c_j}\right) -$$
$$-2\sqrt{3}\frac{Z_1 Z_2}{(c_1 c_2)^{3/2}} \exp\left[-\frac{2\pi}{\sqrt{3}}\left(\frac{d_1 - 2\delta + D/\Delta}{c_1} + \frac{d_2 - 2\delta + D/\Delta}{c_2}\right)\right]. \quad (6.22)$$

Component $\sigma_{ei}(\beta, \delta)$ can be represented as

$$\sigma_{ei}(\beta,\delta) = \sigma_{ei}(\beta,0) + \Delta\sigma_{ei}(\beta,\delta) \quad (6.23)$$

with an additional term

$$\Delta\sigma_{\text{ei}} = \frac{4\pi\Delta}{\beta^3(1+\Delta)} \sum_{j=1,2} n_j^2 a_j d_j (1-\exp(\beta\delta_j)) \times$$

$$\times \exp(-\beta d_j / 2)\text{ch}(\beta r_j) + \sum_{j=1,2} 2\pi n_j^2 a_j d_j \delta_j^2, \qquad (6.24)$$

which takes into account the effects of lattice relaxation of the surface.

As a result, the total interfacial energy $\sigma(\beta, \delta)$ as a function of variational parameters is determined by the sum of the contributions given by expressions (6.7) and (6.22) and (6.23), with an energy density of the inhomogeneous electron gas $w[n(z)]$ (6.5). The values of variational parameters β, δ, used to determine the interfacial energy of metals, were found from the condition of its minimality. As a result, values of β and δ acquire a clear functional dependence on the parameters n_j, Z_j, d_j, c_j, r_j, characterizing the electronic properties of metals, their symmetry and orientation of surface facets along the interface, and the implicit dependence on each other. In [32] we conducted self-consistent calculations of the parameters δ of the displacement of the surface ion planes of the contacting metals Pb–Al with respect to their volume in dependence on the size dielectric gap $2D$ and permeability ε. The calculation results are presented in the form of graphs in Fig. 6.11. It is seen that the displacement δ taken maximum value at $\beta D/\sqrt{\varepsilon} \approx 1$. The values of δ in the position of the maximum increase with increasing values of ε. Subsequent reduction of the δ values with increasing size of the gap $2D$ is characterized by the asymptotic tendency of the parameters δ to the values characteristic for solitary metals (with $\varepsilon = 1$ and $\beta D \gg 1$ the value δ_j coincide with the values in the case of metals with free surfaces (see Table 5.2.)).

Analysis of the influence of surface displacement ion planes on the adhesive characteristics of the contacting metal showed that the effect of possible values for the gap, with and the parameters of which are characterized by a maximum δ_j. At the same time the relative influence of these effects increases with increasing D, due to a rapid decrease in the values of adhesive characteristics than the displacement parameters δ_j. Thus, in the position of the maximum δ_j parameters for a pair of Pb–Al, we have: $\Delta E_a(\delta_j)/E_a(0) \approx 2\%(\varepsilon = 1)$, $6\%(\varepsilon = 15)$; $\Delta F_a(\delta_j)/F_a(0) \approx 0.5\%(\varepsilon = 1)$, $1\%(\varepsilon = 15)$; with $2D = 10$ a.u.: $\Delta E_a(\delta_j)/E_a(0) \approx 7\%(\varepsilon = 1)$, $10\%(\varepsilon = 15)$; $\Delta F_a(\delta_j)/F_a(0) \approx 3\%(\varepsilon = 1)$, $6\%(\varepsilon = 15)$. It should be noted that the inclusion of displacement of the effects of surface ion planes leads to an increase of energy and strength of adhesion.

The identified effects of surface displacement of ion planes, having the character of the expansion for both metals (Pb, Al) when $\delta_j > 0$, can play a significant role in the processes of separation of metals from each other. This is due to the fact that as a result of the exponential nature of the dependence of the overlap integrals of atomic wave functions on the

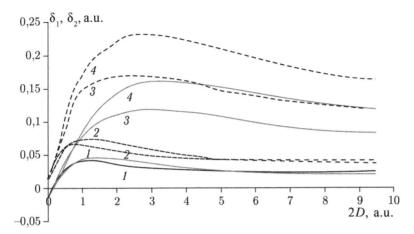

Fig. 6.11. Parameters of displacement δ_1, δ_2 of subsurface ion planes in the interface section of a Pb–Al pair, depending on the size of dielectric gap $2D$ and permeability: $1 - \varepsilon = 1$; $2 - \varepsilon = 4$; $3 - \varepsilon = 10$; $4 - \varepsilon = 15$; dashed lines correspond to Al, the solid lines correspond to Pb.

distance between atoms even a minor ion plane displacement can lead to a noticeable change of the binding energy between the atoms of the surface ion planes. As a result, instead of the rupture fracture, the interface between the metals may show cohesive rupture can occur in one of the metals, characterized by the greatest effects of displacement (extension) of ion surface planes. This contributes to an increase in the energy and force of adhesion of metals, which accompanies these effects of displacement of near-surface ion planes.

In conclusion, we note that the use of the concept of dielectric permittivity for the layer with the thickness of the order of interatomic values variables is not valid and is acceptable only as a model abstraction. Therefore, the results of calculation for $\beta D / \sqrt{\varepsilon} < 1$ can be regarded as estimates. For $\beta D / \sqrt{\varepsilon} \gg 1$ the adhesive properties of the metal–insulator–metal system will be determined by the characteristics of the metal–insulator contact. The inevitable roughness of the contacting surfaces requires the introduction of the vacuum gap between them. In this, account of the vacuum gap in the metal–insulator system leads to a change in the repulsive forces to the attraction force, which was confirmed by similar calculations for a metal–insulator system with a vacuum gap [33]. In addition, for the distance $\beta D / \sqrt{\varepsilon} \gg 1$ along with the electrostatic forces the attraction forces also start to act and become decisive at $D > 100$ Å.

6.5. Adhesion of metal and ionic crystal

Currently, of particular interest is the study of the adhesion interaction of dissimilar bodies in which the internal structures are determined by the chemical bonds of different nature.

The metal–insulator system is such an example. The adhesion properties of such systems were considered in [33, 37]. However, the simplification of the dielectric properties in these system by setting a single parameter – dielectric permittivity – does not allow to take into account the specifics of the microscopic properties of various classes of dielectrics, influencing, of course, also their adhesive characteristics. As a result, there is a need for considering specific classes of dielectrics. The adhesion properties of the metal and the ion crystal with the NaCl lattice were studied in [72] and in the works cited in [220]. However, the general drawbacks of this work include the lack of accurate description of the metal (metal is described within the 'jelly' model, without taking into account the effect of discreteness of the structure and gradient corrections to the inhomogeneity of the electron density at the metal boundary), and also the lack of consistency in the procedure for determining the total energy of interaction between the metal and ionic crystal.

In this section, in accordance with our studies [195, 196], we consider amendments to the 'jelly' model due to the discreteness of the distribution of the positive charge of metal ions, and also propose to take into account in the energy density of the composite the gradient corrections to the exchange correlation energy in the Vashishta–Singwi approximation (VS). To describe the generalized electron density of the composite we used the approach developed in [33, 37], thus applying the subsequent self-consistent description of the energy state of the metal – ionic crystal system.

The density functional method, used in this study, consists in solving the variational problem of finding a minimum energy of the inhomogeneous electron system considered on the background of the given charge of the ions. As the test functions of electron distribution we selected the solutions of the linearized Thomas–Fermi equation and the variational parameter is the inverse screening length β.

Consider a system consisting of a semi-infinite metal and an ionic crystal, separated by a vacuum gap with the size $2D$. We choose the origin of the coordinates midway between the contacting surfaces in the direction of the z-axis in the direction of the ion crystal. The full interfacial interaction energy of the system per unit of surface area, has the form

$$\sigma(\beta, D) = \sigma_0(\beta, D) + \sigma_{ii}(D) + \sigma_{ei}(\beta, D) + \sigma_{int}(\beta, D). \tag{6.25}$$

where

$$\sigma_0(\beta, D) = \int_{-\infty}^{\infty} \left\{ \omega\left[n(z)\right] - \omega\left[n_0 \Theta(-z)\right] \right\} dz \tag{6.26}$$

represents the interfacial energy of the electron gas of the metal within the 'jelly' model. In this approximation, the bulk density energy of the of the inhomogeneous electron gas is represented as a gradient expansion

$$w[n(z)] = w_0[n(z)] + w_2[n(z), |\nabla n(z)|^2] + O(\nabla^4), \tag{6.27}$$

where the energy density of the homogeneous electron gas is

$$w_0[n(z)] = 0.5\varphi(z)n(z) + 0.3(3\pi^2)^{2/3}n^{5/3}(z) -$$
$$-0.75(3/\pi)^{1/3}n^{4/3}(z) - 0.056\frac{n^{4/3}(z)}{0.079 + n^{1/3}(z)}, \tag{6.28}$$

a the term

$$w_2[n(z), |\nabla n|^2] = \frac{1}{72}\frac{|\nabla n|^2}{n(z)} + \frac{A(n)B^2(n)|\nabla n|^2}{3^{4/3}\pi^{5/3}n^{4/3}}, \tag{6.29}$$

$$\begin{cases} A(n) = 0.4666 + 0.3735k_F^{-2/3}(n), \\ B(n) = -0.0085 + 0.3318k_F^{1/5}(n), \end{cases}$$

$k_F = (3\pi^2 n)^{1/3}$ is the Fermi wave vector, and takes into account the gradient of the second order corrections to the inhomogeneity of the electron gas for the kinetic energy in the Weizsacker–Kirzhnits approximation and the exchange–correlation energy in the Vashishta–Singvi approximation.

The potential of the electrostatic field φ in (6.28) should correspond to the generalized distribution of electron density $n(z)$ and ionic charges of the metal – ionic crystal system. We do not take into account the periodicity of the field φ in the plane (x, y), created by ions of the ionic crystal, and also due to the 'corrugation' of the electron density distribution created by these ions. Assuming that for the distances $2D \geq \beta^{-1}$, where β is the inverse screening length, the electron density is homogeneous with respect to the coordinates x, y (as noted in [220], this is performed with an accuracy of better than 10%), we use the one-dimensional distribution for both the electron density $n(z)$ and potential φ. In addition, we believe that for $2D \geq \beta^{-1}$ the effect of the distribution of charges in an ionic crystal on $n(z)$ can be approximately described by regarding the ionic crystal as a medium with permittivity ε. Then the test functions of the electron distribution in the metal $n(z)$, according to [33], are determined as the solutions of the linearized Thomas–Fermi equation and for the considered system are as follows:

$$n(z) = \begin{cases} n_0 - 0.5n_0\exp[\beta(z+D)](1+\eta e^{-4\beta D}), & z < -D; \\ 0.5n_0 e^{-\beta D}\{e^{-\beta z} - \eta\exp(\beta(z-2D))]\}, & |z| \leq D; \\ \dfrac{n_o}{1+\gamma}e^{-2\beta D}\exp(-\beta(z-D)/\gamma), & z > D, \end{cases} \tag{6.30}$$

where $\gamma = \sqrt{\varepsilon}; \eta = (\gamma - 1)/(\gamma + 1)$.

Given the Poisson equation, we have the relation $\varphi(z) = -4\pi n(z)/\beta^2$.

For amendments to the interfacial energy due to interaction between the metal ions we used the interpolation formula [308]

$$\sigma_{ii} = \frac{\sqrt{3}Z^2}{c^3} \exp\left(-\frac{4\pi d}{\sqrt{3}c}\right), \tag{6.31}$$

where Z is the valence of the ions; c is the distance between the nearest neighbours in the plane parallel to the metal surface; d is the interplanar distance.

The amendment to the interfacial energy, taking into account the interaction of the electron gas with discrete positive ions in the metal, is calculated using the Ashcroft pseudopotential, averaged over the crystal planes parallel to the surface of the metal, and, according to [33], has the form

$$\sigma_{ei} = \frac{2\pi n_0^2}{\beta^3}\left(1 + \eta e^{-4\beta D}\right)\left(1 - \frac{\beta d e^{-\beta d/2}}{1 - e^{-\beta d}} \operatorname{ch}(\beta r_c)\right), \tag{6.32}$$

where r_c is the Ashcroft pseudopotential cutoff radius.

Component $\sigma_{int}(\beta, D)$ of the interfacial energy of interaction that characterizes the energy of the electrostatic interaction between the electrons of the metal and the ions with the crystal structure such as NaCl, we define as follows. As the functions $n_{+(-)}(r)$, approximating the electron density of positive (negative) ions in an ionic crystal, we use the Wang–Parr piecewise exponential distribution. In this case, as shown in [220], for the range of the adhesion distances it is sufficient to consider only the contribution from the electrons of the last shell

$$n_i(r) = A_i e^{-a_i r}, \tag{6.33}$$

where $i = +, -$ is the index corresponding to the positive or negative ions of the crystal; $A_i = a_i^3/(8\pi)$ is the normalization factor, and the value of a_i, which characterizes the rate of decrease of the density of the valence electrons in the ion crystal is easily identified by comparing the theoretical and experimental values of the ionization energy [56]. In the approximation (6.33): $a_i = 2\sqrt{-2E_i}$, E_i are the experimental values of ionization energy.

Given the Poisson equation, the density (6.33) corresponds to the electrostatic potential

$$\varphi(r) = -4\pi A_i e^{-a_i r}\left(\frac{1}{a_i^2} + \frac{2}{a_i^3 r}\right). \tag{6.34}$$

Averaging this potential along the m-th ion plane parallel to the crystal surface, we obtain

$$V_{im}(z_m') = \pi n_s \int_0^\infty \varphi_i (\sqrt{\rho^2 + z_m'^2})\rho d\rho = \frac{-4\pi^2 A_i n_s}{a_i^4}(3 + a_i \,|\, z_m' \,|)e^{-a_i|z_m'|},$$

(6.35)

where $z_m' = D + a_s(2m-1)/2 - z$, a_s is the lattice constant of an ionic crystal, n_s is the concentration of ions in the crystal in a plane parallel surface. The region of applicability of this approach corresponds to the condition for the applicability of the uniform distribution of the electron density in the plane (x, y), i.e. for $2D \geq \beta^{-1}$.

As a result, the energy of the electrostatic interaction of electrons of the metal with the ions of the crystal can be written as follows:

$$\sigma_{int}(\beta, D) = \sum_{i=+,-}\sum_{m=1}^{\infty} \int_{-\infty}^{\infty} V_{im}(z_m')n(z, \varepsilon = 1)dz,$$

(6.36)

where m is the index of summation of the ionic planes parallel to the crystal surface.

Carrying out the elementary integration of this expression, we obtain

$$\sigma_{int}(\beta, D) = \frac{-\pi n_s n_0}{2} \sum_{i=+,-}\sum_{m=1}^{\infty}\left\{\left(\frac{4}{a_i^2} + \frac{h_m}{a_i} - \right.\right.$$
$$\left. - \frac{3 + a_i h_m}{a_i^2 - \beta^2} - \frac{a_i^2 + \beta^2}{(a_i^2 - \beta^2)^2}\right)e^{-a_i h_m} + \left(3 + \frac{a_i^2 + \beta^2}{a_i^2 - \beta^2}\right)\frac{e^{-\beta h_m}}{a_i^2 - \beta^2}\right\},$$

(6.37)

where $h_m = 2D + a_s(2m - 1)/2$ is the distance from the metal surface to the m-th ion plane of the crystal.

In accordance with the method of the density functional the value of the variational parameter β is found from the requirement of the minimum total interfacial energy

$$\frac{d\sigma(\beta, D)}{d\beta}(\beta_{min}) = 0.$$

(6.38)

The solution of this equation defines the parameter β_{min} as a function of the gap $2D$. The result of solving the variational problem is the total interfacial energy of the system $\sigma(\beta(D), D)$. Knowing this energy, it is easy to find the energy of adhesion of the system as the work required to remove the metal and ionic crystal from each other to infinity, i.e. $E_a(\beta_{min}, D) = \sigma(\beta_{min}, \infty) - \sigma(\beta_{min}, D)$.

To the above procedure was used for numerical calculations of the adhesion characteristics of a number of composites based on alkali metal and aluminum with alkali–halide crystals. The original data used are presented in Tables 6.3–6.5. For small values of the gap we obtained $\beta_{min} \cong 2$ a.u. The range of applicability of calculated adhesion characteristics is specified by the gaps $2D \geq 0.5$ a.u. However, given that in the geometry of our model the surface ion plane for both the metal and for the ionic crystal is situated at half the interplanar distance from the surface of the corresponding material, this analysis is valid also for smaller values of the gap.

Figures 6.12 and 6.13 show plots of the calculated dependence of the adhesion energy on the gap between the contacting materials for aluminum in contact with the alkali–halide crystals. As seen from the dependence, the adhesion strength increases in the transition to composites containing ionic crystals with lighter ions. This result is in accordance with both with the experimental data for composites of this kind [220] and with simple physical considerations. Indeed, in ionic crystals with lighter ions, the lattice period is smaller and this increases the energy of electrostatic interaction of the metal electrons and ions of the crystal (6.36)–(6.37), both due to increasing concentration n_s of ions in the crystal along the contact surface, and due to the close functional relationship $\sigma_{int}(\beta, D)$ on the lattice spacing a_s. In addition, the intensity of the electron–electron repulsion in ionic crystals is reduced during the transition to lighter ions. This means a reduction in repulsive forces between the tail of the electron density of the metal behind its boundary and the electronic shells of ions in the crystal. As a result, the surfaces of the contacting materials come closer to the distance at which the electrostatic forces of attraction between the metal ions and the generalized electron density of the composite start to exert a strong effect in accordance with the electron–ion component (6.32) of the interfacial interaction energy.

It should be noted that the results of the calculation of $E_a(D)$ for Al– NaF and Al–NaCl composites, given in [72] (Table 6.6), show the inverse

Table 6.3. The values of input parameters for the metals used in the calculations

Metal	Z	n_0, a.u.	r_c, a.u.	d, a.u.	c, a.u.
Al(111)	3	0.0269	1.11	4.29	5.25
Li(110)	1	0.0069	1.30	4.67	5.72
Na(110)	1	0.0038	1.73	5.65	6.92
K(110)	1	0.0021	2.25	6.89	8.55
Rb(110)	1	0.0017	2.42	7.47	9.14

Table 6.4. Baseline data for the ions used in the calculations

Ion	Li^+	Na^+	K^+	Rb^+	F^-	Cl^-	Br^-	I^-
E_i, eV	75.64	47.29	31.63	27.29	3.40	3.62	3.37	3.06
a_i, a.u.	4.72	3.73	3.05	2.83	0.99	1.03	0.99	0.95

Table 6.5. Baseline data for ionic crystals, used in the calculations

Crystal	NaF	NaCl	NaBr	NaI	LiCl	KCl	RbCl
a_i, a.u.	4.38	5.33	5.63	6.12	4.86	5.95	6.22
ε	5.05	5.90	6.28	7.28	11.95	4.84	4.92

relationship of the reduction of the adhesion interaction for the composites with the lighter ions. Their incorrectness was mentioned in [220].

Figure 6.12 shows, for the Al–NaF composite, a sharp increase in the adhesive interaction of the metal with the fluorine-bearing crystals. From our point of view this is also explained by the lower values of the lattice constants of fluorine-bearing crystals in comparison with other alkali–halide crystals and, therefore, significantly higher values of the energy of electrostatic interaction of electrons of the metal with ions of the crystal.

Analysis of the results obtained for the forces of adhesion interaction (inset in Fig. 6.13) also confirms the conclusion that the maximum force of adhesion is observed for the composites with lighter ions. In particular, among the compounds represented in the inset, the maximum adhesion force $(F_a(2D = 0.8$ a.u.$) = 16.1 \times 10^{10}$ N/m^2) has Al–LiCl composite, and the minimum $(F_a (2D = 1.2$ a.u.$) = 4.1 \times 10^{10}$ N/m^2) – the Al–RbCl composite.

Calculations showed that the contribution to the energy of electrostatic interaction of electrons of the metal with ions of the crystal (6.37) already from the second ion plane does not exceed 5% of the first ionic plane for the NaF – a crystal with the smallest lattice constant, and for other ionic crystals this contribution is even smaller.

Figure 6.14 shows graphs characterizing the adhesion energy of the NaCl crystal with a number of alkali metals. It is seen that in this case the higher energy of adhesion have lighter metals. A similar effect was observed in the study of other families of metals of the same valence. This result

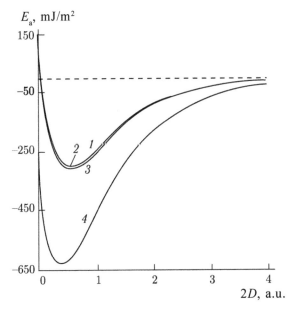

Fig. 6.12. The energy of adhesion depending on the size of the gap between the contacting surfaces of composites: *1* – Al–NaBr; *2* – Al–NaCl; *3* – Al–NaI; *4* – Al–NaF.

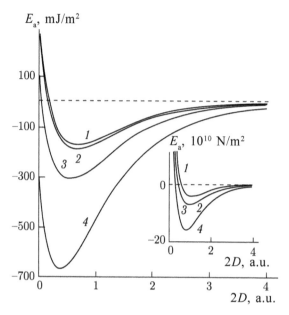

Fig. 6.13. The energy of adhesion (inset – the force of adhesion), depending on the gap between the contacting surfaces of composites: *1* – Al–RbCl; *2* – Al–KCl; *3* – Al–NaCl; *4* – Al–LiCl.

is explained by the fact that in the transition to heavier metals the lattice spacing increases and the bulk electron density n_0 decreases. The decrease of electron density for the heavier metal of the same number valence also causes primarily their lower adhesion to other materials.

Noted in [33], the effect of increasing adhesion with increasing dielectric permeability of the ionic crystal remains valid in the present study for the examined composites. However, for some ionic crystals (see Fig. 6.12), in which the increase in the dielectric permeability is accompanied by an

Table 6.6. The calculations based on adhesion characteristics of the various works

Study	Composite	$2D_0$, a.u.	$E_a(2D_0)$, erg/cm²
This paper	Al–NaCl	0.6	303
	Al–NaF	0.4	620
[33]	Al–NaCl	0.8	154
	Al–NaF	0.8	120
[72]	Al–NaCl	1.8	780
	Al–NaF	2.8	230
[220]	Al–NaCl	3.3	140
	Al–NaF	2.8	260
Experiment	Al–NaCl	–	334
$2D_0$ – equilibrium distance between contacting substances ($F_a(2D_0) = 0$)			

increase in the lattice constant, this effect is suppressed by the already mentioned reverse effect of reducing the magnitude of the electrostatic interaction between the electrons of the metal and ions of the crystal with increasing lattice constant which most often is determining and specifies the reduction in energy and adhesion force of attraction of the contacting materials. Among the composites considered in Fig. 6.12, only in the Al–NaI contact a higher dielectric constant of NaI leads to higher values adhesion energy in comparison with the Al–NaBr and Al–NaCl composites.

Table 6.6 compares the values of adhesion energy of the Al–NaF and Al–NaCl composites obtained in different studies, with results of the present work. The data provided by the table suggests that our results are in better agreement with the experimental value of the adhesion energy of the Al–NaCl composite, equalling 334 erg/cm^2 and presented on [220].

With regard to the data in [33], in which the dielectrics were defined only by their permittivity, the difference between the results of this work and the data in [33] is due to considering the gradient corrections for the exchange–correlation energy in the energy density of the composite and the energy of electrostatic interaction of the metal electrons and ions of the crystal. These contributions partly compensate each other.

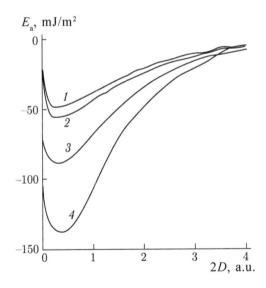

Fig. 6.14. The energy of adhesion, depending on the size of the gap between the contacting surfaces: *1* – Rb–NaCl; *2* – K–NaCl; *3* – Na–NaCl; *4* – Li–NaC.

7

CALCULATION OF ADHESION CHARACTERISTICS OF METALS, SEMICONDUCTORS AND COMPLEX COMPOUNDS BASED ON DIELECTRIC FORMALISM

Introduction

One of the most urgent needs of modern engineering is the improvement of technological methods to improve performance, durability and longevity of components of friction pairs, as well as reducing the power loss due to friction in them. This requires a detailed accounting of factors affecting the magnitude of the interaction of dissimilar materials, and the factors influencing directly the adhesion strength. Knowledge of these factors makes it possible already in the technological preparation stage to select the most appropriate study materials, coatings, and to predict their longevity. In connection with this the main purpose of current research in the field of surface science is the development of methods of analytical description of the friction process, based on the laws of physics, chemistry, mechanics, and development on the basis of these laws of calculation methodologies of the adhesion characteristics.

As shown in previous chapters, in the description of the adhesive properties of materials particularly effective methods are the density functional methods or their modifications [101, 181, 220, 299, 361]. It should be noted that the density functional method allows to obtain a fairly good agreement between the calculated values of the surface and adhesion characteristics and with the experimental results only for monatomic metals, i.e. one-component systems, but cannot be used to describe complex alloys and compounds, as well as to take into account the band structure of semiconductors.

Thus, the density functional method, despite its undoubted importance in the general theoretical formulation of basic principles and concepts of adhesive interaction of metals, does not always makes it possible to obtain reliable calculated values of the characteristics of the adhesive interaction due to the lack of data on a number of empirical parameters which, as a rule, are not known for complex substances, such as: the binding energy of the atoms, the electron work function from the surface of the material, the elastic constants in the thin surface layer, etc. As the density functional method is based on the local characteristics of the electron density distribution in the material, which define a set of correction terms in the density of the surface energy, is not universal for all materials, this leads to cumbersome and computational complexity, as well as to the use of too large a set of (variational) parameters and the corresponding set of empirical data.

The above difficulties make it necessary to create approaches, although less theoretically justified than in the density functional method, however, to ensure receipt of reliable results with the minimal input source data. In this chapter, to describe the adhesive properties of metals and semiconductors, we use an approach based on the dielectric formalism. The use of model approximations for the dielectric functions of these materials makes it possible to determine their adhesion characteristics based only on the concentration of valence electrons and the band gap width.

The possibilities of this approach when applied to the calculation of molecular (van der Waals) forces of interaction of the surfaces of various bodies, are shown, for example, in [11, 83]. van der Waals forces determine the interaction of bodies for sufficiently large values of the gap h between their surfaces ($h \geq 100$ Å) and are associated with the correlation effects of the interaction through the fluctuating electromagnetic field caused by fluctuations of the induced dipole moments of atoms and molecules. At smaller values of the gap, along with the correlation energy of interaction interaction, we must take into account the fluctuation component of the exchange interaction energy of the electrons with the exchange–correlation holes. The combined influence of these exchange–correlation effects and the interaction of the electrons is determined primarily by the energy adhesion of various bodies for both small and sufficiently large values of the gap h to $h_0 \simeq 10^{-5} \div 10^{-4}$ cm, where the correlation energy of interaction of bodies should take into account the effects of the delay. In this paper, we neglect the influence of the delay effects, i.e. assume that $h \ll h_0$. The main theoretical ratios for the exchange–correlation interaction of fluctuations of the electron densities of the various bodies are discussed in the long-wave approximation.

In the long-wave approximation, the single-particle effects of the electron system are superimposed with the collective effects of excitations of the system which have the character of high-frequency plasma oscillations. Therefore, in the long-wave approximation, the interaction energy of contacting condensed media will be determined primarily by the energy of surface plasma waves excited in the surface layers. The relationship of the energy of surface plasma oscillations with the surface energy of metals was followed by several authors (see references in [270]). According to the ideas developed in these studies, the surface energy of the material can be expressed through the energy difference between the 'zero' ($T = 0$ K) plasmons on the surface and in the volume. In [171, 172], based on the concept of surface plasmons, the authors obtained formulas and calculations of surface energy and adhesion energy with the direct contact of the surfaces ($h = 0$) for a number of metals and semiconductors.

This chapter presents the results of the generalization [26–28] of concepts and results of the studies [11, 171] for the case of the arbitrary size of the gap h between the contacting materials, and the methods of calculation of the energy and strength of adhesive interaction between metals and semiconductors as a function of the gap h.

In describing the phenomenon of adhesion the authors of the monograph developed [27, 28, 64] theoretical methods and carried out the calculations of the adhesion characteristics of the contact of different metals, semiconductors and complex compounds. The factors affecting the adhesion properties of the contacting materials were determined. The developed methods have found application in the adhesion theory of dry friction of metal surfaces developed by the authors [61–68], allowing the calculation of contributions to the coefficient of friction due to electrodynamic effects of energy dissipation and the effects of adhesive attraction of surfaces of friction pairs. Principles and methods of selecting the optimum material pairs for non-lubricated friction pairs, taking into account the nature of their adhesion interaction.

7.1. Theoretical principles for determining the adhesion characteristics of the contact surfaces

An approach based on dielectric formalism [149, 171] allows, using the model approximations for the dielectric functions of metals, to determine the law of dispersion of collective excitations in the system – plasmons. The dispersion law of plasmons strongly depends on whether they are on the surface or inside the material. Collective effects in solids, which include the adhesion phenomenon, are defined by the energies of the plasmons in the long-wavelength limit.

Consider the interaction between two semi-infinite materials, located at $T = 0$ K and occupying regions $z < 0$ and $z > h$. Neglecting retardation effects in the interaction of bodies, in the Maxwell equations we can formally put $c \to \infty$ and thus use the electrostatics equation for the potential of the electrostatic field ϕ in the system:

$$\Delta\phi = 0, \quad \phi = \phi(z)e^{i(k_x X + k_y Y - wt)}, \tag{7.1}$$

and get

$$\frac{d^2\phi(z)}{dz^2} - k^2\phi(z) = 0, \quad k^2 = k_x^2 + k_y^2. \tag{7.2}$$

We are interested in solutions having the character of the collective oscillations localized at the surface (disappears when $z \to \infty$). As a result from (7.2) we have:

$$\phi(z) = \begin{cases} Ae^{kz}, & z < 0; \\ Be^{kz} + Ce^{-kz}, & 0 < z < h; \\ De^{-kz}, & z > h. \end{cases} \tag{7.3}$$

'Sewing' together solutions (7.3) at the interface ($z = 0, h$) from the conditions of continuity of the tangential component of the strength of the electric field and the normal component of electric induction (these conditions are equivalent to the continuity of $\phi(z)$ and $\dfrac{d\phi}{dz}$), we obtain as the condition for the existence of a non-trivial solution the following dispersion equation for surface waves in the system:

$$D(\omega, k, h) = \frac{\left[\varepsilon_1(\omega, k) - 1\right]\left[\varepsilon_2(\omega, k) - 1\right]}{\left[\varepsilon_1(\omega, k) + 1\right]\left[\varepsilon_2(\omega, k) + 1\right]} e^{-2kh} = 1. \tag{7.4}$$

The roots of this equation are of interest to us as the eigenfrequencies of surface vibrations. To find them, we must specify the explicit form of the functions of dielectric permittivity $\varepsilon_n(\omega, k)$ for both materials within the framework of the accepted model of the interaction of these environments. According to [11], the interaction energy associated with the presence of interfaces of two semi-infinite bodies at a distance h (per unit area) is equal to

$$E(h) = \frac{\hbar}{2} \iint \frac{d^2 k}{(2\pi)^2} \frac{1}{2\pi i} \int_{-i\infty}^{i\infty} \ln D(\omega, k, h) d\omega, \tag{7.5}$$

where the function $D(\omega, k, h)$ is given in (7.4). The function $D(\omega, k, h)$ is analytic everywhere except a finite number of poles $\omega_{si}(k, \infty)$, corresponding to the frequencies of surface waves when $h \to \infty$. The zeros of $D(\omega, k, h)$, equal to $\omega_{si}(k, h)$, correspond to the frequencies of surface waves with arbitrary but finite h. According to the principal argument of the complex variable function theory, the integral $\dfrac{1}{2\pi i} \int_{-i\infty}^{i\infty} \ln D(\omega, k, h) d\omega$ is the difference between the total number of zeros and poles of $D(\omega, k, h)$. As a result, we obtain

$$E(h) = \hbar/2 \sum_{i=1,2} \iint \left[\omega_{si}(k, h) - \omega_{si}(k, \infty)\right] \frac{d^2 k}{(2\pi)^2}. \tag{7.6}$$

This formula has a simple physical meaning: the interaction energy is the energy difference between the 'zero' surface plasma fluctuations when the surfaces of the contacting bodies are, respectively, at a distance h and at an infinite distance from each other.

We are interested in collective excitations of the electronic system of solids due to the interaction of electrons with the exchange–correlation holes. Consider an electronic system in the 'jelly' model, where the charge of the electron of each material is offset the homogeneous positive background. The exchange and correlation effects in the interaction of electrons are taken into account in the Hartree–Fock approximation. The collective excitations of the electronic system – plasmons – in the long-wave approximation are mathematically consisten twith the use of the approximation of the random phases. Depending on the nature of the interacting materials (metals, semiconductors, dielectrics), the dielectric permeabilities of the materials $\varepsilon_i(\omega, k)$ in the random-phase approximation have different expressions [171, 182], thus determining from the solutions of the dispersion equation different dispersion laws of the connected surface plasmons.

However, the calculations should take into account that surface plasmons $\hbar\omega_{sj}$ at a certain critical of the wave vector k_c, determined by the condition

$$\omega_{sn}(k_c) = k_c v_{fn} + \frac{\hbar k_c^2}{2m}, \tag{7.7}$$

break up, transferring their energy and momentum to single Fermi electrons [189]. This means that when $k > k_c$ the plasmon cannot exist as a coherent motion of all electrons, i.e. becomes almost unobservable.

Each of the plasma modes for different materials is activated when $h \to \infty$, and is characterized by its critical value of the wave vector k_{cn}. In this connection, it is necessary in the calculation of the formula (7.6) to carry out integration over the wave vectors $k < k_c^{(min)}$, where $k_c^{(min)}$ corresponds to the minimum of the values of the critical wave vectors k_{cn} of the materials in question. Thus, only the contribution of the collective states to interaction $E(h)$ is taken into account. The adhesion energy of two different materials us directly associated with the interaction energy $E(h)$. Thus, $E_a(h) = -2E(h)$ [507], and, consequently, the final formula for the adhesion energy of materials, separated by a gap h, takes the form

$$E_a^{(12)}(h) = \hbar \sum_{i=1,2} \int_0^{k_c^{(min)}} \left[\omega_{si}(k,\infty) - \omega_{si}(k,h) \right] \frac{k\,dk}{2\pi}. \tag{7.8}$$

The surface energy can be expressed through the energy of adhesion of the same materials by the following relation:

$$\sigma = 1/2 E_a^{(11)}(0). \tag{7.9}$$

The strength of adhesive interaction of various materials such as function of the gap h between them can be obtained by differentiating the adhesion energy $E_a(h)$ for h, i.e.

$$F_a(h) = -\frac{dE_a}{dh}. \tag{7.10}$$

7.2. Adhesion of metals

Consider the case of contact between two metals separated by a vacuum gap with width h. To obtain the dispersion laws of the coupled surface plasmons, we have to write the expression for the dielectric constants of metals $\varepsilon_i(\omega, k)$. In the approximation of the random phase, they are [171, 182]:

$$\varepsilon_n(\omega, k) = 1 + \frac{\omega_{pn}^2}{\omega_{kn}^2 - \omega^2}, \tag{7.11}$$

$$\omega_{pn}^2 = \frac{4\pi e^2 N_n}{m}, \quad \omega_{kn}^2 = \mu v_{fn}^2 k^2, \quad \mu(\omega) = \frac{7}{18} + \frac{19}{90}(\omega / \omega_{pn})^2.$$

Here ω_{pn}, v_{fn}, N_n are respectively, the plasma frequency, the Fermi velocity and the concentration of valence electrons of the n-th metal; $\mu(\omega)$ is the adjustment factor, numerical coefficients in which provide the standard dispersion of the bulk plasma frequency [149]

$$\varepsilon[\omega_p(k), k] = 0, \quad \omega_p^2 = \omega_{p0}^2 + \frac{3}{5}v_f^2 k^2. \tag{7.12}$$

To determine the law of the dispersion of the frequency of the intrinsic surface plasma oscillations $\omega_s(k)$ the dielectric constants of the material $\varepsilon(\omega, k)$ should be substituted from (7.11) to the dispersion equation

$$\varepsilon(\omega, k) + 1 = 0. \tag{7.13}$$

which yields

$$\omega_{sn}(k) = \frac{1}{\sqrt{2}}\left(\omega_{pn}^2 + \frac{89}{90}v_{fn}^2 k^2\right)^{1/2}. \tag{7.14}$$

Solving the equation (7.4) in the approximation when using relations (7.11), we obtain the following dispersion relations for surface plasma waves in metals whose surfaces are separated by distance h:

$$\omega_{si}(k, h) = \frac{\omega_{pl}}{2}\{1 + \Delta + 2(\gamma_{1i} + \gamma_{2i}) \pm [(1 - \Delta +$$
$$+ 2\gamma_{1i} - 2\gamma_{2i})^2 + 4e^{-2kh}\Delta]^{1/2}\}^{1/2}, \tag{7.15}$$

where the notation $i = 1, 2$, corresponding in (7.15) to the signs $\pm$ ($\omega_{p1} \geq \omega_{p2}$) are used

$$\Delta = (\omega_{p2}/\omega_{p1})^2, \quad \gamma_{11} = v_{1F}^2 k^2 (89 + 19\Delta)/180\omega_{p1}^2,$$

$$\gamma_{21} = v_{2F}^2 k^2 (89 + 19\Delta^{-1})/180\omega_{p1}^2, \quad \gamma_{12} = 7v_{1F}^2 k^2/18\omega_{p1}^2,$$

$$\gamma_{22} = 7v_{2F}^2 k^2/18\omega_{p1}^2.$$

To calculate the adhesion energy, the expressions (7.14) and (7.15) should be substituted into (7.8) and integrate over the wave vectors.

In accordance with the methodology of determining the adhesive properties of different materials presented in section 7.1, numerical calculations of surface energy and adhesion energy and force for a number of metals were carried out.

Using the above relations allows us to calculate the values of the surface energy for metals using only one empirical parameter N – the concentration of valence electrons, since $v_f = \hbar(3\pi^2 N)^{1/3}/m$. The values in Table 7.1, taken from different literature sources [182, 189, 225], give the plasmon energies, experimental values of surface energy and our calculated values of the surface energy of various metals. As the table shows, the agreement between the calculated and experimental values of surface energies is good. On the one hand, it is surprising, since such an agreement is reached within the framework of a simple electron–ion model of the crystal lattice – the 'jelly' model. For comparison, one should consider the fact that the values of the surface energy of metals, calculated using the density functional

Table 7.1. The energies of the plasmons and surface energies for different metals

Element	Z	$\hbar\omega_p$, eV	$\sigma_{(theor)}$, mJ/m^2	$\sigma_{(exp)}$, mJ/m^2
K	1	3.72; 3.87	143	103 ± 8 (liquid.)
Na	1	5.71; 5.25	273; 284	280 ± 5
Mg	2	10.6	767	790
Zn	2	13.0	1078	1020
Be	2	18.9	2011	–
Al	3	15.3	1414	1140 ± 200
Cu	3	17.8	1820	1750 ± 90
Pb	4	14.0	1219	560 ± 4
Fe	4	21.2	2435	2170 ± 330
Ni	4	22.1	2610	1820 ± 180
Cr	4	21.0	2397	2200 ± 250
Ti	4	17.6	1786	2050
V	5	22.0	2591	1950 ± 50
Mo	6	22.2	2630	2600 ± 200
W	7	23.5	2892	2800 ± 280

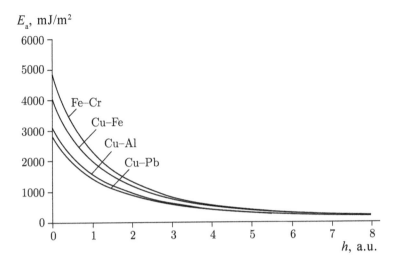

Fig. 7.1. Adhesion energy E_a of metals depending on the size of the vacuum gap between the surfaces of materials.

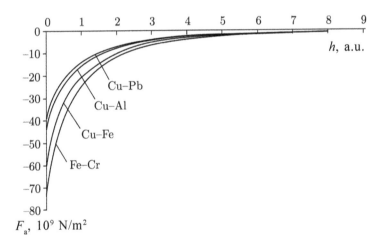

Fig. 7.2. Adhesion force F_a of metals depending on the size of the vacuum gap between the surfaces of materials.

method within the 'jelly' model, have non-physical negative values. This suggests that the surface properties of metals are largely determined by long-wavelength collective excitations of the electron–ion system.

Figures 7.1 and 7.2 show the results of the calculation of the energy and adhesion force for a number of simple and transition metals, depending on the gap. The adhesion force, obtained within the dielectric formalism, is attractive in the whole range of change of h in the range of applicability of the theory $a \ll h \ll \lambda_0$, where a is the interatomic

Table 7.2. The values of adhesion energy E_a, mJ/m² and the forces of adhesion F_a, 10^9 H/m² various metals in the gap size $h = 0$

Value	Cu–Fe	Cu–Al	Cu–Pb	Fe–Cr	Pb–Al	Pb–Cr	Cu–Cr
E_a, mJ/m²	4030	3091	2798	4828	2573	3043	4009
F_a, 10^9 N/m²	−59.4	−43.6	−38.7	−73.7	−35.2	−42.4	−59.1

distance, and λ_0 is the radiation wavelength characteristic for atoms (almost $\lambda_0 \sim 10^{-5} \div 10^{-4}$ cm).

At $k_c h \gg 1$ electronic systems of two materials are separated, exchange effects are negligible, and the remaining correlation effects of the interaction of electronic systems correspond to van der Waals forces of interaction without retardation effects. Thus, within the dielectric formalism the van der Waals interaction forces are automatically taken into account in the above formulas, while when using the electron density functional method only the electrostatic component of the interaction of electronic systems is taken into account.

Table 7.2 shows the results of the calculation of the energy and force of adhesion of a number of metals for the gap size $h = 0$. If we compare the values of adhesion force at $h = 0$, calculated using the dielectric formalism and the method of the electron density functional, one can note the good agreement of these values (see Table 6.2).

7.3. Adhesion of semiconductors

Consider the case of contact between two semiconductors separated by the vacuum gap of size h. To do this, the expression (7.11) for the permittivity $\varepsilon_n(\omega, k)$ is modified, taking into account the band gap of the semiconductor E_{gn} separating the valence band and the conduction band. Thus, according to [171, 182], in the random phase approximation the expressions for the dielectric permittivity of the semiconductors can be expressed as

$$\varepsilon_n(\omega, k) = 1 + \frac{\omega_{pn}^2 - \omega_{gn}^2}{\omega_{kn}^2 + \omega_{gn}^2 - \omega^2}, \quad \omega_{gn} = E_{gn} / \hbar, \tag{7.16}$$

and other parameters have the same expression and the same meaning as in (4.11). From the solution of the dispersion equation

$$\varepsilon_n(\omega, k) + 1 = 0$$

it is easy to obtain the dispersion law for intrinsic plasma oscillations on the surface of the semiconductor

$$\omega_{sn}(k, \infty) = \frac{1}{\sqrt{2}} \left(\omega_{pn}^2 + \omega_{gn}^2 + \frac{89}{90} v_{fn}^2 k^2 \right)^{1/2}. \tag{7.17}$$

The solution of the dispersion equation (7.4) in the case of arbitrary h leads to the following expressions for the surface plasma frequencies $\omega_{sj}(k, h)$:

$$\omega_{sj}(k,h) = \frac{1}{2}\omega_{pl}\{1+\Delta+\beta_1+\beta_2+2(\gamma_{1j}+\gamma_{2j})\pm$$

$$\pm [(1-\Delta+\beta_1-\beta_2+2\gamma_{1j}-2\gamma_{2j})^2 +$$

$$+ 4\exp(-2kh)(1-\beta_1)(\Delta-\beta_2)]^{1/2}\}^{1/2}, \qquad (7.18)$$

where the notation $j = 1, 2$, corresponding in (7.18) to the signs $\pm$ are used,

$$\Delta = \left(\frac{\omega_{p2}}{\omega_{pl}}\right)^2, \quad \omega_{pl} \geq \omega_{p2}, \quad \beta_1 = \left(\frac{\omega_{g1}}{\omega_{pl}}\right)^2, \quad \beta_2 = \left(\frac{\omega_{g2}}{\omega_{pl}}\right)^2,$$

$$\gamma_{11} = \frac{v_{f1}^2 k^2}{180\omega_{pl}^2}(89+19\Delta), \quad \gamma_{21} = \frac{v_{f2}^2 k^2}{180\omega_{pl}^2}(89+19/\Delta),$$

$$\gamma_{12} = \frac{7v_{f1}^2 k^2}{18\omega_{pl}^2}, \quad \gamma_{22} = \frac{7v_{f2}^2 k^2}{18\omega_{pl}^2}.$$

To calculate the interaction energy of the materials, the expressions (7.17) and (7.18) for $\omega_{sj}(k, h)$ and $\omega_{sj}(k, \infty)$ must be substituted into (7.8) and integrated over the wave vectors.

These relations were used for numerical calculations of the surface energy and adhesion energy and force for a number of semiconductors using the experimental values of plasma frequency. Table 7.3 shows the values of the energies of the plasmon and the calculated values of surface energy for a series of semiconductors. Unfortunately, we do not know the experimental values of surface energy for semiconductors. But because similar calculations of the surface energy of metals are in good agreement with experimental values, we can assume that good agreement can also be obtained for semiconductors.

Figures 7.3 and 7.4 show the results of calculations of the adhesion force for a number of semiconductors, depending on the size of the gap h. From the graphs it can be seen that over the entire range of h in the range

Table 7.3. The energies of the plasmons and surface energies for a number of semiconductor compounds

Elements	Z	$\hbar\omega_p$, eV	$\sigma_{(theor)}$, mJ/m^2
Si	4	16.9	1664
Ge	4	16.0	1524
InSb	4	12.0	943
PbSe	5	15.0	1368
ZnS	4	17.0	1686

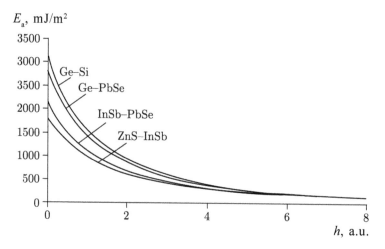

Fig. 7.3. Adhesion energy E_a of number of semiconductors, depending on the size of the vacuum gap between the surfaces of materials.

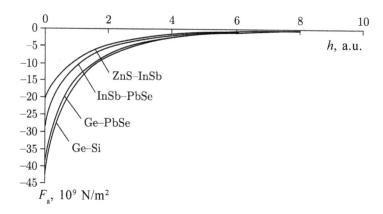

Fig. 7.4. Adhesion force F_a of a series of semiconductors, depending on the size of the vacuum gap between the surfaces of materials.

of applicability of the theory the adhesion force of semiconductors, as well as metals, is of the attraction type. In this case, analysis shows that the best adhesion characteristics are shown by materials with a greater surface energy, and the adhesion of metals is usually much greater than the adhesion of semiconductor materials.

7.4. Calculation of the adhesion characteristics of contact of metals and semiconductors with a dielectric

Consider the case of the interaction of a metal or semiconductor and a dielectric with permittivity ε_0, separated by a vacuum gap with size h. In this system, there exists a unique mode of surface plasma oscillations localized on the surface of the metal (semiconductor). The dispersion law for it can be obtained by solving the equation (7.4) by replacing $\varepsilon_2(\omega, k)$ for ε_0.

As a result,

$$w_s(k,h) =$$

$$= \frac{1}{\sqrt{2}} \left(\omega_{gl}^2 + \omega_{pl}^2 + \frac{(54+35\varepsilon_0)}{45(1+\varepsilon_0)} v_{fl}^2 k^2 + (\omega_{gl}^2 - \omega_{pl}^2) \frac{(\varepsilon_0-1)}{(\varepsilon_0+1)} e^{-2kh} \right)^{1/2}.$$

(7.19)

The energy of adhesion of the metal (semiconductor) and the dielectric as a function of the distance h between the contacting surfaces can be calculated using the relation

$$E_a(h) = \hbar \int_0^{k_c} \left[\omega_s(k,\infty) - \omega_s(k,h) \right] \frac{k\,dk}{2\pi},$$

(7.20)

where for $\omega_s(k, \infty)$ we use the expression

$$\omega_s(k,\infty) = \frac{1}{\sqrt{2}} \left[\omega_{gl}^2 + \omega_{pl}^2 + \frac{(54+35\varepsilon_0)}{45(1+\varepsilon_0)} v_{fl}^2 k^2 \right]^{1/2},$$

(7.21)

for $\omega_s(k, h)$ the expression (7.18), and to determine the critical wave vector k_c the expression (7.7). The strength of adhesive interactions of the metal (semiconductor) and an insulator as a function of the gap h between them can be obtained by differentiating adhesion energy $E_a(h)$.

Figures 7.5–7.8 show the results of the calculation of the adhesion characteristics of the contact of copper and germanium with the materials having the dielectric properties with permittivities $\varepsilon = 4, 10, 15, 80$. The graphs for the adhesion energy of materials, presented in Figs. 7.5 and 7.7, show that for small values of the gap h the adhesion energy is characterized by a maximum whose position shifts with increasing ε to larger h. This feature reflects the change of the nature of the adhesion interaction from repulsion to attraction that occurs at small h. In this case, the repulsion effects at small values of the gap h increase with the increase of the dielectric permeability of the material and the size of the gap at which the nature of the interaction changes also increases.

The graphs show that as ε increases the absolute energy and adhesion force also increase, and the rate of their decay with an increase in the gap decreases. These effects have been found by us also in the description of the adhesion of metals and insulators in the framework of the density functional method. They were physically interpreted in section 6.4. However, in contrast to the results obtained in the density functional method, which for large values of the gap show repulsion of the metal and the dielectric, in this approach by taking into account the van der Waals forces at large

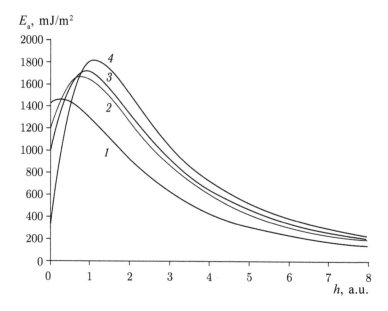

Fig. 7.5. The adhesion energy of the metal (Cu) – a dielectric with permittivity: $1 - \varepsilon = 4$, $2 - \varepsilon = 10$, $3 - \varepsilon = 15$, $4 - \varepsilon = 80$, depending on the size of the vacuum gap.

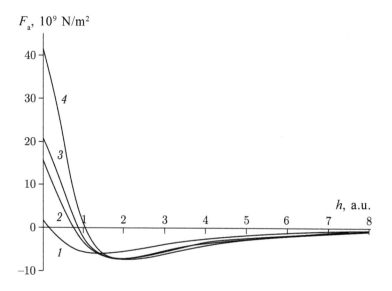

Fig. 7.6. The adhesion strength of the metal (Cu) – a dielectric with permeability: $1 - \varepsilon = 4$, $2 - \varepsilon = 10$, $3 - \varepsilon = 15$, $4 - \varepsilon = 80$, depending on the size of the vacuum gap.

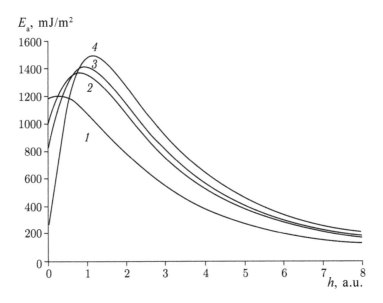

Fig. 7.7. The adhesion energy of the semiconductor (Ge) – a dielectric with permittivity: $1 - \varepsilon = 4$, $2 - \varepsilon = 10$, $3 - \varepsilon = 15$, $4 - \varepsilon = 80$, depending on the size of the vacuum gap.

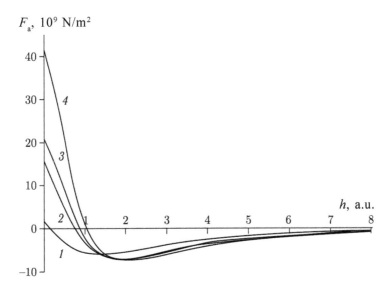

Fig. 7.8. The adhesion strength of the semiconductor (Ge) – a dielectric with permittivity: $1 - \varepsilon = 4$, $2 - \varepsilon = 10$, $3 - \varepsilon = 15$, $4 - \varepsilon = 80$, depending on the size of the vacuum gap.

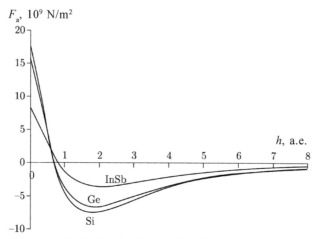

Fig. 7.9. The strength of adhesive interaction of a number of semiconductors with dielectric constant $\varepsilon = 10$, depending on the size of the vacuum gap.

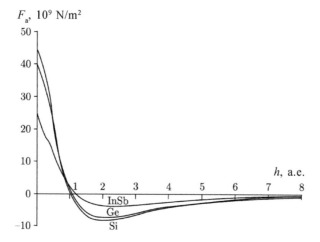

Fig. 7.10. The strength of adhesive interaction of a number of semiconductors with dielectric constant $\varepsilon = 80$, depending on the size of the vacuum gap.

values of the gap the effects of the van der Waals attraction of materials become dominant when compared which the correlation effects of repulsion.

Figures 7.9 and 7.10 give the values of the adhesion force for contact of semiconductors with different dielectrics with the permittivity $\varepsilon = 10$ and $\varepsilon = 80$. From these figures it is clear that the semiconductors with a greater surface energy show a sharper minimum which corresponds to the maximum possible value of the adhesive attraction force of these materials with the dielectrics. Moreover, for all materials the maximum attractive force is reached at the gap size $h \approx 2$ a.u. and with increasing ε is only slightly shifted toward higher values of the gap.

Table 7.4. The maximum values of the attractive force F_a, 10^9 N/m^2 of different materials with the dielectric constant $\varepsilon = 10$ and $\varepsilon = 80$

Permeability	Cu	InSb	Ge	Si
$\varepsilon = 10$	−8.4	−3.5	−6.7	−7.5
$\varepsilon = 80$	−9.3	−3.9	−7.3	−8.3

Table 7.4 shows the calculated maximum values of forces attraction for a number of materials in contact with the dielectric permittivity $\varepsilon = 10$ and $\varepsilon = 80$.

7.5. Adhesion of complex compounds

This section presents the results of the calculation of the adhesion characteristics of contact of various coatings of wear-resisting and self-lubricating materials used in microcryogenic techniques for modifying the surfaces of friction pairs in order to give them high tribological properties, characterized by high wear resistance and low friction coefficient [61–68].

In joint work [61–68], the authors and scientists and technologists of emerging technologies of the Sibkriotekhnika' (Omsk) (currently, LLC 'NTC Cryogenic Technology') company developed methods for calculating the adhesion component of the friction force in metal friction pairs under dry friction conditions. The equation derived for the first time in [199] for the friction coefficient of metallic pairs directly contained the dependence on the adhesion force of the materials that make up a friction pair. There was a problem of calculating the adhesion force, and then the friction coefficient for a series of wear-resisting materials based on nitrides and carbides of transition metals.

In microcryogenic technology, the force of interaction of the rubbing surfaces at the dry friction sites is reduced by self-lubricating materials – graphite, plastics, molybdenum disulphide (MoS_2), etc. Nitrides and carbides of transition metals are mainly characterized by the metallic properties, although some of these with deviations from stoichiometry can acquire semiconducting properties. Self-lubricating materials with electrophysical properties are dielectrics (teflon, plastics) or semiconductors (graphite, MoS_2, etc.). In metallic dry friction pairs with coatings of wear-resisting and self-lubricating materials, we are faced with the need to calculate the adhesion and tribological properties for metal–metal, metal–semiconductor, and metal–dielectric pairs. The methods developed for calculating the adhesion characteristics of different materials based on the dielectric formalism make it possible to solve this problem better than any currently available theoretical methods.

From the formulas given in the preceding sections, we see that to calculate the adhesion characteristics of each metal–metal pair it is enough to have information only on experimental plasma frequencies $\omega_p^2 = 4\pi e^2 n/m$ and the Fermi velocities of electrons $v_F = \hbar(3\pi^2 n)^{1/3}/m$. But these relations show that in fact we need to know only the values of the concentration n of valence electrons of the metal, which can be determined by the valence Z of the metal ions, the molar mass M, mass density ρ and Avogadro's number N_A, i.e. $n = ZN_A\rho/M$. Comparison of the theoretical calculated and experimental values of the plasma frequency ω_p shows a good agreement only for simple metals (K, Mg, Be, Al, etc.) that do not have the d-band, and some semiconductors (Si, Ge, ZnS, PbSe, etc.). This agreement breaks down in metals and alloys with the d-band near the Fermi surface, such as Cu, Ag, Au, and in most transition metals. Therefore, the most desirable experimental parameter for the calculation of the adhesion properties of metals and alloys is the plasma frequency. Unfortunately, it is not available for some materials. In particular, for the nitrides and carbides of transition metals in the determination of ω_p and subsequent calculating the adhesion energy and adhesion strength of interaction it was necessary to use the experimental values of the Hall coefficients R [87, 205] or the electron density of states $N(\varepsilon_F)$ on the Fermi level [87].

For semiconductors, the additional parameter compared to metals is the band gap E_g. In the case of adhesive interaction of a metal with an insulator the characterization of the latter in the present model requires only permeability ε_0.

We computed adhesion characteristics of the contact of steel 30Kh13 with a wide range of nitrides and carbides of transition metals (TiN, Mo_2N, Cr_2N, TiC, Mo_2C, WC, Cr_3C_2), as well as with teflon and molybdenum disulphide MoS_2, i.e. a set of materials widely used in tribotechnology. The values of input parameters, used in the calculation of the adhesion characteristics of these materials, are shown in Table 7.5, the results of calculation of the adhesion force at $h = 0$ – in Table 7.6.

The subsequent calculation of the friction coefficient of nitrides and carbides of the transition metals on a steel using the data on the adhesion force revealed that among the considered set of materials the most suitable tribological characteristics are those of the molybdenum nitride coating Mo_2N.

However, when choosing the coating material for the friction pairs we must take into account a number of fundamental contradictions. They are related to the fact that the antifriction properties of a coating are achieved largely due to its low adhesive properties with respect to the materials contained in the friction pair, but at the same the time it is necessary to ensure that the antifriction coating is efficiently bonded with the substrate material. Typically, the materials of the working body in a friction pair and the substrate material are steels with respect to which the material of the antifriction coating has similar adhesion properties. In the case of molybdenum nitride this means that, as an excellent material because of its tribological properties in a friction pair with steel grade 30Kh13, Mo_2N

Table 7.5. The values of the lattice constants a, the plasma frequency $\hbar\omega_p$, the band gap E_g, the dielectric constant ε, the Hall constant R, the effective carrier density n^*, the effective charges Z^* and ion conductivities for a number of metals and compounds

Compound	a, Å	$\hbar\omega_p$, eV	E_g, eV	ε	R, 10^{-10} m^3/C	n^* $10^{28}\ \text{m}^{-3}$	Z^*, e	σ 10^6 $(\text{ohm·cm})^{-1}$
TiC	4.327	–	–	–	–14	0.44	0.045	0.005
Cr_3C_2	5.553	–	–	–	–0.47	13.2	1.2	0.0133
Mo_2C	4.321	–	–	–	–0.85	7.30	0.91	0.0141
WC	2.906	–	–	–	–21.8	0.29	0.03	0.0522
TiN	4.249	–	–	–	–0.55	11.2	1.19	0.04
CrN	4.149	–	–	–	–264	0.024	0.0022	0.0127
Mo_2N	4.169	–	–	–	–2.83	2.21	0.32	0.0051
ZrN	–	–	–	–	–1.44	–	–	–
NbN	–	–	–	–	0.52	–	–	–
ZrC	–	–	–	–	–0.4	–	–	–
NbC	–	–	–	–	–1.05	–	–	–
30X13	2.94	14.4	–	–	–	–	4	0.13
MoS_2	–	21	1.2	10	–	–	6	–
Teflon-4	–	–	–	2.2	–	–	–	–

Table 7.6. The values of adhesion force $F_a(0)$, 10^9 N/m² for a number of transition metals, nitrides and carbides of transition metals in contact with steel 30Kh13 and molybdenum nitride Mo_2N, when the value of $h = 0$ the gap between the surfaces of materials

Mate-rials	Mo	Ni	Cr	Zr	Ti	Cr_3C_2	NbN	TiN	Cr_2N	NbC	ZrN	Mo_2N
Mo_2N	9.92	9.91	9.78	9.55	9.29	8.48	8.31	8.23	7.82	7.11	6.56	–
Steel 30Kh13	45.5	45.4	44.2	42.1	39.9	32.2	30.2	29.1	24.7	18.5	14.8	8.69

coating can be poorly bonded to the steel surface of the friction pair due to the low adhesion properties.

The way out of this contradiction can be found by applying an intermediate sublayer between the surface of the component and the Mo_2N coating. The requirement, which the material of the intermediate layer must meet, is the following: the strength of adhesion of the material with both the surface of the component and with the molybdenum nitride coating must be greater than that of Mo_2N with the surface of the component.

When searching for the material for the intermediate layer the following sets of transition metals were examined: Mo, Ni, Cr, Ti, Zr – and nitrides and carbides of transition metals: Cr_3C_2, TiN, TiC, Cr_2N, ZrC, ZrN, NbC, NbN. A computer program was developed for calculating the energy and adhesion force of the above transition metals and compounds with 30Kh13 steel and molybdenum nitride Mo_2N. The calculation results are also shown in Table 7.6. It can be seen that the greatest adhesion strength to both steel and molybdenum nitride is obtained for the transition metals in the sequence Mo, Ni, Cr, Zr, Ti, followed by a sequence of compounds Cr_3C_2, NbN, TiN, Cr_2N, NbC, ZrN. The adhesion strength of these materials with the steel exceeds the adhesion strength of the same materials with nitride molybdenum about 4 times. This sequence of materials, ordered by the value of adhesion strength, is also retained at h not equal to zero. This is due to a monotonic decrease in the adhesion force modulus (which is attractive and negative in sign) with increasing size of the gap h between the contacting surfaces of materials. This is confirmed by plots of energy and adhesion force, shown in Figs. 7.11 and 7.12 for the series materials in contact with the steel grade 30Kh13.

Given that the adhesion strength of molybdenum nitride to steel at $h = 0$ is equal to $F_a(0) = -8.69 \cdot 10^9$ N/m², the adhesion strength of the selected material with the molybdenum nitride should not be lower than the given value. Of the materials for which calculations were carried out, this criterion is satisfied only by the transition metals. The best from this point of view are molybdenum and nickel. In view of the fact that nickel is magnetic, complicating its deposition by ion-plasma methods, the most preferred material for the intermediate layer is molybdenum

The results of calculations of adhesion of steel 30Kh13 with a number of self-lubricating materials, playing the role of solid lubricant and facilitating the process of burnishing materials in dry friction units, showed that the best adhesion properties (highest energy and strength of adhesive interactions) are shown by molybdenum disulphide MoS_2.

To test the predictions of the theory, Sibkriotekhnika company conducted tribological tests to determine the friction coefficient of pairs based on 30Kh13 steel and materials with coatings of Mo_2N, Cr_2N, TiN. The test results showed good agreement of the values of the friction coefficient

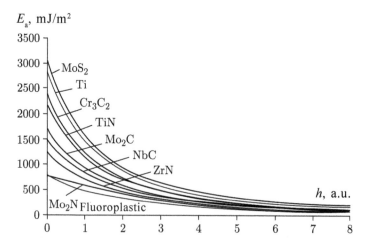

Fig. 7.11. The energy of adhesion for a number of transition metals, nitrides and carbides of transition metals, as well as some self-lubricating materials in contact with the steel grade 30Kh13.

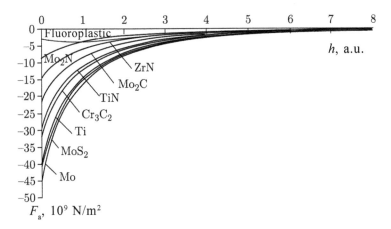

Fig. 7.12. Adhesion force for a number of transition metals, nitrides and carbides of transition metals, as well as some self-lubricating materials in contact with the steel grade 30Kh13.

with the results of the calculation, confirming a lower friction coefficient on steel of the sample coated with Mo_2N [63, 68].

To verify the calculation results in the adhesion characteristics, the AO 'Sibkriotekhnika' Company studied the adhesion strength of wear-resisting coatings of TiN, Cr_2N and Mo_2N on a substrate of steel 30Kh13 by the introduction of an indenter (the method is described in section 7.7.1).

The obtained values of the adhesion strength of coatings σ_a (200.8 $\mp$ 8,3 MPa for coatings of TiN, 163.2 $\mp$ 8,1 MPa for Cr_2N, 120.4 $\mp$ 9.6 MPa for Mo_2N) indicate higher adhesion properties of the coatings of TiN and Cr_2N, while the coatings of Mo_2N are characterized by a significantly lower adhesion. These findings are in complete agreement with the results of theoretical calculations of the adhesion properties of these coatings (Fig. 7.12). The numerical values of the adhesion strength of the coatings correspond to the results of theoretical calculations with the effective size of the gap $h \cong 10$ a.u. $\cong 0.5$ nm.

To increase the adhesion strength of the Mo_2N coatings, which have remarkable tribological properties, experiments were carried out with the method of ion implantation of the surface of a steel sample with Mo ions with a dose of $\sim 10^{16}$ cm^{-2} followed by spraying of Mo_2N. Measurement of adhesion strength of the film by the indenter gave $\sigma_a = 263 \pm 28$ MPa, i.e. showed more than doubling of the adhesion strength of the coating.

In order to test and implement the recommendations on the optimal choice of materials and coatings for combined friction sections and the use of the deposition technology of multilayer coatings in series production, the Sibkriotekhnika company designed and manufactured equipment for surface treatment of machine parts. The system makes it possible to carry out multiple types of surface treatment of parts and tools, including multilayer deposition of ion-plasma coatings with CIB (condensation method in vacuum with ion bombardment), magnetron sputtering and ion implantation in a single-vacuum cycle. The resulting coatings are characterized by high adhesion with the component and uniformity in thickness across the treated surface. The use of this equipment to modify the friction surfaces in the component–sublimator combined friction section allowed to increase the product life by 1.5 times.

7.6. Description of the method of calculation of the adhesion properties of diamond-like coatings

In recent years, experts have developed a number of technological methods for deposition of diamond-like coatings (hereinafter DLC) with stable, controlled and repeatable characteristics [36, 81, 198, 207, 222]. The great interest in the DLC is due to their unique physical, chemical and tribological properties, which can be varied by the technology of formation of coatings and doping [242, 340–342, 473, 518]. A number of the properties of diamond-like coatings and their applications will be characterized:

– The thickness of the coating, while maintaining the uniformity of properties and structure of the material, may vary in the range from 0.005 to 100 μm;

– High hardness is controlled by the deposition conditions in the range $6.0 \div 20.0$ GPa;

– The coatings are sufficiently elastic and can withstand well repeated bending;

– The friction coefficient of the metal–coating is low;

– The durability of coatings is very high because of the hardness, smoothness and low friction coefficient;

– The coatings are resistant to many inorganic and organic etchants in a wide temperature range;

– The unalloyed coatings can withstand many hours of heating to 300°C in air and oxygen-free environments up to 1200°C;

– The coatings doped with some metals can withstand hours of heating in air to a temperature of 900°C;

– The coatings have good biocompatibility with the biochemical medium of the human environment;

– The resistivity of the coatings by doping with various metals can vary from 10^{14} to 10^{-4} ohm·cm;

– Undoped dielectric coatings withstand the electric field strength of up to $3 \cdot 10^6$ V/cm;

– The coatings are opaque for the ultraviolet range;

– In the visible range the coatings are transparent in thin layers up to a thickness of $1 \div 2$ mm;

– In the infrared range the coatings are transparent to wavelengths from 1 to 14 μm.

The variety of the properties of DLC results in a wide area of application. Here are some of the areas of application [200]:

– Friction nodes (pairs) used in machine construction, automotive and aerospace industries;

– Cutting tools made of high-speed steels and a hard alloy for processing (cutting, milling, cutting thread) of non-ferrous and light metals and their alloys, such as aluminum alloys with high silicon content, increasing resistance $2 \div 10$ times;

– Components of dies (working, blanking, deep drawing) working on non-ferrous and light metals and their alloys;

– Blades for cutting foil;

– Honing tool based on a thick ($10 \div 20$ mm) film with high abrasive properties;

– Elements of valves and distribution equipment for the petrochemical industry;

– Protection of elements and structures of chemical and electrochemical equipment;

– Protection of audio and video heads, the production of high quality acoustic membranes;

– Manufacture of medical technology (implants, bioprosthesis);

– Protection of the optical elements against the atmospheric effects such as wind erosion;

– Coatings;

– Dies for wire drawing;

– Detectors of nuclear radiation;

– Powerful compact heaters with operating temperature of $100 \div 500°C$, and miniature IR emitters with operating temperature up to $900°C$ and high light output;

– Separating dielectrics with high thermal conductivity.

One of the important parameters of the DLC is its phase composition, i.e. the ratio of diamond and graphite phases, since this parameter determines most of the properties of DLC. The ratio of diamond and graphite phases determines the wear resistance, residual stress, adhesion to the substrate, the friction coefficient and other characteristics of the coating. It is therefore important to be able to theoretically predict the properties of the coating, based on the knowledge of its phase composition. In accordance with this information about the unique properties of diamond-like coating it was necessary to develop methods of calculation of the adhesion characteristics for the various DLC materials used in industry as modifying surfaces.

To describe the adhesive interaction of DLC with different materials, the authors of the monograph [16, 29–31] have used an approach based on the dielectric formalism and the use of representations of surface plasma oscillations. The approach allowed model approximations for the dielectric

functions of materials to determine their adhesion characteristics. Basics
of the method and some of its applications are set out in previous sections
of this book. As already noted, within the approximations of this approach
the calculation of the adhesion properties of different materials can be
carried out using only the information about the concentration of valence
electrons in them, and additionally the width of the band gap in the case
of the semiconductor properties of the material.

On the basis of model representations of the diamond coating as a
semiconductor solid solution consisting of a dielectric phase (sp^3) of
diamond with a relative fraction c of carbon atoms in it and the metallic
phase (sp^2) of graphite with a relative fraction of $1 - c$ carbon atoms we
can obtain the functional dependence of the physical parameters of DLC
needed to calculate the adhesion characteristics on its phase composition.
To determine the plasma frequency ω_p for DLC, we used the expression
$\omega_p = (4\pi e^2 N/m)^{1/2}$, where the concentration N of valence electrons in the
DLC was the sum of the concentrations of valence electrons in the diamond
and graphite phases:

$$N = cN_{\text{diam}} + (1-c)N_{\text{graph}}. \tag{7.22}$$

The concentration of valence electrons in each of the phases were
determined from the relation

$$N = Zn_{\text{cell}} / v_{\text{cell}}, \tag{7.23}$$

where Z is the valence of the atoms of the material, n_{cell} is the number
of atoms in the unit cell, v_{cell} is the volume of the unit cell. We used the
following empirical evidence.

For the diamond: the unit cell – a face-centered cubic, consists of
8 atoms, the lattice constant $a_{\text{diam}} = 0.3566$ nm, the unit cell volume $v_{\text{cell}} = a_{\text{diam}}^3$,
$Z = 4$, the width of the forbidden zone is equal to $E_g = 5.48$ eV. As a result,
$N_{\text{diam}} = 3.716 \cdot 10^{25}$ cm^{-3}. For graphite: the unit cell – hexagonal, contains
four atoms, the lattice constant $a_{\text{graph}} = b_{\text{graph}} = 0.24612$ nm, $c_{\text{graph}} = 0.6708$
nm, volume the unit cell $v_{\text{cell}} = 0.86a_{\text{graph}} \cdot c_{\text{graph}}$, $Z = 4$, $E_g = 0$. As a result,
$N_{\text{graph}} = 4.541 \cdot 10^{23}$ cm^{-3}.

Changes of the band gap for the DLC with the change of the phase
composition were found from the relations of the theory of semiconductors
for the electron density distribution, given that the diamond is a proper
semiconductor, and the graphite phase is a donor for it. After appropriate
transformation and normalization the following expression for the change
in the width of the band gap of DLC on the diamond phase concentration
c and temperature T was obtained:

$$E_g(c) = -2k_B T \ln\left[1-c+c\exp\left(-\frac{E_g(1)}{2k_B T}\right)\right], \tag{7.24}$$

where $E_g(1)$ is the value of the band gap of pure diamond.

Figure 7.13 is a plot of the dependence of the band gap of the DLC on the concentration of the diamond phase, calculated for a temperature $T = 300$ K.

The correctness of the calculation method and the selected parameter values were tested by calculation of the surface energy of diamond and graphite, using the relations (7.7)–(7.9) and (7.17), (7.18) and comparing the calculated values with experimental ones. Thus, the surface energy calculated by the dielectric formalism method, for diamond was found to be $\sigma_{theor} = 4348$ mJ/m², experimental – $\sigma_{exp} = 4550 \div 5000$ mJ/m².

The layered structure of graphite leads to strong differences in the values of surface energy depending on the orientation of the surface edge. Thus, the measured value of surface energy along the natural layers of graphite is $\sigma_{exp}^{\parallel} = 300 \div 500$ mJ/m², and at the same time, the surface energy $\sigma_{exp}^{\perp}$ for the faces perpendicular to the layers is approximately 5000 mJ/m². The calculated value of the surface energy of graphite is $\sigma_{theor} = 3440$ MJ/m². Here it should be noted that the adhesion characteristics are calculated using the 'jelly' model in which the anisotropy of the material does not occur. Therefore, the calculated values can be compared only with the experimental values averaged over the faces. Thus, the surface energy of graphite, averaged over the faces, is $\sigma_{exp}^{mean} = 3400 \div 3600$ mJ/m², which is in good agreement with the calculated value.

The good agreement between the calculated and experimental values of the surface energy of diamond and graphite suggests that the technique allows to reliably calculate the adhesive characteristics of DLC using the given approximation method. Table 7.7 shows the values of the physical parameters determined by us for the DLC with different phase composition, and then used to calculate the surface and adhesion characteristics.

Figures 7.14–7.16 show plots of the calculated dependences sof the adhesion force as a function of vacuum gap size h for the contact of

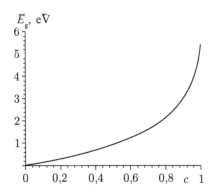

Fig. 7.13. The dependence of the band gap of the DLC on the concentration of the diamond phase for the temperature $T = 300$ K.

Table 7.7. The calculated values of physical parameters for DLCs with different phase composition

Concentration of diamond phase c, %	Plasma frequency, $\hbar\omega_p$, eV	Band gap E_g, eV	Lattice constant a, Å
0 (graphite)	25.12	0	6.708
50	28.32	0.95	5.137
70	29.50	1.63	4.509
83	30.25	2.36	4.100
95	30.92	3.75	3.723
100 (diamond)	31.19	5.48	3.566

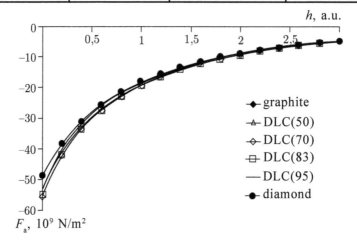

Fig. 7.14. The strength of adhesion of aluminum with DLC at various contents of the diamond phase.

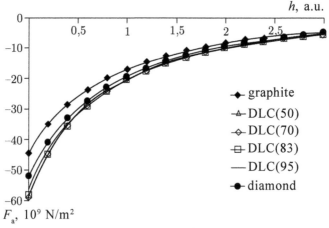

Fig. 7.15. The strength of adhesion of germanium with DLC at various contents of the diamond phase.

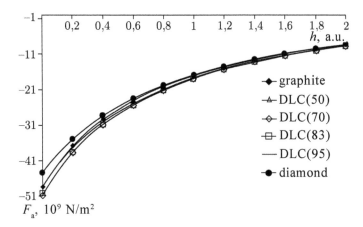

Fig. 7.16. The strength of adhesion of 30Kh13 steels with of DLC at various contents of the diamond phase.

Table 7.8. The values of adhesion force $F_a(0)$, 10^9 N/m^2 of the diamond-like coating with a number of simple and transition metals, semiconductors and steel 30Kh13 at the gap $h = 0$ between the surfaces of materials

Materials	W	Mo	Fe	Cr	Zr	Ti	Si	Ge	Al	30Kh13
F_a	107	96.2	89.5	88.3	81.2	75.6	62.9	58.9	55.7	50.8

aluminum Al, germanium Ge and 30Kh13 steel coated with a diamond-like coating with various contents of the diamond phase, and also with diamond and graphite. Analysis of the effect of phase composition showed that the best adhesion properties have DLCs with 70 ÷ 80% content of diamond phase, although in the range 50 ÷ 80% the differences in the adhesion properties of the DLC are small and this demonstrates high adhesion in this range. Thus, at $h = 0$ the adhesion force takes values $F_a = (50 ÷ 60) \cdot 10^9$ N/m^2 for the contact of the considered material with the DLC with the phase composition optimal for adhesion. Table 7.8 gives the calculated adhesion strength values of the diamond-like coating with the phase composition optimal for the adhesion properties, with a range of materials with the gap $h = 0$ between the surfaces of materials. The calculations have shown that diamond-like coatings are characterized by high values of the normal component of the adhesion force F_a. Thus, the adhesion force of metals such as tungsten, molybdenum, chromium, zirconium, titanium, with a DLC is 1.5 ÷ 2 times higher than the adhesion force of the same metals with 30Kh13 steel.

But it is well known that the contact of dissimilar materials together results in the formation of tangential components the interaction forces in addition to the normal forces. They can occur due to differences between the values of the temperature coefficients of linear expansion α_i of the coating and the substrate, and also because of the temperature difference and the deposition temperature and the temperature at which the coating operates formed during coating. The magnitude of the tangential stresses occurring at the interface of materials is proportional to difference in the coefficients of linear expansion $\Delta\alpha$ and the difference in the deposition temperature and the operating temperature of the coating ΔT:

$$\sigma_\tau = \frac{E}{1-\mu}\Delta\alpha\Delta T, \tag{7.25}$$

where E is the modulus of volume expansion, μ is Poisson's ratio. For DLC: $E = 7.8 \cdot 10^{11}$ N/m^2, $\mu = 0.3 \div 0.4$.

The coefficient of linear expansion of a DLC with the two-phase composition can be found as $\alpha = c\alpha_{diam} + (1 - c)\alpha_{graph}$. Consequently, to obtain high-quality coatings, it is necessary to select a pair of contact materials not only on the basis of the maximum of their adhesive characteristics but also the minimum difference between their temperature coefficients of linear expansion. However, the feature of the diamond-like coatings is the low value of the temperature coefficient of linear expansion: $\alpha \approx (2\div3)\cdot10^{-6}$ K^{-1} (for the phase composition with the optimal adhesion properties). This leads to rather stringent requirements for the selection of material on which the DLC is deposited, or leads to the necessity of introducing the intermediate sublayer with the value of the coefficient α as close as possible to the value of the temperature coefficient of linear expansion of the diamond-like coating. Moreover, we must remember that when the diamond-like coating is deposited on metals the surface layer can show the formation of the corresponding metal carbide phase, the presence of which affects both the adhesion strength and the tangential stress. Our calculations show that the carbides of transition metals are characterized by lower strength of adhesion to diamond-like coatings than the transition metals.

If we consider that the threshold strength of the DLC is characterized by the value $\sigma_B = 1.28$ GPa, and assume that the temperature change in spraying and use of the coating $\Delta T = 200$ K, then from (7.25) we can determine the maximum difference between the temperature coefficients of linear expansion of the substrate and the DLC $\Delta\alpha_c$ at which the tangential stresses at the interface of materials do not yet lead to the destruction of the coating. Thus, the calculations showed that $\Delta\alpha_c \approx 5.47\cdot10^{-6}$ K^{-1}. Taking into account the value of α for the DLC we get that the values of the temperature coefficient of linear expansion of the substrate material must not exceed the values $\alpha_c \approx (7.68 \div 8.42)\cdot10^{-6}$ K^{-1}. These include a series of transition metals: tungsten $- \alpha \approx (4.1 \div 4.5)\cdot10^{-6}$ K^{-1} at $T = 200 \div 400$ K, molybdenum $- \alpha \approx (4.57 \div 5.45)\cdot10^{-6}$ K^{-1}, zirconium $- \alpha \approx (5.71 \div 6.65)\cdot10^{-6}$ K^{-1}, chromium $- \alpha \approx (5.18 \div 8.3)\cdot10^{-6}$ K^{-1}, titanium $- \alpha \approx (7.23 \div 8.82)\cdot10^{-6}$ K^{-1}.

7.7. Methods and tools for measuring the adhesion of thin films

The phenomenon of adhesion between surface layers of dissimilar materials and coatings, brought into contact, is determined by the forces of interatomic interaction. However, the value of adhesion strength of coatings depends not only on the characteristics of the relationship between bodies, brought into contact, but also on the method of measurement as well as the method of separation.

Methods of non-destructive testing of adhesion strength, which give reliable results, so far do not exist yet. The results of measuring the adhesion strength can affect the stress state of the interface between the film and the substrate due to shrinkage or thermal effects in materials. All this points to the difficulty of obtaining reliable experimental values of the adhesion strength of coatings. In this section we describe the traditional method and the methods of indentation and of the contact potential difference and their application to the determination of the adhesion strength of wear-resisting coatings on the surface of specimens of steel 30Kh13.

7.7.1. Application of the indentation method for determining the adhesion strength of coatings

The indentation method allows to obtain semi-quantitative information on the values of the force of adhesion interaction of the coating to the substrate, so that the resultant values can be used to compare the coatings on the basis of the adhesive interaction. This method allows to determine the effect of quality of the surface layer of the substrate (surface preparation method) on the adhesion strength. It can be used to determine the adhesion strength of coatings deposited by ion-plasma, plasma-arc, magnetron, electroplating methods.

The method is based on the introduction of a conical indenter through the coating into the substrate sequentially increasing fixed loads prior to the formation of a concentric crack in the coating or coating delamination around the indentation and then define the geometric parameters of the indentation and of the deformed zone of the coating with the substrate.

Indentation into the substrate results in complex elastoplastic deformation, leading to the buckling of the substrate material around the indentation and the formation of a bead. The strains and stresses formed at the bead disrupt the adhesive bond between the coating and the substrate. The hole formed in the process of deformation of the coating and the substrate has the form shown in Fig. 7.17.

The relationship between the adhesion strength of coatings, the force on the indenter, which leads to cracking of the coating, and the geometric characteristics of the indentation is expressed by the dependence

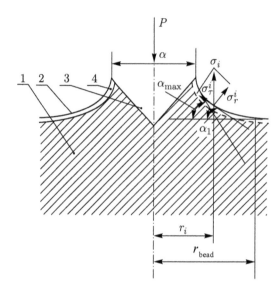

Fig. 7.17. Standard form of the indenter (indentation) and the deformed region [64]: *1* – substrate, *2* – coating, *3* – conical indenter, *4* – cylinder.

$$\sigma_a = \frac{P_a}{d^2} \cdot \sin\alpha_{max}, \qquad (7.26)$$

where σ_a is the adhesion strength of the coating; P_a is the load on the indenter leading to a violation of the adhesive bond; d is the diameter of the indentation made by the indenter; α_{max} is the maximum angle between the tangent to the profile of the deformed area and the original surface.

Thus, the measured parameters in the implementation of the method are the load on the indenter, the indentation diameter, profile of the deformed area on which the calculation method determined α_{max}. The following information about the devices and methods of measuring the adhesion strength of coatings is based on a book [64].

Instruments for studying adhesive interaction forces and measurement procedure

Experimental objects include flat cylindrical specimens or parts with dimensions that allow them to placed on the MII-4 interferometer and 252 profilograph–profilometer. In developing the coating process or surface preparation modes prior to coating it is most appropriate to use the flat samples.

The cracking force is determined by a TP 2090 device for hardness measurements by the Rockwell method (hereinafter the device 2090 TP).

Cracking of the coating after indentation is recorded by a PMT-3 microhardness tester or any microscope with a magnification of ×250.

The profile of the deformed part is constructed using a Linnik microinterferometer MII-4 and, if necessary, with the help of a model 252 profilometer.

A device for measuring Rockwell hardness (TP 2090)
The 2090 TP device is designed to measure the hardness of metals and alloys by the Rockwell method. The device is calibrated in units of kgf (1 kgf = 9.81 N). The device performs measurements at three fixed loads: 60, 100, 150 kgf (respectively 588, 981, 1471 N). The design feature of the device is automated application of the pre-load of 10 kgf (98 N). Because the interval between the applied loads large (40 ÷ 50 kg), there is a need for refining and calibrating the device. Refinement of the device allows loading from 10 to 152.5 kgf with an interval of 10 kgf.

Determination of indenter loads of 10 to 50 kgf is made directly using exemplary type of dynamometer DOSM-3-0.2 with the measurement range 20 ÷ 200 kgf. The principle of operation of this device is based on the possibility of additional (intermediate) loading of the levers to create loads of 52.5 to 152.5 kgf with an interval of 10 kgf. The device is calibrated using calibration data (Table 7.9).

Before measuring the samples were carefully degreased The object table of the 2090 TP device (with the dynamometer) rises to achieve the desired value of load (10 kg), which is determined by Table 7.10.

With loads of 10 ÷ 50 kgf (with an interval of 10 kg) the sample surface is indented (after 2 ÷ 3 mm). The microscope of PMT-3 equipment is used to determine the minimum force at which the coating cracks or concentric cracks appear. If the surface is not disrupted at low loads, pricks

Table 7.9

Load P, kgf	Preload, units of instrument scale	Position of load switch, units of instrument scale	The resulting load, kgf
0	10	588	52.5
0.680	10	588	62.5
1.330	10	588	72.5
0	50	981	82.5
0.680	50	981	92.5
1.330	50	981	102.5
1.330	70	1471	112.5
0.680	100	1471	122.5
1.330	140	1471	132.5
1.880	180	1471	142.5
2.380	210	1471	152.5

Table 7.10

P, kg	10	20	30	40	50
Values of indicator of dynamometer, units	1.31	1.62	1.94	2.24	2.55

are produced at high loads (52.5 ÷ 1025 kgf) at intervals of 10 kgf. At a certain force on the indenter, resulting in a violation of the adhesive bond (P_a), at least three pricks are made in the sample surface. The diameter of the craters (d), obtained at the cracking force, is measured by the Linnik microinterferometer MII-4 or the model 252 profilograph–profilometer in accordance with their instructions for use.

The scatter of the data determined by the formula $\Delta_i = \dfrac{\sigma_{a,mean} - \sigma_i}{\sigma_{a,mean}} \cdot 100\%$.

The data for which the spread is greater than 25% are discarded, and redetermined $\sigma_{a,mean}$. The measurements are repeated if the variance for all measurements is greater than 25%.

7.7.2. Contact potential difference method for determining the condition of the modified surface

The quality of coatings with different technologies of their application includes many parameters (reliability and longevity, strength, manufacturability, adhesion strength, etc.), which are largely determined by the state of the surface before coating.

The most important characteristics of the surface, determining its energy state, are the surface energy and the electron work function from the surface. Experimental determination of the surface energy of materials is quite a complicated task. On the other hand, the work function is much like the surface energy in terms of the functional dependence on the parameters of the material [146–148], and is experimentally determined significantly easier.

The work function, as discussed in chapters 4 and 5, is determined as the minimum energy required to remove an electron of the solid and placing it near the surface. 'Near' here means the distance a large on the atomic scale, but small compared with the linear dimensions of the crystal. The physical nature of the work function consists in the fact that on the metal surface there is a dipole potential barrier D determined by the interaction between the ions and electrons of the metal lattice. Due to this potential barrier, the electrons are confined in the metal, even in the case when the metal is in a vacuum. The energy of this potential barrier must be less than the energy corresponding to the Fermi level, because at room temperature, the electrons do not detach from metal and higher temperatures are required to observe an appreciable thermionic emission. The origin of the potential barrier can be explained as follows: due to tunneling of electrons outside the crystal, the value of electron density varies smoothly in the surface region and this is accompanied by the formation of the dipole (double) electric layer the charges of which would prevent further 'leakage' of electrons.

In addressing the issue of work function W, it must be borne in mind that we study the clean surface of the metal placed in a vacuum. Contamination of this surface, the adsorption of foreign substances, and the distortion of the lattice by foreign atoms will change the properties of the metal surface and thus affect the value of the work function. The surface of the metal can adsorb for a variety of reasons both positive and negative ions, neutral atoms, molecules, representing rigid or induced dipoles, as well as molecules affected from the metal surface by the van der Waals forces. The adsorption of positive ions on the surface of the metal causes the formation of an electrical double layer and the work function W strongly decreases. The adsorption of negative ions leads to an increase in W. The adsorption of molecules having a permanent or induced electric dipole moment is also accompanied by the formation of the electrical double layer which greatly lowers the work function. In addition, many atoms can be adsorbed on the metal surface by chemical means to form single- and polyatomic layers, such as oxygen in the oxide layers. These layers on the surface can be polarized. In many cases, the atoms or molecules of these layers are arranged in some preferred areas identified by the lattice structure. This causes a significant change in the values of W. Since the metals tend to adsorb atoms and molecules, the measurement of the work function requires high-vacuum techniques and intensive study of surface treatment of metals.

Knowing the value of the work function for a given material makes it possible to control the purity of its surface by direct measurement of the work function.

We now consider two metals M_1 and M_2 connected in such a way that the electrons can move freely from one of them to another. Let the Fermi energy of one of them be higher than the energy of the other metal $(E_{F1} > E_{F2})$. Then in this system for some time there will be the predominant electron transfer from metal M_1 to metal M_2. At the same time, a positive charge appears on the metal M_1 and the negative charge on the metal M_2. An electric field forms between the metals perpendicular to the plane of contact and will be localized in a layer with the thickness of the order of 10^{-8} cm. The directional flow of the electrons from metal M_1 to metal M_2 ceases at the moment when the Fermi levels for both metals are the same.

The potential difference established in the equilibrium state in the outer space between the surfaces of the contacting metals is called the *external contact potential difference*. Its value is associated with the work function of the metals M_1 and M_2 and is given by

$$\varphi = \varphi_2 - \varphi_1 = \frac{W_2 - W_1}{e}. \qquad (7.27)$$

Through the contact potential difference we can, according to [243], characterize the value of the interfacial energy of the interaction of metals, and thus the value of their adhesion energy.

From the relation for the contact potential difference (CPD), it follows directly that one way of measuring the work function of the metal can be the measurement of the contact potential difference between this metal and the metal of the known work function.

The principle of measuring contact potential difference
The existing methods of measuring the contact potential difference can be reduced (with various modifications) to the four basic methods [226, 232]: ionization, photoelectric, condenser, thermal electrons.

The most common method of the dynamic capacitor (compensation method), which is applicable in any environment, provides sufficiently accurate direct indications and is used in the case of surfaces. One disadvantage of this method is the need for careful screening.

When using the method of the dynamic capacitor [66], plates of the materials M_1 and M_2 are placed at some distance from each other. One of them is fixed motionless and the other one vibrates (using a special device) with a frequency of a few tens of hertz and an amplitude of several tenths of a millimeter. Between the plates there are installed CPD and the electric field. The inner surface of each plate has the electric charge $q = C\varphi$, where C is the capacitance between the plates. An alternating electric current $i = \varphi dC/dt$ forms in the circuit, and the alternating voltage on the load resistance $R - Ri = R\varphi dC/dt$.

Contact potential difference φ can be offset by the voltage of the opposite sign from a battery. In this case, the current through the resistance R will cease. This voltage from the load resistance (an amplifier) is fed to an oscilloscope. Changing (with a divider) the voltage from the battery, the voltage fluctuations on the load resistance can be interrupted. In this case, the voltmeter indicates the desired contact potential difference.

Figure 7.18*a* is a block diagram of the stand for the measurement of the contact potential difference using the method of the dynamic capacitor [64]. The measuring container contains a measuring circuit element – the dynamic capacitor formed by the vibrating probe – the sample pair. The vibrator of the probe is shown in Fig. 7.18*b*.

The power gain and compensation unit (LFA) of the amplifier has signals from the generator G3-118 and the CPD compensation scheme. The amplified signal from the G3-118 is fed to the coil of the vibrator and causes fluctuations of the probe with a frequency of 120÷160 Hz. The optimal frequency of the probe is selected individually for each vibrator from the maximum amplitude of longitudinal oscillations observed through a magnifying glass or microscope. The vibrator is placed in a copper casing and attached to the holder in the rack. During measurement the test sample is placed on the plate of the foiled plastic and secured to the table. Vibrator and the sample are placed on the movable elements of the rack that allows them to move smoothly relative to each other when measuring the CPD.

The compensation circuit includes element O76 with a nominal voltage of 1.4 V and adjustable resistances of 4.7 and 22 ohms, respectively, for coarse and fine compensation for the CPD. Directly counting of CPD is

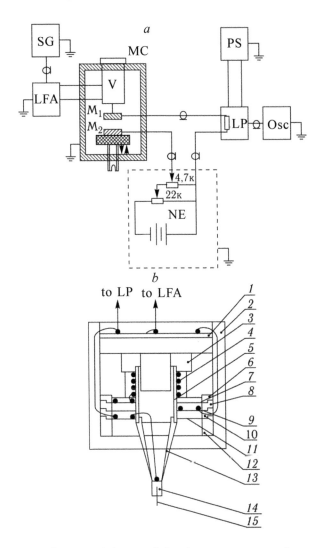

Fig. 7.18. Block diagram of the stand for the measurement of contact potential difference (*a*) and probe vibrator (*b*): SG – sound generator type GZ-118 [64], V – vibrator; M_1 – probe; M_2 – the sample; LFA – power low amplified, power 6 W; PS – power supply unit; LP – a low-noise preamplifier; Osc – Oscilloscope; NE – a normal element; MC – measuring cell; *1* – top cover, *2* – ebony tube, *3* – magnetic, *4* – Coil, *5* – tube tracing, *6, 8, 10, 12* – hard rubber puck, *7, 9, 11* – membrane, *13* – foam cone, *14* – tube, *15* – probe.

carried out by a digital readout device ShCh-301. The measured value of the compensation voltage is the value of the CPD between the probe and the studied surface.

7.7.3. Studies of adhesion characteristics of wear-resisting coatings of surfaces of friction pairs

In studies conducted in [64, 66] the adhesion strength of coatings was determined using one of the traditional methods – the method of penetration of the indenter through the coating to the substrate, as well as the method of the contact potential difference (CPD) not developed for these purposes. The method of pressing a conical indenter is a destructive method and due to a number of methodological features, which were mentioned above, gives only semi-quantitative values of the force of adhesive interactions. Nevertheless, its numerical values can be used to compare the adhesion strength of coatings and test the predictions of the theory of adhesive interaction. As regards the CPD method, it is not a direct method of determining the adhesion properties. The work function of an electron from the surface of the material determined by this method allows to determine the microscopic parameters of the theoretical model, on the basis of which we can calculate the surface and adhesion characteristics of different materials. However, the application of the CPD method puts into the hands of experimentalists not only the non-destructive method, but and in particular the development of the rapid method for controlling the adhesion strength of coatings.

Investigation of adhesion strength of wear-resisting coatings by the introduction of the indenter
The above procedure establishes a detailed sequence of measurements.

In accordance with (7.26), which specifies the relationship of the adhesion strength of coatings with the force on the indenter, which leads to cracking of the coating, and the geometric characteristics of the indentation, the measured parameters are: the load on the indenter, the diameter of the indentation, the profile of the deformed part used to calculate α_{max}.

Test samples were prepared in the form of flat disks of steel 30Kh13 with HRC 46–50 with dimensions that allows to place them on the interferometer MII-4 and profilograph–profilometer model 252. Of all the available materials based on the nitrides of transition metals used to create wear-resistant antifriction coatings for machine parts, according to the results the calculation of tribological characteristics we chose TiN, Mo_2N and Cr_2N. Before applying the coating, the surfaces of the sample disks were ground and finished by an abrasive paste based on TiC to the roughness $R_a < 0.1$ μm, and then subjected to washing in gasoline B-90 and wiping with ethyl alcohol rectified to GOST 5962-67.

Coatings were deposited by the ion-plasma method in equipment NNV-6/6-I1 equipped with a 'Radikal' gas ion source and a pulsed arc source of metallic ions with an accelerating voltage of 35 kV and 75 of the 'Diana-1 type'. The coating method consisted of the following:

a) Surface cleaning by Ar ions with energy $E = 1 \div 3$ keV;

b) Heating and purification by ion bombardment of the deposited material by arc evaporation at an accelerating voltage $U = 800 \div 1500$ V depending on the type of ions;

c) Modification (implantation) of the surface with high-energy ions using a pulsed source Diana-1;

d) The deposition of ion-plasma coatings.

Experience in the research of the adhesion strength of coatings shows that this method of measurement of both the initial measured parameters meters and the final value of the adhesion strength depends on the coating thickness. Therefore, the modes and deposition time were selected so that at the high homogeneity of the coating (no droplet phase), the thickness of coatings on all samples was $3 \div 5$ mm.

To determine the cracking force, an improved instrument was used for measuring the Rockwell hardness (TR 2090), which allowed for loading from 100 to 1500 N at intervals of 50 N. Injections were made through the surface of the sample of $2 \div 3$ mm. The microscope of PMT-3 equipment was used to determine the minimal force resulting in a violation of the adhesive bond of the coating in the form of swelling, radial or concentric cracks. At a certain force on the indenter P_a, leading to a violation of the adhesive bond, at least five pricks were made in the sample surface.

The diameter d of the imprint of the craters, obtained at the cracking force, and the profile of the deformed area were measured using the Linnik microinterferometer MII-4 and the 252 profilograph–profilometer.

The obtained values of the input parameters, their mean values and the calculated values of the adhesion strength of coatings are in Table 7.11. The values of the adhesion strength of coatings indicate higher adhesion of coatings of TiN and Cr_2N, whereas the Mo_2N coatings are characterized by much smaller adhesion. These findings are in complete agreement with the results of the theoretical calculations of the adhesion properties of these coatings, presented in section 7.5.

The numerical values of the adhesion strength of coatings are consistent with the theoretical calculation with the effective value of the gap $h \approx 10$ a.u. ≈ 0.5 nm.

To increase the adhesion strength of the Mo_2N coatings, which have remarkable tribological properties, experiments were carried out using the method of ion implantation of the surface of a steel sample by Mo ions with a dose of $\sim 10^{16}$ cm^{-2} followed by spraying of the Mo_2N coating. Measurement of the adhesion strength of films by the indentation method gave $\sigma_a = 263 \pm 28$ MPa, i.e. showed doubling of the adhesion strength of the coating.

Table 7.11

Sample no.	Coating	$\sin \alpha$	P_a, N	d, mm	σ_a, MPa
1	TiN	0.03931		0.435	207.7
		0.03883		0.453	189.2
		0.03564		0.445	180.0
		0.03842	1000	0.431	206.8
		0.3761		0.419	214.2
	average	–		–	199.6 ± 12.9
2	TiN	0.03938		0.443	210.7
		0.03893		0.435	216.0
		0.04064		0.452	208.9
		0.03908	1050	0.434	217.9
		0.03917		0.440	212.4
	average	–		–	213.2 ± 3.9
3	TiN	0.03923		0.448	205.2
		0.04015		0.473	188.4
		0.04108		0.492	178.2
		0.04063	1050	0.481	184.4
		0.03957		0.465	192.2
	average	–		–	189.7 ± 9.1
4	Cr_2N	0.02897		0.281	146.8
		0.03771		0.316	151.1
		0.03173		0.304	137.3
		0.03284	400	0.309	137.6
		0.03326		0.312	136.7
	average	–		–	141.9 ± 6.0
5	Cr_2N	0.03387		0.287	185.0
		0.03671		0.303	179.9
		0.03478		0.294	181.1
		0.03526	450	0.306	169.5
		0.03305		0.292	174.4
	average	–		–	178.0 ± 5.6
6	Cr_2N	0.03273		0.297	167.0
		0.03566		0.309	168.1
		0.03473		0.286	191.1
		0.03518	450	0.312	162.6
		0.03195		0.303	156.6
	average	–		–	169.1 ± 11.9
7	Mo_2N	0.01551		0.225	107.2
		0.01975		0.229	131.8
		0.02307		0.256	123.2
		0.02115	350	0.232	137.5
		0.02034		0.223	143.2
	average	–		–	128.6 ± 12.5

		0.01657		0.213	109.6
8	Mo_2N	0.01801		0.225	106.7
		0.01707		0.206	120.7
		0.01695	300	0.199	128.4
		0.01481		0.188	125.7
	average	–		–	118.2 ± 8.7
9	Mo_2N	0.01735		0.221	106.6
		0.01708		0.217	108.8
		0.01815		0.212	121.2
		0.01532	300	0.204	110.4
		0.01623		0.197	125.5
	average	–		–	114.5 ± 7.6

Using the CPD method for determining the adhesion strength of wear-resisting coatings

The CPD method is one of the ways of measuring the work function of the metal surface, determined by measuring the contact potential difference between the surface and the reference metal of the known work function.

The contact potential difference was measured by the method of the dynamic capacitor. The principles for determining the CPD by the dynamic capacitor were already described and the block diagram of measurement of the CPD was presented. The measuring dynamic element of the circuit – dynamic capacitor – is formed by the vibrating probe–the sample pair. The probe was a gold wire with the work function $W = 4.89$ eV. The contact potential difference between the probe and the sample surface was measured to determine the electron work function from the surface of the sample.

As already noted, the measurement of the work function requires high-vacuum technology and intensive surface treatment study of metals, to avoid the influence of atoms and molecules adsorbed on the metal surface, on the value of the work function. Therefore, to measure the CPD using the method of the dynamic condenser, a special measuring stand was constructed on the basis of the VUP-5 universal vacuum post, with the measuring cell placed in its vacuum chamber. In addition to traditional methods of cleaning the surfaces of the samples, glow discharge cleaning in vacuum was also used.

In order to study the adhesion properties of wear-resisting coatings, friction section were used for measurements of the CPD and the values of the work function were determined for the coatings of TiN, Mo_2N, Cr_2N and 30Kh13 used as a substrate for these coatings. Thus, $W_{TiN} = 3.82 \pm 0.1$ eV, $W_{Mo_2N} = 2.97 \pm 0.1$ eV, $W_{Cr_2N} = 3.45 \pm 0.1$ eV, $W_{30Kh13} = 4.54 \pm 0.1$ eV. The work function is largely similar to the surface energy of the functional dependence on the parameters of the metal. This may allow the experimental values of the work function to be used to determine the parameters of the metal and then the surface energy. This idea was realized in [64, 66] using the method of the density functional. The calculation of the adhesion characteristics yielded the dependence of

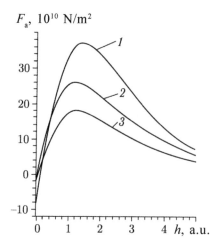

Fig. 7.19. The dependence of the strength of adhesion interaction were F_a on the magnitude of the gap h between the substrate of steel and coatings 30Kh13: 1 – TiN; 2 – Cr$_2$N; 3 – Mo$_2$N, calculated the experimental data on the electron work function from the surface material [64].

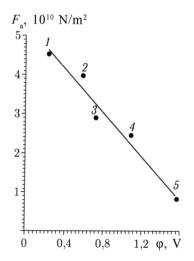

Fig. 7.20. The strength of adhesive interaction F_a of various coatings: 1 – Mo; 2 – Ti; 3 – TiN; 4 – Cr$_2$N; 5 –Mo$_2$N to the substrate steel 30Kh13 as a function of the CPD [64].

the adhesion force of coatings based on nitrides of transition metals with the substrate made of steel 30Kh13 as a function of the gap $2D$ (Fig. 7.19).

Figure 7.19 shows that the maximum adhesion is shown by the coating of TiN, while the lowest – by the coating of Mo$_2$N. Figure 7.20 shows the dependence of the calculated values of the adhesion strength of coatings on the value of the measured contact difference between the substrate and the coating of molybdenum, titanium, and nitrides of titanium, chromium, and molybdenum. It is seen that the smaller CPD value between the coating and the substrate, the higher is the adhesion force.

Thus, the developed method for calculating the surface energy and adhesion characteristics for various metals according to the work function of an electron from a metal surface makes it possible to use the method of the contact potential difference as a method for predicting the adhesion strength of coatings and determining the state of the surface of various materials.

8

FRICTION OF SURFACES OF SOLIDS WITH NO LUBRICATION (DRY FRICTION)

Introduction

Raising the technical level, quality, reliability and durability of products is the main direction of development of modern engineering. The main conditions for a successful solution to this problem are: the creation of friction units of machines and mechanisms, which have high durability, as up to 90% of cars fail due to wear of parts, creating assemblies of machines that have low friction, because friction losses reach of 40% power consumption.

The problem is acute for products friction units of which operate in extreme conditions, i.e. under conditions of elevated speeds and loads, high and low temperatures, corrosive and inert media in vacuum and weightlessness, etc. These products include systems for microcryogenic technology. Cylinder–piston groups with precision friction pairs define the service life of products.

The use of traditional methods of improving the wear resistance of friction units of machines is limited by the inability to use any lubricant, including natural gas (for example, in vacuum cryopumps), and also by severe restrictions on the type of materials used and the weight of the product (cryopumps in space vehicles, submarines, tanks). When using cryopumps in space vehicles there is an urgent need for repair-free use of non-lubricated moving friction sections in the product with an abnormally long life, as assessed on the basis of the terms of the use of the spacecraft as a whole.

The problem of increasing the service life of friction units operating in extreme conditions can be solved by the development and application of new processes and methods of modifying the surfaces of friction pairs with the desired properties. Existing technologies do not always provide the necessary quality of the working surfaces of the guide cylinder and surfaces of the metal–metal friction pairs, thus leading to low wear resistance of friction units. An equally important issue is to reduce the consumption power, which can be partially solved by reducing power losses due to friction. All this requires detailed research and taking into account factors affecting the magnitude of the interaction of heterogeneous materials that make up a pair of friction, as well as taking into account the factors that influence directly the friction force. Knowledge of these factors is used to select, already in the technology development stage, the materials and coatings for friction units, characterized by the lowest friction coefficient, and to predict their longevity. Therefore, the main purpose of current research in the field of friction science is the development of analytical methods for the description of the friction process, based on the laws of physics, chemistry and mechanics.

8.1. Types of friction and dry friction laws. Wear

The friction of solids is a complex phenomenon that depends on many processes occurring at the interface in the areas of actual contact and in the thin surface layers of these bodies at their relative tangential movement.

The complexity and multifactor nature of the processes accompanying friction and wear of materials, as well as the vital importance of proper accounting of these processes in the development of friction units, designed for continuous operation in service stress conditions, led to extensive research of tribological problems and the emergence of a large variety of theories and hypotheses in this area.

In friction, mechanical, thermal, electrical and chemical processes occur at the same time. Friction strengthens or softens the material, damages the surface of materials in the friction pair, followed by an increase in friction and wear or is a self-organizing process in which the transfer of material from one surface to another causes, on the contrary, decreased wear and friction with the improvement in the overall tribological characteristics of the section [47].

Research and development of the theory of friction and wear cover a long period of time (the history of studies of friction dates to the days of Aristotle and Leonardo da Vinci). Despite the fact that the word 'friction' is known to everyone, even the experts often have difficulty in explaining this phenomenon.

By the nature of the relative motion there is sliding friction and rolling friction. Sometimes both types of friction occur together, when rolling is accompanied by sliding, for example, in gear–screw transmissions or between wheels and rails.

It is necessary to distinguish between the internal friction as the resistance of the same body to relative displacement, and the external friction of solids, or simply friction, which will now be considered.

From everyday experience we know that if two bodies are in contact with each other, their relative displacement requires some tangential force. When this force increases from zero up to a certain magnitude, the contact zone shows having micro-displacements called *preliminary displacement*.

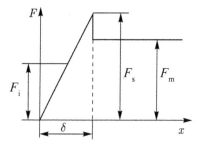

Fig. 8.1. The transition of static friction to the friction of motion [166].

Preliminary displacement, due to deformation of the body, primarily deformation of the irregularities of their surfaces under the influence of shearing force, preceeds the relative macrodisplacement of bodies. But the bodies themselves remain motionless until the tangential force does not reaches the limiting value, called *the static friction force* F_s (Fig. 8.1). In this preliminary displacement reaches its limiting displacement δ. An intermediate value of the applied tangential force is *the incomplete friction force* F_i [166].

Thus, between two contacting and fixed bodies relative to each other there is a static friction force, which prevents the disruption of their state of rest, translational displacement of one body relative to another. This fundamental fact, according to of B.V. Deryagin's proposal [79], should be regarded as a fundamental law of friction. In order to avoid violations of the laws of the numbering adopted in the literature about friction, the law may be called the *zero law*. Its importance is great, despite the simplicity of the formulation. The presence of static friction is the hallmark of sliding friction which allows to distinguish it from the internal friction formed at any infinitesimal tangential load.

After the body has moved off the ground, the friction force usually decreases, but does not disappear completely and remains approximately constant, if the movement conditions do not change. This resistance to motion is called *friction of movement* and the resistance force by the *friction force of movement* F_m. The ratio of the friction force (static or movement) to the normal load is called *the friction coefficient* (respectively, static or motion). Leonardo da Vinci (1452–1519) believed that for smooth surfaces the friction coefficient is constant and equal to 0.25, and this view was held this in science for a very long time. G. Amonton (1663–1705) believed that the friction coefficient is equal to 1/3 of the surfaces of iron, leather, lead and wood investigated by him. In fact, the friction coefficient can be in a fairly wide range from about 0.001 to easily loaded ball bearings for tens of units for carefully purified similar metals in contact in vacuum. Under normal conditions of friction in air, the friction coefficient generally changes in a relatively narrow range of 0.1 to 1. But here the value of the friction coefficient is not important and it is its constancy, which, in essence, is the content of the first law of friction, often referred to as the *Amonton law*.

Thus, the first law of friction is formulated as follows: the friction force is directly proportional to normal load. If f denotes the friction coefficient, which is constant and independent of the normal load, the first law of friction can be written as

$$F = fP. \tag{8.1}$$

But the founders of tribology already understood that this ratio is true for certain load–speed conditions, combinations of friction materials and other friction conditions. Amonton wrote that the friction force is a complex function of normal pressure, sliding velocity and time. However, in boundary friction and for a number of non-lubricated friction materials, the first law of friction is satisfied with sufficient accuracy.

The most paradoxical is the second law of friction: the friction force is independent of the size and shape of the surface area corresponding to the nominal contact area. Under the *nominal contact area* we mean the surface area of the site on which the bodies would make contact if they were perfectly smooth. Even Leonardo da Vinci experimentally showed that the effort to move the body of the plane does not depend on which of its different-sized facets friction takes place. No one was interested in the cause of this phenomenon until Amonton showed in his experiments that the friction does not depend of the nominal contact area of the rubbing bodies. This statement was the essence of the second Amanton friction law, and its interpretation anticipated the modern notion of discrete contact of solids. The essence of the seeming paradox of the second Amanton law lies in the fact that the solid bodies do not make contact over the whole surface but only on individual spots because of the roughness of real surfaces. The total area of these spots (the actual area of contact) is very small relative to the nominal area. If the nominal area is defined, then, as shown by the studies, the actual area increases linearly with increasing load. In turn, the friction force is proportional to the actual contact area. With a sample with different facets, we retain the value of actual contact, which depends only on the load. This shows the independence of the friction of the nominal contact area.

A third law, attributed to Ch.O. Coulomb (1736–1806), is sometimes added to the above laws: the friction force is independent of sliding velocity. This law is substantiated less than others. By the way, Coulomb himself discovered in experiments both increasing and decreasing the speed dependences of the friction force. However, if the contact temperature during sliding is changed slightly and the properties of the contact area are almost unchanged, then in some speed range this law can be justified.

It should be noted that the laws of friction are still good empirical rules that allow to navigate in situations that arise in practice.

Depending on the presence of the lubricant we distinguish dry friction (friction without lubrication) and the friction with the lubricating material. Friction with the lubricant in this book will not be considered except for the description of friction using solid lubricants. This is due to the intention of this book to describe phenomena in which the physical properties of the surface are most evident. In the process of dry friction the characteristics of the surface properties of solid contact bodies and their interfacial interactions are decisive.

Dry friction with the lubricant and contamination of the surfaces is carried out in the brakes, friction gear, machines at the sites of the textile, food processing, chemical industries, where the lubricant is not allowed to prevent damage to product or for security reasons, as well as in sections of machines operating at very high or very low temperatures when no lubricant is suitable.

Friction has a molecular-mechanical nature [119]. On the planes of actual contact surfaces there are molecular attraction forces (van der Waals forces), which appear at distances tens of times larger than the interatomic distance in crystal lattices, and increase with increasing temperature as they are fluctuating in nature. Molecular forces cause adhesion to the contact points. Adhesion, as shown above, can form between the metal and oxide films. It can also be due to electrostatic forces. These interactions result in the resistance to relative displacement of the surfaces, called the *molecular* (*adhesion*) *force component of friction*. The adhesion forces are directly proportional to the actual contact area. The applied pressure affects these forces indirectly through the actual contact area [120].

The adhesion forces, being normal to the surface, it would seem should not work at relative tangential movement of the surfaces. In fact, the relative displacement of the surfaces at adhesion interaction is accompanied by shear deformation which due to imperfect elasticity of the material requires the expenditure of energy in an irreversible manner. Even larger tangential force must be applied if the bond between the bodies is not broken along the boundary surface but at some depth from the surface. A stronger manifestation of adhesion forces is grasping of the surfaces. In the zones of actual contact of solids forming the friction pair the asperities of the harder body penetrate into the surface of the softer counterbody. This is explained by the difference in the mechanical properties of bodies, their heterogeneity in some areas and by differences in the geometric contours of the contact sites. Therefore, the sliding of one body relative to another results in deformation of the softer surface layer by embedded asperities. The resistance to deformation of the surface layers during sliding is called the *deformation* (*mechanical*) *component of the friction force*. It can be calculated by knowing the mechanical properties of surface layers, the geometric shapes of asperities and the stress state in the contact zone, with the main provisions of continuum mechanics [166].

The development and comparison of different theories of friction and wear [20, 79, 119] allowed to determine the most common tribological law, confirming the two-term dependence of the friction force in mutual displacement of two solids, which consists of adhesion and deformation components:

$$F_{fr} = F_a + F_{def}. \tag{8.2}$$

Note that the additivity of the adhesion and deformation components of friction forces can only be used as a first approximation, since in general they are mutually related. Depending on the type and conditions of friction, as well as the structure of the body and bonds in them these friction force components may increase or decrease or even disappear altogether. Thus, in extreme cases in internal friction the adhesion component is close to zero, while in external friction of the perfectly smooth surfaces the deformation component would be zero. However, in real cases, both components are nonzero. In dry friction both the adhesion and deformation components of friction forces depend mainly on the nature and mechanical properties of materials of the rubbing surfaces as well as on their microgeometry.

Many researchers believe that the deformation component of friction force is usually very small (only a few percent cent of the total friction force). Thus, the friction of metal surfaces in vacuum is accompanied by a high friction coefficient (more the unity). If air is supplied to the vacuum chamber, in a very short period of time the friction coefficient is reduced by several times. On this basis, we can conclude that the adhesion component of friction is the cause of the high friction coefficient in vacuum. Note that in rolling friction the adhesion component has a relatively little effect on friction.

The static friction force depending on the duration of fixed contact increases to a certain limit. The friction force of movement depends on the sliding speed of surfaces, and the friction coefficient may increase monotonically, decrease and pass through a maximum or minimum.

Friction without lubrication (dry friction) is accompanied by discontinuous (jump-like) sliding of the surfaces accompanied, for example, by vibration of the car when the clutch is engaged, 'twitching' during braking, 'squeal' of brakes, vibration of cutting tools during cutting and violation of smooth operation of slow-moving parts. One can mention some ways of dealing with 'jumps' in friction: increasing rigidity of the system, increasing the sliding speed, the selection of friction pairs for which the friction coefficient depends weakly on the duration of the fixed contact and at higher speeds does not pass through a minimum [47, 119].

Oxide films, moisture and dirt on metal surfaces affect the friction coefficient in two ways. The forces of adhesion interaction between them can be hundreds of times smaller than in the case of the interaction of metals in pure contact. In addition, the strength of the oxides is usually lower than the strength of base metal, so resistance to 'gouging' and cutting of particles at movement is significantly reduced, and the friction coefficient decreases. On the other hand, the thick oxide films have a lower hardness and their presence leads to an increase in the actual contact area. If the increase in the actual contact area is faster than the decrease in the mechanical component of the friction force, this may lead to an increase in friction force.

One of the main problems of tribotechnology is the prediction of the durability of friction sections. To solve this problem, it is necessary to develop methods of calculation of machine parts for wear, taking into account the physico-mechanical characteristics of the materials of the moving parts, operating modes in friction, external friction conditions and design features of friction units. However, the development of these methods is a much more complex matter than the development of methods for calculating strength. This is due to the following reasons:

1) the amount of material receiving the load is not constant, it varies depending on the pressure, the roughness of the rubbing surfaces and the various films formed on the friction surfaces;

2) the actual contact of solids is discrete, and deformation takes place in material microfragments to which the provisions of the classical classical theory of elasticity of the homogeneity and isotropy of the element the body do not apply;

3) in the calculation of wear we evaluate the characteristics of the fracture process, in contrast to the calculations of strength which examine the conditions of non-failure of bodies;

4) in friction the properties of materials change significantly, and it is very difficult to determine in advance the extent of these changes and hence the new destruction conditions.

These difficulties to some extent explain the diversity of proposed, at different times, analytical dependences for calculating the components and materials for wear. A large part of them relates to the calculation of the degree of wear of the friction material wear and do not take into account features of construction of the friction section.

The choice of materials for friction units of cryogenic technology is based on analysis of existing workloads and specific operating conditions. The main criteria for selection are the physico-mechanical and tribological properties at operating temperatures and technological properties of the materials.

Although the laws of friction have been studied rather thoroughly, no satisfactory quantitative laws of wear have as yet been proposed. In general, if it is safe to say is that the wear increases with the time of relative movement of the contacting bodies and that the wear of hard surfaces is smaller than the wear of softer surfaces. However, in the process of wear of materials, there are many exceptions, and the dependence of wear on the load, the nominal contact area, speed, etc., in general terms has not been accurately defined. This is due to the abundance of the factors influencing the wear and tear, and small changes can completely change the role of each factor or the nature of their impact. A typical and serious complication in the study of wear is the inverse transfer of material: wear particles can move from one surface on another, and then return to the first surface.

The theory of wear in comparison with the theory of friction was formulated relatively recently, in about the last 60–70 years. Development of the types of wear was the subject of papers of I.V. Kragelskii [119] M.M. Khruschov [230], B.I. Kostetskii [118], and other scientists.

When analyzing the theory of wear, it is important to note the importance of the classification of wear. The most extensive and proven fundamental research can be considered the classification of B.I. Kostetskii in which all kinds of fracture surfaces divided into acceptable and unacceptable. Acceptable forms of wear include: mechano-chemical oxidation wear, the mechano-chemical wear of non-oxygen origin, mechano-chemical form of abrasive wear. Unacceptable forms of wear are: bonding of type I, bonding of type II, fretting process, mechanical form of abrasion wear, fatigue in rolling and other types of damage (corrosion, cavitation, erosion etc.).

In 1957, I.V. Kragel'skii proposed the fatigue theory of wear, which has been developed most extensively and is widely uses. It is based on the concept of the need for repeated effect on areas of actual contact surfaces [119, 120].

The discovery in Russia (the authors of D.N. Garkunov and I.V.Kragel'skii) of selective transfer led to a new direction in the development of the science of friction and wear related with the study of the mechanism of selective transfer and development of new methods for fighting wear on the basis of the established phenomenon.

Scientists have found that in the friction of copper alloys on steel in boundary lubrication conditions, eliminating the oxidation of copper, the selective transport of copper from the solid solution of the copper alloy on steel and back from steel to the copper alloy takes place. This process is accompanied by a decrease of friction to liquid friction and leads to significantly reduced wear of the friction pair [46, 47].

8.2. The principles of choosing optimal pairs of materials for non-lubricated moving friction sections

Reduction of power losses due to friction and increase of the wear resistance of rubbing parts of machines are the main problems in constructing various machines, especially high-precision machines working in extreme conditions. These machines include components for microcryogenic engineering, friction units of which operate without lubrication or in boundary friction conditions in a medium of dry inert gases. In this regard, the optimal choice of materials and coatings not only defines the service life, but also reduces the friction coefficient and hence the power loss in friction.

The interaction of the materials that make up a pair of friction is a consequence of many factors and events. In the approach to the problem of optimal selection of materials for friction pairs, fundamental to tribotechnology, we must first determine the range of factors that determine the friction processes and wear of materials under friction. Also, it should be remembered that even the best selection of materials for friction pairs does not mean that these materials will be used to produce parts of friction units. Anti-friction and wear-resisting materials are used only as a coating of friction parts or a solid lubricant.

When choosing materials and coatings for friction parts it is necessary to take into account a number of fundamental contradictions. They are related to the fact that the antifriction properties of a coating are achieved largely due to its low adhesion properties with respect to the material included in the friction pair, but at the same time it is necessary that this antifriction coating bonds well with the substrate material. Typically, the working material of the body in a friction pair and the substrate material are certain brands steel with respect to which the material of the antifriction coating has similar adhesion properties. To solve this problem, it is essential develop a methodology for the choice of contacting materials with a mandatory assessment of their adhesion and tribotechnical properties.

The phenomena and the factors determining friction and wear of different materials will be reviewed. This will allow the selection of the most important of them and raise the question of their optimal values.

The research results show a close relation of the friction and wear characteristics with the physical and mechanical properties the surface layer of material with the physical and chemical nature of the forces of interaction of atoms and molecules in the lattice, with the electrodynamic and thermophysical properties of materials.

The properties of materials are significantly affected by ambient temperature. Lowering the temperature to cryogenic helps clearer expression of the relationship between the crystal structure, adhesion and tribological properties of solids, as in rapid cooling the effect of adverse factors on the contact interaction decreases, and the reduction of thermal energy leads to a relative increase in the influence of the interatomic bonding forces.

Factors determining the wear rate can be divided into the following functional groups:

1) external friction conditions – specific load (nominal pressure);

2) the mechanical properties of the wearing material – modulus of elasticity and strength characteristics of the material;

3) the microgeometry of the wearing surface – indicators of roughness and waviness of the surface;

4) the friction characteristic – friction coefficient.

The quality of the machined surfaces of metals and alloys has a significant impact on many of the performance characteristics of the components and, in particular, on their wear resistance. The quality of the surface is a complex of the physical and mechanical properties of the surface layer: roughness and microrelief, microhardness, residual stresses, etc. In the process of technological processing of the surface, as well as in running-in in sliding thin surface layers significantly change their structure and properties. To date, a large amount of experimental data on the influence of methods of surface treatment of metals on their mechanical properties has been collected [64].

Sliding velocity also has a significant influence on friction and wear of materials. The dependence of the frictional properties of materials on sliding velocity has not been studied enough. Sliding velocity determines the duration of a single frictional bond and, consequently, the rate of deformation of materials. The sliding velocity determined the heat generation power and temperature at the contact. Heating of the surface layers of rubbing bodies leads to a change in their mechanical and frictional properties.

The energy approach to the theory of friction and wear, developed in recent years, has opened the prospect of further theoretical and experimental research in this area. Thus, it is known that the type and parameters of the crystal structure largely determine the energy state of the metal surface. Therefore, changes in the surface layer during friction are directly reflected in the surface energy of the material. It was found that the energy state of the metal surface undergoes substantial changes in work in friction pairs. There is reason to believe that the surface energy as a more complete physical description of the surface condition can replace microhardness, taken usually as the main parameter characterizing the wear resistance of materials.

In the zones of actual contact of solids forming the friction pair friction, harder microasperities penetrate into the surface of the softer body. This is explained by the difference in the mechanical properties of bodies, their heterogeneity in individual areas and by differences in the geometric shapes of the contact sites. Therefore, the sliding of one body relative to another is accompanied by the deformation of the less rigid surface layer by the embedded irregularities. Along with the deformation of the surface layers during friction in the areas of surfaces with relatively small distance between them ($1 \div 10$ Å), there are also noticeable interatomic interactions. As noted in the previous section, the development and comparison of different friction theories has been used in establishing the most common tribological law, claiming the binomial dependence of the friction force in the mutual displacement of two solids. Thus, the friction force is made up of adhesion and deformation components (8.2).

In dry sliding friction both the adhesion and deformation components of the friction force depend mainly on the nature of and mechanical properties of materials of the rubbing surfaces, as well as their microgeometry. In the friction of solids, especially metals, a decrease in the surface roughness reduces deformation component of the friction force and increases the adhesion component. With some differences from metals, similar phenomena on the surface are observed for polymeric materials. Consequently, there exists an optimal value of the surface roughness at which the friction coefficient is minimal.

In many metal friction pairs the adhesion component of the friction force is much more important than the deformation component, so that in friction of metal on metal the contribution of the deformation component can be neglected. According to modern views on the nature of friction, in the case of two perfectly smooth surfaces of solids friction is determined by the adhesion component which is due to the effects of intermolecular interactions between the contacting bodies. According to the theory of B.V. Deryagin [78, 79, 82], the adhesion component of the friction force is represented as

$$F_{ad} = f(P + P_{ad}),$$ (8.3)

where f is the friction coefficient, P is the pressure due to external load, P_{ad} is the pressure caused by the adhesion attraction of the surfaces.

According to B.V. Deryagin, the friction coefficient f is the ratio of the tangential components of the forces of adhesion repulsion to the normal components of the repulsive forces and at a given location of the atoms is constant for a given friction pair.

Since adhesion forces act only at the contact points with the actual area S_f, we can write

$$P_{ad} = p_{ad} S_f,$$ (8.4)

where p_{ad} is the force of adhesion attraction per unit area of actual contact of the surfaces.

If the surface roughness is modelled by a system of spherical segments of radius r, which is most commonly practiced for surfaces with low surface roughness height, then for the actual contact area in accordance with the Hertz theory we can write the expression

$$S_f = \alpha r^{2/3} P^{2/3},$$ (8.5)

where α is the angle contracting the segment.

As a result of the expression (8.3), for the adhesion component of the friction force we can write

$$F_{ad} = f(P + \alpha r^{2/3} p_{ad} P^{2/3}).$$ (8.6)

This expression implements the binomial law of the dependence of the adhesion component of the friction force on the external load P. The second term in equation (8.6) is determined by the specific force of adhesion attraction which is calculated by the methods for calculating the adhesion characteristics of various metals discussed in the previous chapters. In addition, the expression (8.6) shows that the adhesion component of the friction force increases with increasing magnitude of adhesion of the metals that form the friction pair. Therefore, while taking into account the mechanical properties for selecting the friction pairs with a minimum value of the friction force, it is important to select metals with the smallest mutual adhesion.

The revealed multifactor nature of the processes of friction and wear makes it necessary to include in the procedure for optimizing a large number of parameters, but all of them can be divided into two subsystems: 1) the parameters determining the frictional properties of materials of friction pairs, and 2) the parameters characterizing the wear of materials. As a result, the principle of selecting the best friction pair for the non-lubricated friction unit can be formulated as follows: the optimum pair should be regarded as the one which is characterized by a set of structural, energy and electrophysical parameters providing the minimum friction coefficient between them and the maximum wear resistance of each materials in the given external loading and speed conditions.

A set of parameters that determine the frictional properties of materials depends on their nature, i.e. on whether they are metals, semiconductors or insulators. This is due to first of all to the electrodynamic nature of the adhesion component of the friction force. Thus, at relative motion of the bodies the ordered motion of charges in the crystal lattice causes emission of electromagnetic waves and the absorption effects of these waves also determine friction. As noted above, the adhesion component of the friction force depends on the adhesion properties of the pair of materials. Therefore, a set of parameters that determine the frictional properties, should also include parameters that define the surface energy of materials and the energy and strength of their adhesion.

The nature of wear in dry friction of materials is determined mostly by rupture of intermetallic bonds [20]. If the rupture of the metal bonds occurs along the interface of the metals, this type of wear is determined by the force of adhesion interaction of metals. If the rupture of metal bonds takes place inside one of the metals, this type of wear is determined by the cohesive strength of the interaction between the atoms of the metal. As the cohesive interaction is a special case of adhesion, it may be possible to describe the process of wear of metals on the basis of the theory of adhesion. However, this problem requires a more detailed study.

At present there is no physically consistent theory of determining the adhesion component of the friction force. This is due to the complexity of non-equilibrium processes in the areas of direct contact of materials and accompanied by the release and dissipation of energy, as well as by the indeterminacy of the physical and mechanical parameters of the surface layers.

Since the friction is due to processes occurring in very thin surface layers, the friction force depends on the physico-mechanical properties of these layers, which differ in their properties from the layers located at depth. The reason is that the binding forces of the atoms in the surface layer are not symmetric, and the atoms can not occupy a position corresponding to the minimum value of energy within the material. The distortion of the structure of the surface layers also manifests itself in the machining of the surface and in the process of friction under the influence of deformation of these layers and temperature changes. Therefore, the internal energy of the surface layers will be higher than that inside the metal layers. The molecules of the surrounding environment are adsorbed on the solid surface due to chemisorption and form a film of chemical compounds with the solid. In the simplest case oxide films form. Thus, in general, the near-surface layers have a distorted structure and contain oxide films.

Very often, lubrication is used in order to reduce the friction force of interaction of the bodies. However, in machines in microcryogenic technology friction units operate without lubrication or in boundary lubrication conditions in an environment of dry inert gases. The force of interaction of the rubbing surfaces in dry friction is reduced using solid lubricants – graphite, molybdenum disulphide (MoS_2), plastics. Unfortunately, the solid lubricants are characterized by low wear resistance, and, therefore, to increase the longevity of units of dry friction materials with wear-resistant coatings based on nitrides and carbides of transition metals are used.

8.3. The method and calculation of the adhesion component of friction force of metals under dry friction conditions

The force interaction of solids in friction is explained in most cases by the widespread adhesion–deformation theory of friction, according to which the friction force in mutual displacement of two solids is made up of adhesion and deformation components (8.2). In the case of two perfectly rigid bodies with perfectly smooth surfaces the friction force between them is determined by the adhesion component and is due to interatomic interaction effects. Under normal conditions, in many metal pairs the adhesion component of friction force has a much greater impact than the deformation component.

[212] describes the calculation of the tangential component of the force of interaction of the ionic lattice with a moving dielectric medium. To solve this problem, the author considers the following model: the crystal lattice (such as NaCl), contacting with a body moving along the boundary with velocity v, is considered as a continuous medium with a given dielectric permittivity $\varepsilon(\omega)$. The task is to calculate the forces which act on the moving medium per unit area of the crystal lattice. The force acting on the crystal at a 'lattice–environment' tangent is defined as the summation of forces acting on each of the ions:

$$F_{\text{tang}} = -\sum_{n_i} (-1)^{n_1 + n_2 + n_3} Q \, \text{grad} \, \mathbf{V}(\mathbf{r}_{n_i}, \omega) \big|_{\omega=0}, \qquad (8.7)$$

where Q is the charge of the ions forming the lattice; n_i are the indices of lattice nodes; $\mathbf{V}(\mathbf{r}_{n_i}, \omega)$ is the total potential of the ions at the point $\mathbf{r}_{n_i}$.

By the last expression the problem is reduced to finding the potential generated by the charge $Q(t) = Qe^{i\omega t}$ in the presence of the moving dielectric medium. Neglecting retardation effects in the propagation of the field for the Fourier-image of the potential, an expression as the sum of the potentials of the charge and its electrostatic image was derived in [208]

$$V(\mathbf{q},z,z',\omega) = -\frac{2\pi Q}{q}\left(\frac{\varepsilon(\mathbf{q},\omega)-1}{\varepsilon(\mathbf{q},\omega)+1}\cdot e^{-q(z+z')} - e^{-q|z-z'|}\right). \qquad (8.8)$$

Equation (8.8) for the potential has two components: the anisotropic part, which depends on the direction of the vector $\mathbf{q}$, and the isotropic part, which does not depend on it. From (8.7) it follows that the contribution to the force tangent to the boundary is provided only by the anisotropic part of the potential whose existence is associated with the presence of anisotropic spatial dispersion caused by the motion of the medium. Using this expression and taking into account the symmetry properties of generalized susceptibilities with respect to the operation of inversion of the wave vector:

$$\text{Re } \varepsilon(\mathbf{q},0) = \text{Re } \varepsilon(-\mathbf{q},0), \quad \text{Im } \varepsilon(\mathbf{q},0) = -\text{Im } \varepsilon(-\mathbf{q},0), \qquad (8.9)$$

we obtain an expression for the tangential force acting on unit area:

$$F'_{tang} = \frac{8\pi Q^2}{a^4}\sum_{i}^{(+)}\frac{C_i}{C_i}\text{Im}\left(\frac{1}{\varepsilon(C_i,0)+1}\right)\left(\frac{e^{-C_i h}}{(1+e^{-C_i a})}\right)^2. \qquad (8.10)$$

Sign (+) at the sum means that the summation is carried out only over vectors C_i, having a positive projection on the direction of the speed. Due to the influence of the exponential factor, the contribution to the sum comes from the terms corresponding to the minimum values of the vector C_i, i.e., we can confine ourselves to considering the contribution of the vectors $\{\pm\pi/a, \pm\pi/a\}$ with the modulus $C = \pi\sqrt{2}/a$, where a is the lattice constant of the ionic crystal. From (8.10) it follows that the dependence of friction on the speed determined by the dispersion law of the medium. The imaginary part $\varepsilon(\omega)$ describes the absorption of electromagnetic waves by the medium, thus the friction is related to the dissipative properties of the medium. At real speeds of the bodies, the absorption regions at the characteristic frequencies of optical and plasma oscillations give a negligibly small contribution, and only the contribution from the absorption at low frequencies due to conduction of the medium remains. In this case, it suffices the following approximation for $\varepsilon(\omega)$:

$$\varepsilon(\omega) = \varepsilon_0 + i\frac{4\pi\sigma}{\omega}, \qquad (8.11)$$

where ε_0 is the static dielectric constant; σ is conductivity of the medium.

Using the fact that for the medium moving at non-relativistic speeds $\varepsilon(\mathbf{q}, \omega) = \varepsilon(\omega - \mathbf{q}, \mathbf{v})$ is the Doppler effect, we obtain using (8.11) and at $C_i = \pi/a$

$$\mathrm{Im}\,\frac{1}{\varepsilon(-\pi v/a)+1}=\frac{4\left[v/(\sigma a)\right]^{2}}{16+(\varepsilon_{0}+1)^{2}\left[v/(\sigma a)\right]^{2}}. \qquad (8.12)$$

As a result, the following expression was obtained in [212] for the friction force between the ionic crystal and a dielectric medium acting on unit area:

$$F'_{\mathrm{tang}}=\frac{8\pi\sqrt{2}Q^{2}\exp(-2\sqrt{2}\pi h/a)}{a^{4}[1+\exp(-\pi\sqrt{2})]^{2}}\cdot\frac{4v/(\sigma a)}{16+(\varepsilon_{0}+1)^{2}\left[v/(\sigma a)\right]^{2}}. \qquad (8.13)$$

The results of [212] were developed in [62, 64, 67–69, 199] for the case of friction of metal surfaces. Thus, to calculate the adhesion component of friction force by the formula (8.6) we must determine the value of the friction coefficient f for the contacting friction pairs. In the friction coefficient we distinguish two components:

$$f = f_{\mathrm{el}} + f_{\mathrm{d}}. \qquad (8.14)$$

One component – electrodynamic f_{el} – is linked to the effects of absorption of electromagnetic waves in metals at relative movement of the charges forming the metal lattice. This component can be expressed in terms of the characteristic of the contact of metals with similar lattice constants a, moving relative to each other with relative velocity v.

Another component of the friction coefficient – dislocation f_{d} – is due to the fact in contact of metals with different lattice constants networks of edge dislocations form along the interface and reduce the adhesion energy of the contacting materials causing at the same time a reduction of shear stresses at the interface [72]. The relative displacement of the metals leads to the formation of a component of the friction force which prevents changes in the dislocation structure.

To determine the electrodynamic part of the friction coefficient f_{el} it is assumed that the contacting metals (1 and 2) have the same lattice constants along the interface: $a = (a_1 + a_2)/2$. We then generalize the theory of molecular friction forces between the ionic lattice and the dielectric medium, presented in [212], for the case of friction of metal surfaces.

Metals are a system of equally charged ions embedded in an electron gas. Therefore, in contrast to [212], we have replaced the ionic crystal lattice with opposite charged ions by the lattice with positively charged ions. With regard to the contribution of the electron gas to the friction force, taking into account its uniform distribution in the plane parallel to the interface, the electron gas does not contribute to the tangential component of the friction force between the surfaces but determines the strength of adhesion between the contacting metals and thus through the additional pressure P_{ad} of the forces of adhesion attractive of the surfaces also the adhesion component of the friction force of metal surfaces. In addition, for the metals we must use the approximation

$$\varepsilon(\omega) = 1 + i\frac{4\pi\sigma}{\omega}. \qquad (8.15)$$

As a result, using the procedure similar to [212], the following expression was obtained for the friction coefficient f_{el} associated with electrodynamic effects of energy dissipation and the effects of adhesion attraction of the surfaces:

$$f_{el} = \frac{4\pi\sqrt{2}\exp(-2\sqrt{2}\pi h/a)}{p_{ad}a^4[1-\exp(-\pi\sqrt{2})]^2}\left((Z_1e)^2\frac{(v/\sigma_2 a)}{4+\left[v/(\sigma_2 a)\right]^2}+\right.$$

$$\left.+(Z_2e)^2\frac{v/(\sigma_1 a)}{4+\left[v/(\sigma_1 a)\right]^2}\right), \qquad (8.16)$$

where Z_i is the valence of metals, σ_i is the conductivity of metals, h is the gap between the surfaces; $p_{ad}(h)$ is the force of adhesion attraction per unit area. Thus, the expression (8.16) allows for values of Z_i, σ_i, fixed for the given friction pair, to get the dependence of f_{el} on the relative speed v and the size of the gap h between the surfaces.

The formula (8.16) was used as a basis for developing a computer program to calculate the friction coefficient for any metal friction pair and a subprogram for calculation of the dependence of force p_{ad} of adhesion interaction of the metal surfaces on the size of the gap h. An example of calculation results for the TiN–W, TiN–Mo, TiN–Fe, TiN–Cr friction pairs using the tabulated values of Z_i, σ_i, a_i are shown in Fig. 8.2a in the form of the dependence of f_{el} on the size of the gap h at a fixed speed of relative movement of the surfaces $v = 1$ m/s. Figure 8.2 shows a monotonically decreasing dependence of the electrodynamic part of the friction coefficient f_{el} on h with a maximum f_{el} value at $h = 0$. In this case, the above-mentioned frictional pairs are listed in order of decreasing values of the friction coefficient. The maximum value of the friction coefficient is obtained for the TiN–W friction pair, and the minimum – TiN–Cr pair. It should be noted that when increasing the gap to 5 a.u. and more the f_{el} value can be neglected.

The dependence of the friction coefficient f_{el} on the speed shown in Fig. 8.2b is characterized by vanishing at $v = 0$ and a maximum at $v \approx (\sigma_1 + \sigma_2)/2a$ (a very low rate of relative movement of the bodies) and by the monotonic decrease with increasing speed at $v > (\sigma_1 + \sigma_2)/2a$. This is due to the fact that at the actual speeds of the bodies the absorption regions at the characteristic frequencies of the optical and plasma oscillations give a negligible contribution to the mechanism of energy dissipation, and only the contribution from the absorption at low frequencies due to the conductivity of the medium remains. The intensity of the absorption effects decreases with increasing frequency; since the absorption frequency $\omega \sim v/a$, then the

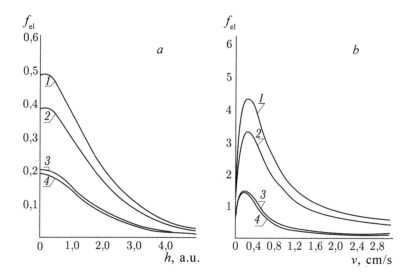

Fig. 8.2. The dependence of the electrodynamic part of the friction coefficient f_{el}: a – on the gap between the surfaces at $v = 1$ m/s, and b – on the relative speed of the surfaces for friction pairs: 1 – TiN–W; 2 – TiN–Mo; 3 – TiN–Fe; 4 – TiN–Cr [64].

intensity of the effect of electromagnetic absorption decreases and hence the friction coefficient also decreases with increasing speed in accordance with the hyperbolic law $f \sim v^{-1}$. Since the electrodynamic friction does not occur at $v = 0$, at intermediate frequencies (speeds) the friction coefficient should be maximum. As can be seen from Fig. 8.2, this maximum is observed at relative speeds of the order of $0.2 \div 0.4$ cm/s. The friction coefficient at the maximum point and the corresponding value of the adhesion component of friction force can serve as an estimate of the static friction force F_s.

Note that in calculating the tabulated data on the conductivity should be used with caution because the friction force in the electrodynamic model is actually determined by the motion of induced charges in a narrow surface layer the conductivity of which may differ from the conductivity inside the body, for example, due to the presence of the oxide film. In addition, the temperature in the contact area of friction pairs can vary greatly from the ambient temperature, which also affects the conductivity.

Another component of the friction coefficient is associated with the fact that the contact of metals with different lattice constants results in the formation of networks of edge dislocations along the interface which lower the energy of adhesion of the contacting materials causing at the same time a reduction of shear stresses at the interface. The relative displacement of the metals is accompanied by the formation of the component of the friction force which prevents changes in the dislocation structure. The contribution of this component to the total friction coefficient was assessed in [62, 64, 199]. The friction coefficient f_d in the form of the ratio of the energy of the network of dislocations per unit area E_d of the surface of the boundary to the adhesion energy was introduced:

$$f_d = \frac{E_d}{E_a},$$ (8.17)

where f_d is the dislocation component of the friction coefficient; E_d is the energy of the dislocation network per unit area of the interface; E_a is the adhesion per unit area of the interface.

The energy of the dislocations E_d is determined by the discrepancy in the lattice constants $\Delta a/a$, the density and geometry of the dislocations, and the elastic constants of the boundary layer.

Calculation of energy E_d of the system of edge dislocations was based on results of [70, 71]. Let the Burgers vectors of the dislocation network be located in the most favorable directions (for close packed {111} plane in the fcc lattice, they lie along the direction $\langle 1\bar{1}0 \rangle$, $\langle 10\bar{1} \rangle$ and $\langle 01\bar{1} \rangle$, in the {110} plane in the bcc lattice – along the directions $\langle 1\bar{1}1 \rangle$ and $\langle \bar{1}11 \rangle$. The network of such dislocations has a diamond-shaped cell, but we can consider separately one of the systems of parallel dislocations. The x axis is directed along the Burgers vector, and the axis z – perpendicular to the boundary. On the basis of [71] the following set of parameters canbe introduced for determination of E_d for contacting pairs of materials

$$\gamma_{1,2}^2 = \frac{1}{2c_3c_5}\left(\lambda \pm \sqrt{\lambda^2 - 4c_1c_3c_5^2}\right),$$

where

$$\lambda = c_5^2 + c_1c_3 - (c_5 + c_8)^2;\ b_j = \frac{c_5\gamma_j^2 - c_1}{(c_5 + c_8)\gamma_j};\ \lambda_j = c_3b_j\gamma_j + c_8;\ j = 1,2;$$

$$v = \frac{b_1}{\lambda_1} - \frac{b_2}{\lambda_2};\ \mu = c_5\left(\frac{\gamma_1}{\lambda_1} - \frac{\gamma_2}{\lambda_2} - v\right);\ \alpha = \alpha_1 + \alpha_2 = \frac{1}{\mu}\left(\frac{1}{\lambda_2} - \frac{1}{\lambda_1}\right).$$

Here $\gamma_{1,2}$ are the coefficients in the exponents describing the decrease in displacements and stresses with distance from the interface ($\gamma_{1,2}$ characterizes the anisotropy of the crystal field of the materials, for an isotropic medium $\gamma_1 = \gamma_2 = 1$); b_j is the ratio of the normal and tangential displacements; λ, λ_j, v, μ are auxiliary values; c_1, c_3, c_5, c_8 are elastic constants in the xz system. For cubic crystals they can be expressed in terms of the original constants c_{11}, c_{12}, c_{44} using the transformation formulas for the tensors of the fourth rank when rotating the coordinates.

For the fcc lattice we have:

$$c_1 = c_{11} + \frac{1}{2}H;\ c_3 = c_{11} + \frac{2}{3}H;\ c_5 = c_{44} - \frac{1}{3}H;\ c_8 = c_{12} - \frac{1}{3}H.$$

For the bcc lattice:

$$c_1 = c_{11} + \frac{2}{3}H; \ c_3 = c_{11} + \frac{1}{3}H; \ c_5 = c_{44} - \frac{1}{3}H; \ c_8 = c_{12} - \frac{1}{3}H;$$

$$H = 2c_{44} + c_{12} - c_{11}.$$

For the hexagonal system:

$$c_1 = c_{11}; \ c_3 = c_{33}; \ c_5 = c_{55}; \ c_8 = c_{13}.$$

For the joint of two crystals [71] with $a_1 > a_2$, we introduce the parameters:

$$\eta = \frac{a_1 - a_2}{a_2}; \ l = \frac{a_1 a_2}{2(a_1 - a_2)}; \ a' = \frac{2a_1 a_2}{a_1 + a_2};$$

$$c'_{3,5} = \sqrt{c^{\langle 1 \rangle}_{3,5} c^{\langle 2 \rangle}_{3,5}}; \ b = -\frac{\pi a'}{l c'_5} \frac{1}{\alpha_1 + \alpha_2},$$

where η is the degree of lattice mismatch, $2l$ is the distance between adjacent parallel dislocations, a' is the constant of the reference lattice, $c^{\langle 1 \rangle}_3$, $c^{\langle 2 \rangle}_5$ are the elastic constants of the boundary layer.

In the Peierls–Nabarro model, the energy of the dislocation network for unit contact surface, provided that the normal stresses at the interface can be neglected, is defined as follows [71]:

$$E_d = \frac{c'_5 a'}{2\pi^2} \left(1 + b - \sqrt{1 + b^2} - b \ln(2b\sqrt{1 + b^2} - 2b^2) \right). \tag{8.18}$$

Then the energy of adhesion is

$$E_a = E_a^{(0)} - E_d, \tag{8.19}$$

where $E_a^{(0)}$ is the energy of adhesion of the contacting metals, excluding the influence of the system of dislocations.

The above formulas were used in developing a program and in [69, 199] calculations were carried out of the energy of the dislocation for a pair of contacting materials, depending on the relative change $\Delta a / a_i$ of the lattice constant of a material. Figure 8.3 shows an example of a plot of dislocation energy for a TiN film on a steel substrate due to the change of the lattice parameter of the substrate. The calculation results are evident for these materials. At the initial values of the lattice parameters ($a_{st} = 2.87$Å, $a_{TiN} = 4.25$Å, for TiN with the NaCl structure in calculations we used $a_{TiN}/2$) the energy of the dislocation is equal to $E_d = 1338$ MJ/m^2, and it is important to consider of the effect of misfit dislocations, since the change in adhesion energy due to dislocations is comparable to the interfacial energy of interaction of coatings with the substrate.

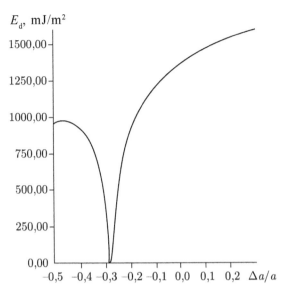

Fig. 8.3. Dependence of the adhesion energy of edge misfit dislocations for a TiN film on a steel substrate on the relative change of the parameter of the substrate lattice $\Delta a/a$ [64].

In [62, 64, 199] the dislocation component of the friction coefficient was assumed to be independent of the speed of relative motion of the bodies. In addition, using the given model, we can not determine the energy dependence of the dislocation E_d on the size of the gap h between the rubbing surfaces. In this regard, although the adhesion energy E_a in the formula (8.17) is a function of h, in fact, the calculation of f_d was conducted only at $h = 0$. The value of E_a ($h = 0$) was calculated in the framework of adhesion interaction, based on the dielectric formalism. The calculations gave the following values for the corresponding friction pairs: TiN–W, $f_d =$ 0.58; TiN–Cr, $f_d = 0.60$; TiN–Fe, $f_d = 0.72$. As a result, the overall molecular friction coefficient, equal to $f = f_{el} + f_d$, for the investigated pairs takes the values: TiN–Cr, $f = 0.79$; TiN–Mo, $f = 0.98$; TiN–Fe, $f = 0.78$; TiN–W, $f =$ 1.2. Comparison of the reference and the experimental values of the friction coefficients of these metal pairs with the values calculated by the above method gave good agreement.

Using the representations of dielectric formalism, in [69] the authors calculated the energy and strength of adhesion interactions of the substrates from corrosion-resistant steels Cr18Ni10Ti and 30Cr13 and various wear-resistant coatings based on nitrides and carbides of transition metals (Cr_3C_2, TiN, Mo_2C, Mo_2N, Cr_2N, CrN, WC, TiC) as a function of the gap h.

Figure 8.4 shows plots of calculated [62, 68, 69] dependences of the friction coefficient f on steel with wear-resistant coated counterbodies on the basis of the above-mentioned nitrides and carbides of transition metals on the size of the gap h at a fixed speed $v = 1$ m/s of relative movement between the surfaces. The dependence of f on h reflects the dependence of

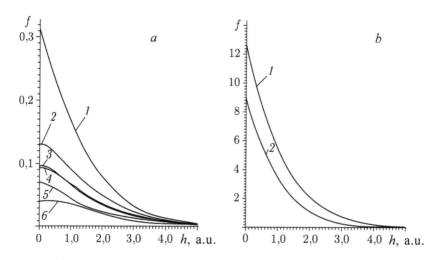

Fig. 8.4. Dependence of the adhesion component of friction force on the size of the gap h for $v = 1$ m/s for the friction pairs based on 30Cr13 steel and wear-resistant materials: a) 1 – TiC; 2 – TiN; 3 – Mo_2S; 4 – Cr_2N; 5 – Mo_2N; 6 – Cr_3C_2; b) 1 – CrN; 2 – WC [64].

friction on the external load. The figure shows the monotonically decreasing dependence of f on h with a maximum value of f at $h = 0$. Comparison of the graphs in Fig. 8.4 for various friction pairs shows that the lowest friction coefficient f on steel is obtained for the movement of the counterbodies coated with Cr_3C_2, Mo_2N, Cr_2N, Mo_2C, TiN, TiC, and the largest – with CrN, WC.

Figure 8.5 shows the dependence of the friction coefficient f of the same friction pairs on the speed v of the relative movement of materials when $h = a = (a_1 + a_2)/2$. Comparison of the graphs $f(v)$ in Fig. 8.5 for various friction pairs at $v > v_{max}$ affirms the above findings of the ratio of the friction coefficients for the given materials with small rearrangements between some pairs.

The calculations also showed that, compared with the Cr_2N coating, the CrN coating has much worse both the adhesion and tribological properties. Therefore, in the process of deposition of the Cr_2N coating it is essential to select deposition regimes in such a way as to avoid the formation of CrN.

Our calculations show that the adhesion component of the friction force depends on a set of the physical properties of materials that make up the friction pair, such as valence, conductivity, on their symmetry and matching of the lattice constants, and on the adhesion characteristics of contacting materials. The procedure for calculating the frictional properties of the nitrides and carbides of the transition metals reflects all the stages of previous studies of the adhesion characteristics of complex compounds.

As for the dependence of the adhesion component of friction force on the relative speed of the materials, it should be noted that the range of

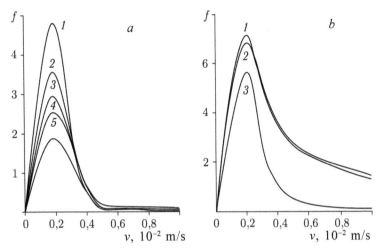

Fig. 8.5. Dependence of the friction coefficient of the adhesion component of the friction force on the speed v of relative motion of bodies at $h = a$ for friction pairs based on 30Cr13 steel and wear-resistant materials: a) 1 – Mo_2N; 2 – Cr_2N; 3 – Mo_2C; 4 – TiN; 5 – Cr_3C_2; b) 1 – CrN; 2 – WC; 3 – TiC [64].

applicability of expression (8.16) for the electrodynamic part of the friction coefficient is limited to low speeds. The dependence of friction forces on the speed is at present among the most difficult issues of tribonics. At the same time, the technical significance of this problem has greatly increased in recent years due to the expansion of the range of speeds used in friction pairs. Experimental investigation of the dependence of the dry sliding friction coefficients on the speed indicates that these dependences have a pronounced minimum. Thus, the minimum for a alundum–steel pair corresponds to a speed of 100 m/s; the silicon carbide – steel pair to 92 m/s; chromium alundum – steel pair – about 80 m/s [102]. At high speeds, friction is determined primarily by the physical properties of sliding metals at elevated temperatures. Under these conditions, the adhesion mechanism is still valid.

Suppose that the friction mechanism, associated with its adhesion part, is based on the transfer of the energy of the adhesion bond released at rupture of the bond, to vibrations of the crystal lattice. At the same time, the duration of activation of lattice vibrations with energy $\sim m\omega^2 a^2/2$ due to rupture of the adhesion bond with the energy ε is determined, with the Boltzmann factor $\exp[-\varepsilon/(kT)]$ taken into account, by the ratio

$$t_{act} \sim \frac{m\omega^2 a^2}{\varepsilon\omega} \cdot e^{-\varepsilon/(kT)}, \qquad (8.20)$$

where ω is the frequency of intrinsic vibrations of the atoms (or the frequency of movement of dislocations), m is the mass of the atom, a is the lattice constant, ε is the energy of the single bond, T is temperature.

With increasing speed v the duration of the physical contact of the surfaces decreases as $t_c \sim v^{-1}$. As a result, the change in energy per unit time, per unit area of contact, due to the rupture of adhesion bonds and the transfer of energy to vibrations of the lattice atoms can be written as

$$\Delta W \sim n_b \varepsilon t_c / t_{act} \sim \frac{n_b \varepsilon^2 e^{\varepsilon/(kT)}}{m \omega v a^2},$$
(8.21)

where n_b is the number of bonds per unit surface area. Equating the energy change of the system to the work of the friction force $F_{ad} v$, we obtain an estimate for the adhesion component of the friction force

$$F_{ad} \sim \frac{n_b \varepsilon^2 S_f e^{\varepsilon/(kT)}}{m \omega v^2 a^2},$$
(8.22)

where S_f is the area of actual contact surfaces.

Thus, the strength of adhesion interaction decreases with increasing speed according to the law $\sim v^{-2}$ and increases with increasing energy of the adhesion bond. However, increasing speed also increases the temperature of the friction surfaces, which increases the pliability of microroughness to plastic deformation and leads to an increase in the actual contact area and hence the friction force. Thus, the minimum of the friction force is associated with the action of speed and temperature factors.

Summarizing, we can make conclusions about the properties of the adhesion component of the friction force:

1. At low speeds of the relative displacement of the rubbing surfaces, when the relations (8.6), (8.14) and (8.16) are valid, the adhesion component of the friction force is characterized by:

a) binomial dependence on the magnitude of the external load P;

b) a linear dependence on the normal component of the adhesion force for a given pair of metals; in this case as the strength of mutual adhesion of the metals increases the relative importance of the adhesion component of the friction force also increases as compared to the deformation component;

c) complex dependence on speed: there is a sharp peak at very low speeds $v \sim (\sigma_1 + \sigma_2)/(a_1 + a_2)$ and a monotonically decreasing dependence of $\sim v^{-1}$ with $v \gg (\sigma_1 + + \sigma_2)/(a_1 + a_2)$.

2. At speeds of $v \geq 10$ m/s the adhesion component is characterized by:

a) the dependence ($\sim v^{-2}$) decreasing with increasing speed, which is then followed by an increasing dependence;

b) the relationship of the minimum of the friction force with the action of speed and temperature factors (in particular, the real contact area S_f increases with temperature due to the effects of plastic deformation of microasperities).

3. A common property of the adhesion component of the friction force is its dependence on the nature of the surface roughness of the friction pair. With the increase of surface roughness (average width of the gap h between the surfaces) the adhesion component of the friction force is reduced.

8.4. The method and calculation of the friction characteristics of metals with solid lubricant materials, oxide and diamond-like wear-resistant coatings

In dry friction sites with solid lubricant materials (fluoroplasticic, plastics, graphite, MoS_2, etc.) friction pairs are usually metal – solid lubrication material pairs. Because of their electrophysical properties, the solid lubricants are usually dielectrics (fluoroplasticic, plastic) or semiconductors (graphite, MoS_2, etc.). Both the tribological characteristics of friction section with solid lubricants and their adhesion properties undergo significant changes when compared with similar characteristics for dry friction units based only on metal friction pairs.

Solid lubricant materials, having good antifriction properties, are unfortunately characterized by low mechanical strength and wear resistance. In recent years much interest in tribotechnology has been attracted by the application in friction sections of materials with oxide wear-resistant coatings based on aluminum and chromium oxides. According to their electrophysical properties, the aluminum and chromium oxides are semiconductors. In this connection, there is an urgent need to establish a method for calculating the adhesion and tribological characteristics of materials and coatings with semiconducting or insulating properties.

To calculate the adhesion characteristics of the adhesion contact of 30Cr13 steel with fluoroplasticic, molybdenum disulphide MoS_2 and Al_2O_3, Cr_2O_3 oxides, the dielectric formalism method outlined in Chapter 7 was applied, and on the relationhips (7.8)–(7.10), (7.17)–(7.21) were used to calculate the dependence of the energy and strength of their adhesion interaction on the size of the gap h between the contacting surfaces (see Fig. 8.6).

Comparison of these results with the values of the adhesion characteristics of 30Cr13 steel with wear-resistant metal coatings, given in the previous section, shows that, excluding the effect of the edge dislocations, the best adhesion properties (the highest energy and the strength of adhesion interaction) in the pair with steel 30Cr13 were obtained for the materials such as Cr_2O_3, Al_2O_3, MoS_2, Cr_3C_2, TiN, and Mo_2C. It may be noted that the estimated worst adhesion to the steel was observed for the coatings characterized by the following materials: CrN, WC, TiC, fluoroplasticics and Mo_2N.

The friction coefficient in the dry friction sections with solid lubricant materials or oxide coatings cannot be calculated using (8.16): these materials are insulators or semiconductors rather than metals for which the relation (8.16) holds. For a friction pair of a metal and a solid lubricant material with dielectric constant ε_0 the formula for the adhesion component of the friction coefficient takes the form [62, 64, 69, 199]

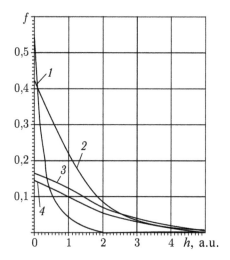

Fig. 8.6. Dependence of the friction coefficient of the adhesion component of the friction force on the size of the gap h for $v = 1$ m/s for friction pairs: 1 – steel 30Cr14 – MoS_2; 2 – 30Cr14–fluoroplasticic, 3 – 30Cr14–Cr_2O_3; 4 – 30Cr14–Al_2O_3 [64].

$$f = \frac{8\pi\sqrt{2}\exp(-2\sqrt{2}\pi h/a)}{p_{ad}a^4[1-\exp(-\pi\sqrt{2})]^2} \cdot \frac{4(Z_1e)^2 v/(\sigma_2 a)}{16+(\varepsilon_0+1)^2[v/(\sigma_2 a)]^2}. \qquad (8.23)$$

Since the conductivity σ_2 of the solid lubricant material or oxide is significantly lower than the conductivity of metals, it immediately follows from (8.23) that the friction coefficient for the metal–solid lubricant material or metal-oxide is much less than for the metal–metal pair.

It should be noted that the temperature in the contact area of the friction pairs may be different from the ambient temperature and this has a very significant effect on the conductivity in the case of semiconductors. Indeed, if the change of metal temperature in the range $100 \div 500$ K leads to a large change in the conductivity (the conductivity of metals decreases with increasing temperature), the increase in temperature for semiconductors, first, leads to an increase in conductivity, and second, can be very significant. Thus, in the references it is noted [225] that in the range $T = 100 \div 300$ K the conductivity of a semiconductor such as GaS, varies in the range $\sigma = 10^{-7} \div 10^{-2}$ ohm·cm^{-1}, for Sb_2S_3 at $T = 100 \div 500$ K in the range $\sigma = 10^{-15} \div 10^{-4}$ ohm·cm^{-1}.

Unfortunately, the literature for anti-friction coatings on the basis of such semiconductor materials such as MoS_2, Al_2O_3, Cr_2O_3, contains no data on changes in the conductivity of these materials at the desired temperature range from room temperature to the melting point. The available information on the conductivity of these materials at room temperature shows that the conductivity values for them are so low that substitution into (8.23) gives the values of the friction coefficient close to zero. This suggests that to obtain the real values of the friction coefficient it is necessary to use higher values of the conductivity of these materials which they can have at temperatures in the friction contact spots close to the melting point of ~2000 K.

The formula (8.23) has been modified [62, 64, 69] to calculate the friction coefficient of metals with dielectrics or semiconducting materials. As a result, the expression for the adhesion component of the friction coefficient takes the form

$$f = \frac{8\pi\sqrt{2}\exp(-2\sqrt{2}\pi h/a)}{p_a a^4 [1+\exp(-\pi\sqrt{2})]^2 \varepsilon_2} \cdot \frac{(Z_2 e)^2 v/(\sigma_1 a)}{4+[v/(\sigma_1 a)]^2}, \qquad (8.24)$$

where σ_1 is the conductivity of the metal in the metal–semiconductor or metal–dielectric friction pair; Z_2 is the valence of the semiconductor or dielectric, ε_2 is its dielectric constant.

This ratio can also be used to calculate the friction coefficient in the metal–solid lubricant or metal–wear-resistant coating pair with the characteristics of the semiconductor or a dielectric (in particular, for diamond-like coatings). Thus, Figs. 8.6 and 8.7 show the results of calculations for corrosion-resistant steel 30Cr14 paired with fluoroplastic, MoS$_2$ and oxides Al$_2$O$_3$, Cr$_2$O$_3$ in the form of plots of the dependence of the friction coefficient f of the size of the gap h at a fixed speed of $v = 1$ m/s relative movement of the surfaces (Fig. 8.6) and on the relative speed v of movement of materials at $h = a = (a_1 + a_2)/2$ (Fig. 8.7).

Figures 8.6 and 8.7 show that:

a) at $h = 0$ the lowest friction coefficient on steel 30Cr14 has Al$_2$O$_3$, followed by Cr$_2$O$_3$, fluoroplastic and MoS$_2$;

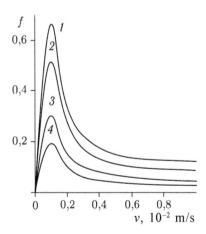

Fig. 8.7. The dependence of the adhesion component of friction coefficient on the speed v of relative motion of bodies at $h = a$ for friction pairs: 1 – steel 30Cr14 – MoS$_2$; 2 – 30Cr14–fluoroplasticic, 3 – 30Cr14–Cr$_2$O$_2$; 4 – 30Cr14–Al$_2$O$_3$ [64].

b) with increasing gap size h the friction coefficient decreases more rapidly for MoS_2; as a result, at $h \geq 2.6$ Å the minimal friction coefficient is obtained for MoS_2, followed by fluoroplastic, Al_2O_3, Cr_2O_3.

Analysis of the dependence of the friction coefficients on speed confirms the above conclusions about the relationship of the friction coefficients for these materials (with minor rearrangements between some pairs).

Our calculations show that the adhesion component the friction force depends on a number of the physical properties materials that make up the friction pair, such as valence, conductivity, on their symmetry and matching lattice constants, and on the adhesion characteristics of contacting materials.

The friction coefficient for diamond-like coatings (DLC) was calculated using the following model representations of the dependence of the lattice constant, the permittivity and the effective ion charge of DLC on its phase composition. Taken into account that a change in the relative share c of carbon atoms, which are in the diamond phase, the lattice constant a of the structure of the diamond-like coating changes in accordance with Vegard's law for solid substitutional solutions $a = a_{diam}c + a_{grap}(1 - c)$. In calculating the friction coefficient the layering of graphite was taken into account and the constant of the structure perpendicular to the graphite layers was used as the process of running-in of the graphite phase is accompanied by the reorientation of the surface faces of graphite along its natural layers.

In accordance with the microscopic theory of the behaviour of dielectrics in electric fields, the dielectric constant for the non-polar dielectrics is given by

$$\varepsilon = 1 + \frac{\sum_i c_i \alpha_i}{1 - \sum_i c_i \alpha_i \beta_i}, \qquad (8.25)$$

where c_i is the concentration of the i-th type of atoms; α_i is their polarizability; β_i is the factor of the internal field, taking into account the interaction of dipoles with each other and determined by the features of the structure of the material.

For most dielectrics with $\varepsilon = 2 \div 8$ there is $\beta \approx 1/3$. In the DLC only the atoms situated in the diamond phase are polarized. The atoms in the metallic phase of graphite are not polarized. Therefore, (8.25) for the case of DLC is simplified

$$\varepsilon = 1 + \frac{c\alpha}{1 - c\alpha\beta}, \qquad (8.26)$$

where c is the fraction of the atoms of the diamond phase, α is the polarizability of the diamond atoms, $\beta \approx 1/3$. Knowing the dielectric constant of diamond $\varepsilon = 5.7$, we can find its polarizability:

$$\alpha = \frac{\varepsilon - 1}{c[1 + \beta(\varepsilon - 1)]}. \qquad (8.27)$$

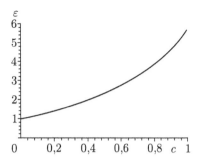

Fig. 8.8. The dependence of the dielectric constant ε on the concentration of the diamond phase c.

Thus, from (8.27) we easily find that the polarizability of carbon atoms in the diamond phase is $\alpha = 1.831$.

With this information, the dependence of the dielectric constant on the concentration of the diamond phase, shown in Fig. 8.8, can be plotted on the basis of (8.26).

The values of the effective ion charge for the diamond and graphite phases were chosen from the condition of matching the calculated and experimental values of the adhesion component of the friction coefficient in vacuum for the friction pairs: steel–diamond with the friction coefficient $f \approx 0.15 \div 0.20$ and steel–graphite with the friction coefficient $f \approx 0.5$. As a result, we obtained the following values: for the effective ion charge of the diamond $Z_{\text{diam}} = 2.0$, for graphite $Z_{\text{graph}} = 0.5$. For DLC the effective charge of the ions was determined by linear approximation:

$$Z = cZ_{\text{diam}} + (1 - c)Z_{\text{graph}}. \qquad (8.28)$$

This procedure was used to calculate the friction coefficients for the diamond-like coating on 30Cr14 steel (Fig. 8.9) in four different phase compositions of the DLC: diamond, 83% of the diamond phase (17% graphite), 70% of diamond phase (30% graphite), 50% for each phase. The friction coefficient was calculated for the relative speed $v = 1$ m/s. The figure shows the decrease of the friction coefficient with increasing content of graphite phase in the range of applicability of the proposed model of DLC. But increase in the graphite phase content is accompanied by an increase in the degree of wear of the coating as a result of the reduction of its strength properties, so there is a phase composition of the DLC with the optimal tribological characteristics. The currently produced DLCs [200], with 83% of diamond and 17% of graphite phases, appear to be optimal.

To compare the tribological properties of the diamond-like coatings with the properties of wear-resistant coatings based on nitrides and carbides of transition metals [29–31], calculations were carried out to determine the friction coefficients of 30Cr13 on these coatings at a speed of relative

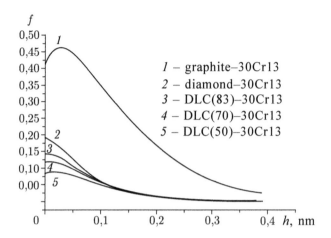

Fig. 8.9. Dependence of the friction coefficient on the size of the gap h for the pair of 30Cr14 steel and diamond-like coating DLC with different phase compositions for $v = 1$ m/s.

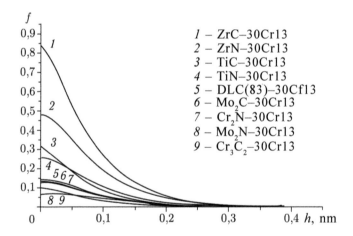

Fig. 8.10. Dependence of the friction coefficient on the size of the gap h for friction pairs on the basis of steel 30Cr14 and wear-resistant materials for $v = 1$ m/s.

movement of the surfaces $v = 1$ m/s (Fig. 8.10), and DLC (83) with the nitrides and carbides of transition metals at $v = 1$ m/s (Fig. 8.11). The calculation results demonstrate that the friction coefficient for the DLC–steel pair is comparable only with the wear-resistant coatings with the best the tribological properties based on the nitrides and carbides of chromium and molybdenum, and in pair with them the DLC show abnormally low friction coefficients.

Calculations of the friction coefficient of the diamond-like coating with materials from transition metals (Ti, Cr, Mo, Fe, W, Cu) with $v = 1$ m/s (see

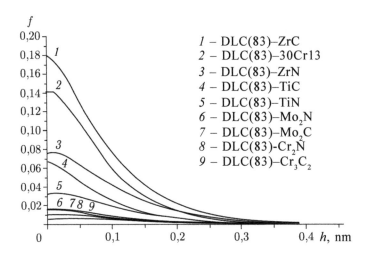

Fig. 8.11. Dependence of the friction coefficient on the size of the gap h for pairs DLC (83) and wear-resistant materials for $v = 1$ m/s.

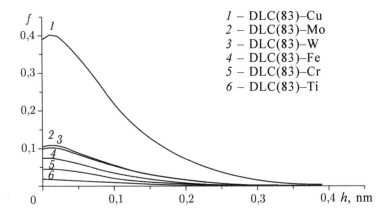

Fig. 8.12. Dependence of the friction coefficient on the size of the gap h for pairs DLC (83) and transition metals with $v = 1$ m/s.

Fig. 8.12) showed [29–31] that the lowest value of the friction coefficient $f \approx 0.02$ is obtained for friction of DLC (83) on titanium. Therefore, titanium alloys UT4, VT6 in pairs with DLC should be characterized by good tribological properties. To counteract the phenomenon of adhesion bonding of titanium-bearing alloys with DLC, we can apply ion nitriding, which leads to hardening of the surface layer of titanium alloys and prevents the transfer of titanium particles on the surface of the DLC. The calculation of the intensity of wear and performance of the cylinder–piston dry friction pair (CPP) of the microcryogenic system with the friction pair of the DLC and the nitrided titanium alloys UT4 and VT6 and showed that the use of these materials results in the service life of the CPP of more than 50 000 h.

8.5. Application of methods for calculating the adhesion and tribological properties to select the optimum friction pairs

Combined friction units are used widely in microcryogenic technology. Thus, the product – a sublimator – contains a friction unit which is a cylinder–piston pair with guide rollers. This composite friction section contains various kinds of friction pairs (metal–metal and metal–wear-resistant coating, metal–self-lubricating material), working in dry friction conditions. To improve the efficiency of the friction section, it is necessary to use antifriction and wear-resistant coatings for guide grooves along which the balls move, and an antifriction coating for the balls. High wear resistance with good antifriction properties are shown by coatings of nitrides, carbides and carbonitrides of transition metals. The example of this combined section will be used to demonstrate the application of the methods for calculating the adhesion and tribological characteristics for selection of the optimum materials for friction pairs.

When selecting a material for an antifriction coating it is necessary to solve the contradicting problem: the coating should have maximum adhesion to the substrate and minimum adhesion to the material working in pair with the coating. This problem can be solved by the following methods: spraying of multilayer coatings which provides a gradual transition from the coating with good adhesion properties with respect to the substrate to the material with good adhesion properties with respect to the antifriction coating; modifying the substrate surface by ion implantation of metal atoms, related to the antifriction coating; use of self-lubricating materials, playing the role of solid lubricant.

When choosing a method for creating antifriction coatings and related materials, we will consider the above-mentioned theoretical methods for determining the energy and strength of adhesion interactions of different materials, the adhesion component of the dry friction coefficient for the metal–metal and metal–solid lubricant, as well as the results of calculations of the adhesion and tribological characteristics of the contact of steel surfaces with some wear-resistant coatings and solid lubricant materials widely used in microcryogenic technology (nitrides and carbides of transition metals, fluoroplastic, molybdenum disulphide). In previous sections we have shown the results of calculation of the adhesion component of the friction coefficient for metal friction pairs based on 30Cr14 steel and nitrides and carbides of transition metals, as well as friction pairs: 30Cr14-fluoroplastic, 30Cr14–MoS_2, and for comparative analysis of the results for different pairs, depending on the speed of the relative displacement of the surfaces and on the distance between the contacting surfaces.

Comparison of the results of calculations for different friction pairs shows that the lowest friction coefficients on steel have Cr_3C_2 and Mo_2N. Analysis of the dependence of the friction coefficient of these materials on the size of the gap h shows (Fig. 8.5) that with increasing h the friction coefficient of Mo_2N on steel decreases faster than the friction coefficient of Cr_3C_2 on steel. In addition, it is well known that Mo_2N has a high wear resistance caused by an abnormally high microhardness of the compound. This makes it possible to recommend the molybdenum nitride as the material possessing the best tribological characteristics with respect to steel.

The calculation of the friction coefficient between the steel surface and solid lubricant materials (fluoroplastic, MoS_2) showed that the best tribological characteristics are shown by the molybdenum disulphide.

Based on these results, the following recommendations can be suggested:
– the best antifriction and wear-resistant material in the friction pair with steel is the molybdenum nitride Mo_2N;
– to further reduce friction and reduce the wear of parts in the running-in period, it is recommended to use a solid lubricant based on molybdenum disulphide MoS_2 which at high tribological properties is chemically inactive with respect to molybdenum nitride.

However, the molybdenum nitride, possessing excellent tribological characteristics in the friction pair with steel, as shown by calculations of its adhesion properties, can not be used directly as a surface coating of guide grooves made of Kh12M steel without prior modification. To improve the adhesion of molybdenum nitride to Kh12M steel, it is necessary to select the material for the intermediate layer, which would have good adhesion properties with respect to both steel and molybdenum nitride.

In [63, 64, 67] the authors calculated the adhesion characteristics of the contact of various materials used in microcryogenic engineering for surface modification of steels, in order to distinguish them from those which would have higher values of the energy and strength of adhesion interaction with steel than the molybdenum nitride and also higher values of the same characteristics with the molybdenum nitride than steel.

Consideration was given to a set of transition metals (Mo, Ni, Cr, Ti, Zr) and nitrides and carbides of transition metals (Cr_3C_2, TiN, TiC, Cr_2N, ZrC, ZrN, NbC, NbN), and the energy and strength of adhesion of these transition metals and compounds with 30Cr14 steel and molybdenum nitride Mo_2N were calculated. The results of calculation of the adhesion force at $h = 0$ are shown in Table 7.6 which shows that the highest values of the strength of adhesion to both steel and molybdenum nitride are obtained for the transition metals in the sequence Mo, Ni, Cr, Zr, Ti, and then for the compounds (Cr_3C_2, NbN, TiN, Cr_2N, NbC, ZrN, Mo_2N). At the same time the adhesion strength of these materials to the steel exceeds the adhesion force of the same materials with molybdenum nitride four times. This sequence of the materials based on adhesion strength is retained at h not equal to zero. This is due to a monotonic decrease of adhesion strength with respect to modulus (adhesion force is attractive and negative in sign) with the increase of the gap h between the surfaces of contacting materials.

Because the value of the strength of adhesion of molybdenum nitride to steel at $h = 0$ is equal to $F_a(0) = -8.69 \cdot 10^9$ N/m², the strength of adhesion of the selected material with the molybdenum nitride should not be less than the above values. In the group of the materials for which calculations were carried out this criterion is only satisfied by the transition metals. The best from this point of view is molybdenum.

Based on the analysis of the adhesion characteristics, molybdenum can be recommended as the material for the intermediate layer between the surface of parts made of steel and antifriction and wear-resistant coating of molybdenum nitride. The adhesion bond of molybdenum with the surface of steel can be improved by modification of the substrate by molybdenum ion implantation.

These recommendations for increasing the adhesion strength of Mo_2N coatings were implemented by ion implantation of the surface of a steel sample with Mo ions with doses of ~10^{16} cm⁻² followed by deposition of the Mo_2N coating. Measurement of the adhesion strength of the film by an indenter gave the value of the adhesion strength of coatings of $\sigma_a = 263 \pm 28$ MPa, while the adhesion strength of coatings on the non-modified surface of the steel sample was characterized by $\sigma_a = 120.4 \pm 9.6$ MPa. Thus, modification of the surface by the Mo ions led to more than doubling of the adhesion strength of the coating.

To test the predictions of the theory, experts at the 'Sibkriotekhnika' company [64, 65] conducted in the dry friction conditions tribological tests to determine the friction coefficient of the pairs based on 95Kh18 steel and materials with coatings of TiN, Cr_2N, Mo_2N and multilayer coatings $Mo–Mo_2N–MoS_2$. Single-layer coatings of nitrides of titanium, chromium and molybdenum were produced by vacuum condensation with ion bombardment (CIB) at NNV-6.6I1 installation, $Mo–Mo_2N–MoS_2$ coatings were produced in a specially constructed apparatus for the deposition of multilayer composite ion–vacuum coatings. This setup allowed to combine in a single vacuum cycle processes of coating by CIB, ion implantation and magnetron sputtering. The resulting coatings are characterized by high adhesion to the workpiece, the uniformity in thickness across the treated surface, high-wear resistance, good running-in and improved antifriction properties.

Tribological studies were conducted on UMT friction installation working by the end technique. A friction pair was composed of a moving sample with the tested coating and a rotating indenter. The sample was a disk with a diameter of 40 mm and 5 mm thick. The indenter was made in the form of a ring with an outer diameter of 25.5 mm, inner diameter of 22 mm and a height of 6 mm. The sample and the indenter were made of steel 95Kh18, heat-treated to HRC $\geq$ 58. The working surfaces of the sample and the indenter were ground and finished to $R_a \leq 0.1$ μm by abrasive paste on the basis of titanium carbide, followed by rinsing in an ultrasonic bath with gasoline and rubbing with alcohol. After depositing coatings the working surfaces were polished to $R_a \leq 0.1$ µm for removal of the drop phase.

Tests were conducted at sliding velocities $v = 0.6$ m/s, 1 m/s and 1.75 m/s in air. In the test, the dependence of the friction coefficient on the applied load created by a loading lever in the range of 0.01 to 0.5 MPa in increments of 0.01 MPa was studied.

The test cycle consisted of the start period (runnning-in) and proper test in a stable mode and the microseizure mode up to seizure (wear) of the coating. The quality of running-in was assessed by two parameters: the appearance of wear tracks on at least 90% of the sample area, and the stabilization of the friction coefficient. The friction surface conditions was estimated visually. Depending on the sliding velocity and the applied load the running-in period was $15 \div 30$ min.

The friction coefficient was recorded every $3 \div 5$ min in the transient mode, and every $10 \div 15$ minutes in the steady mode. The final result was the arithmetic mean of all measurements of the friction coefficient. The confidence interval of determination of the friction coefficient did not exceed 0.05. The test results are shown in Table 8.1 [67].

Analysis of test results shows that in the stable mode under identical test conditions (sliding speed and load), the lowest coefficients of friction against steel 95Kh18 is shown by the combined multilayer Mo–Mo$_2$N–MoS$_2$ coatings and coatings of molybdenum nitride and then by the coatings of chromium and titanium nitrides. These findings are in full compliance with predictions based on theoretical calculations of the friction coefficient of these coatings in a pair with steel. The high tribological properties of multilayer coatings and coatings of molybdenum nitride are also manifested in the fact that they can withstand higher loads before microseizure and bonding.

The calculated friction coefficients of wear-resistant coatings will be compared with the measured values. Thus, for the sliding velocity $v = 0.6$ m/s calculation of the electrodynamic component of the friction coefficient using formula (8.16) gives the following values for the these coatings at the gap size of $h = 0$: $f_{el}(\text{TiN}) = 0.22$, $f_{el}(\text{Cr}_2\text{N}) = 0.15$, $f_{el}(\text{Mo N}) = 0.12$. These values are in good agreement with the measured values of the friction coefficients at low loads on the friction pair, if we take into account the experimental error in measuring the friction coefficient and the approximate theoretical model of the adhesion component of the friction force (the approximation of perfectly flat friction surfaces).

The efficiency of this model is particularly evident in Fig. 8.13, where the calculated dependence of the adhesion component of the friction coefficient for the Mo$_2$N–95Kh18 pair on the sliding velocity is compared with the results of speed tests for the given pair under a load of 0.01 MPa.

With increasing load tests revealed an increase in the coefficient of dry friction (see Figure 8.14). This is due to the fact that with increasing load

Table 8.1.

Coating	Pressure, MPa	Speed m/s	Test time in modes. min		The friction coefficient in the modes of		
			stable	micro-bonding	stable	micro-bonding	seizure
TiN	0.01	0.6	60	–	0.29	–	–
	0.03		60	–	0.36	–	–
	0.05		25	60	0.58	0.88	>1
Cr$_2$N	0.01	0.6	60	–	0.22	–	–
	0.03		–	–	0.30	–	–
	0.05		–	–	0.46	–	–
	0.1		60	–	0.54	–	–
	0.15		35	65	0.48	0.68	~ −1
Mo$_2$N	0.01	0.6	60	–	0.17	–	–
	0.03		–	–	0.24	–	–
	0.05		–	–	0.32	–	–
	0.1		–	–	0.37	–	–
	0.15		60	–	0.42	–	–
	0.2		40	110	0.35	0.48	0.7
Mo$_2$N	0.01	0.6	60	–	0.17	–	–
	0.01	1.0	–	–	0.12	–	–
	0.01	1.7	–	–	0.08	–	–
	0.1	0.6	–	–	0.37	–	–
	0.1	1.0	60	–	0.42	–	–
	0.1	1.7	45	80	0.48	0.54	0.72
Mo–Mo$_2$N–MoS$_2$	0.01	0.6	60	–	0.12	–	–
	0.03		–	–	0.18	–	–
	0.05		–	–	0.23	–	–
	0.1		–	–	0.27	–	–
	0.15		–	–	0.31	–	–
	0.2		60	–	0.35	–	–
	0.25		48	140	0.32	0.42	0.68

micro-roughness smoothed, and the atoms of the surfaces of friction couples closer together. This in turn leads to increase the adhesion component of friction. In the calculation formula (8.16) for the friction coefficient of its dependence on the size of the gap h reflects the friction between surfaces in implicit form the dependence on the load, since an increase in load P is decrease in the gap.

However, during the tribological tests it was revealed that at sufficiently high loads, when the stable friction regime is interrupted by the regime of microseizure and subsequent bonding, the friction coefficient over a time interval corresponding to stable regime is characterized by a smaller

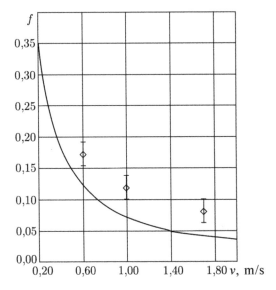

Fig. 8.13. The calculated dependence of the adhesion component of the friction coefficient f on the velocity v and the results of measurements of the friction coefficient for the 95Kh18–Mo$_2$N pair under a load of 0.01 MPa [64].

friction coefficient for a given pair than at lower loads. Perhaps this is due to the fact that at the given load the temperature of the contact spots is so high plastic flow of the material takes place and the hydrodynamic stage of friction comes from a lower friction coefficient starts which then gives way to the microseizure regime with increased adhesion of materials due to the processes of diffusion and sintering.

Tests of the Mo$_2$N–95Kh18 pair also revealed that with increasing load the dependence of the friction coefficient on the sliding speed can be reversed in comparison with the case of low load, i.e. increase with increasing speed. This is due, apparently, to the fact that the temperature of the friction surfaces rises with increasing speed and this increases the pliability of microroughness to plastic deformation and leads to an increase of the actual contact area, and consequently an increase in friction force.

Along with the electrodynamic component of friction, as noted above, it is possible to isolate another – dislocation component of friction coefficient f_d. It is due to the fact that the contact of metals with different lattice constants along the interface is accompanied by the formation of networks of edge misfit dislocations, which lower the energy of adhesion of the contacting materials, causing at the same time a reduction of the shear stress at the interface. The relative displacement of the metals results in the formation of the friction force component which prevents changes in the dislocation structure. The estimated contribution of this component to the total friction coefficient is determined by the ratio of the energy of the

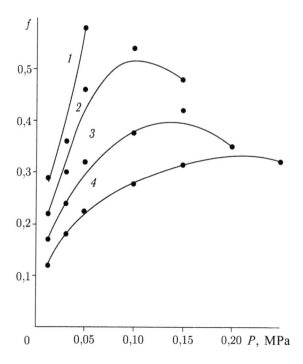

Fig. 8.14. Dependence of the dry friction coefficient f on steel 95Kh18 on load P at the sliding speed $v = 0.6$ m/s for counterbodies coated with: 1 – TiN; 2 – Cr$_2$N; 3 – Mo$_2$N; 4 – multilayer Mo–Mo$_2$N–MoS$_2$ coating [64].

network of the dislocations per surface unit of the boundary E_d per unit boundary surface to the surface adhesion energy (8.17). It is in a highly loaded test mode, especially in the microseizure mode, when the friction surfaces are characterized by the large relative friction area, the dislocation component of the friction coefficient manifests itself to a greater extent.

To calculate the energy of the dislocations in accordance with the formulas of section 8.3 we need the data on the elastic constants of contacting materials. Among the nitrides of transition metals such background data exist only for titanium nitride. Evaluation of the energy of dislocations for TiN–95Kh18 pair, where the elastic constants of iron were used for the steel, gave a value of $E_d = 1338$ MJ/m^2. At the adhesion energy of the pair equal to $E_a = 2188$ MJ/m^2, the dislocation component of the friction coefficient is $f_d = 0.61$. In view of the electrodynamic component of the friction coefficient for the given pair with $v = 0.6$ m/s, equal to $f_{el} = 0.22$, the maximum value of the friction coefficient of the pair is equal to $f = 0.83$. The calculated value of the friction coefficient, surprisingly, is in very good agreement with the measured friction coefficient for the TiN–95Kh18 in the microseizure mode.

The adequacy of the model describing the adhesion and tribological properties of different dry friction pairs is also apparent in the fact that the selection of materials of multilayer coatings based on this model proved to be optimal for achieving high tribological and wear-resistant properties of coatings under dry friction conditions. It is therefore possible to create an engineering methodology for calculating the complex adhesion and tribological properties of alloys and compounds widely used in cryogenics. The application of this methodology and associated software allows already at the stage of technological study of combined friction sections to choose materials and coatings characterized by the optimal adhesion and tribological properties, which provide high durability and performance of friction parts.

Thus, the development of theoretical methods for the determination of the adhesion and tribological properties of different materials and the practice of modifying the surface of various machine parts for microcryogenic technology have shown that the development of technological methods combining several operations of the ion-plasma treatment of surfaces, with the optimal selection of materials based on the developed theoretical methods makes it possible to produce multilayer composite coatings with improved tribological properties thus making it possible to increase the service life of friction units of microcryogenic technology and machinery in general.

9

THEORETICAL MODELS AND METHODS OF DESCRIPTION OF ADSORPTION OF METAL ATOMS ON METALLIC SURFACES

Introduction. Understanding the adsorption process

Investigation of the surface properties of materials and energy characteristics of adsorption of atoms and molecules of different substances is an urgent problem both in terms of fundamental representations of changes in the surface properties of crystals [453] and applied aspects in terms of modifying the properties materials by spraying coatings with the necessary properties [364, 437, 457]. In particular, adsorbates of the alkali metals are used for electrodes with a low effective work function and implementation of heterogeneous catalysis with improved activity and selectivity of catalysts [364, 437]. Thin films of cobalt and iron are widely used in magnetoelectronics [432, 457].

Adsorption of alkali metals has long occupied a central place among the objects of theoretical research to describe adsorption on the surface of crystals [327, 379, 384, 453]. This is primarily due to the simple electronic structure of atoms of alkali metals, as well as the fact that these adsorbates at low temperatures usually do not cause a significant relaxation of the substrate surface and are uniformly distributed in the form of a monoatomic layer with the implementation of the so-called non-activated adsorption [21, 379].

A simple explanation of the interaction between the atoms of the alkali metal and the surface of the metal substrate was proposed I. Langmuir [384]. He suggested that the alkali metal atom completely gives up its valence electron to the substrate to form an ionic bond. In 1935, when a more accurate theoretical description of adsorption of alkali metals was available, Gurney suggested a quantum-mechanical model, applicable to thin coatings [327]. In his model, the adatom of the alkali metal is partially positively charged and induces the density of the negative charge in the substrate, which leads to an increase in the surface dipole moment. With this description it is expected that an increase in the degree of coverage of the coating will increase electrostatic repulsion between the adatoms. To reduce it, some of the valence electrons will move back from Fermi level to the adsorbate. Thus, the surface dipole moment decreases. The effect of depolarization also explains the decrease in the electron work function from the surface [379]. This theoretical description of adsorption of alkali metals on metal surfaces is the analytical basis of many experimental studies (see, e.g. [437]).

In the last decade of XX century as a result of the development of theoretical methods for describing the adsorption and experimental studies it became clear [293, 453] that all the interactions and reactions of metal atoms on metal surfaces are more complex than previously assumed. The calculations, carried out taking into account the full atomic structure, the effects of relaxation of the atoms from their ideal positions, as well as the processes of surface reconstruction [453] show that the traditional view of the adsorption of metals, which is based on the depolarization effect, represents only part of a complete picture of the phenomenon, and the recent experimental data must be taken into account [293]. The following phenomena may occur on the surface:

– surface phase transitions with the formation of 'islands' of adatoms (island adsorption) [257, 274, 416, 454, 474, 477];

– replacement of the surface atoms of the substrate by the adatoms with ejection of the adatoms to the surface (substitutional adsorption) [415, 454, 474, 477];

– mixing the adatoms with surface atoms of the substrate with the formation of surface binary solid solutions (activated adsorption) [273, 457].

In this regard, the role of the theoretical approach to the determination of the structural and energy characteristics of adsorption on metals becomes more important. The theory of adsorption has reached a level where it became possible to calculate the total and free energy as well as adequately describe the electronic and atomic structures of simple adsorption systems with a given accuracy. These calculations, based on quantum-mechanical approach [453], commonly use complex methods and require considerable computing resources. Undoubtedly, the methodological development of this approach to the description of the adsorption of atoms and molecules in the last decade of XX century was impressive, and the further modernization of such methods as well as acceleration of computing processes based on them are still needed. However, at present the theory of adsorption still requires an adequate understanding and explanation of the formation of the adsorption structures and their properties. For example, it is necessary to understand why some adsorbates are able to replace the near-surface atoms of the substrate rather than simply adsorb on the surface, and why the geometry of adsorption may vary with the change of the parameter of the coating Θ [453].

In this chapter, the adsorption of metals on metal surfaces is considered as a special case of a more general phenomenon – the adhesion of metallic films. The density functional method was used for a comprehensive study of the adsorption of metal atoms on the substrates of simple transition and noble metals with different surface orientations of the faces taking into account the gradient amendments to the inhomogeneity of the electronic system and displacements of ionic planes in the interfacial region between the media. The method of self-consistent calculation of the energy of adsorption of metal atoms and electron work function of metal substrates is investigated, taking into account gradient corrections to the heterogeneity of the kinetic and exchange–correlation energy of both the second (alkali metals) and fourth order (transition and noble metals). The model of adsorption of metal atoms, which allows an adequate description of large variety of adsorption structures formed during activated adsorption on the substrates of the simple transitional and noble metal surfaces with different orientations is presented. The dependence of the adsorption energy and the electron work function on the composition of the surface of binary solid solutions is determined.

The material of the chapter is based on the results obtained by the authors of this book in [146, 149–158].

9.1. A multiparameter model of non-activated adsorption of atoms of alkali metals on metal surfaces

For more than a hundred years the adsorption of alkali metals on metal cylindrical surface has been the subject of experimental and theoretical studies [21, 43, 44, 58, 73–75, 106, 116, 125, 139, 161–164, 175–178, 234, 326, 379, 476]. Interest in these systems is due mainly to the technological value of the adsorbates of alkali metals for producing effective electrodes with a low work function and implementation of heterogeneous catalysis with improved activity and selectivity of catalysts [267, 364, 437]. But furthermore, these systems are good models for studying the adsorption of metals and plating of surfaces, accompanied by a record reduction of the emission of electrons from metal surfaces [453, 476].

As a rule, the adsorbed atoms are arranged uniformly on the surface of the adsorbent in the form of a monoatomic layer. However, they may have no mobility and in the process of adsorption are localized at certain points on the surface of the crystal. However, in recent years the study of the distribution of alkali metal atoms on the surface layer of the substrate revealed the effects of 'migration' of adatoms on the surface, depending on the degree of coverage [453, 476] and surface phase transitions with the formation of 'islands' of atoms of an alkali metal [296, 415, 416, 419].

At low temperatures and low coverages Θ the adatoms of the alkali metal are homogeneously distributed on the surface of the adsorbent in the form of a monoatomic layer. As the degree of coverage increases the equilibrium distance of the layer from the substrate gradually increases, i.e., the length of the substrate–adsorption bond increases which leads to its weakening. Simultaneously, the distance between the alkali metal atoms in the surface layer decreases, resulting in an increase of the intensity of interaction between the adatoms. This mutual competition between the adsorbate–adsorption and substrate–adsorption may lead to the destruction of the adsorbed film and its breakdown to separate islands [416, 476].

This section presents the methodology and results of calculation of the interfacial energy, adsorption energy and the electron work function of the metal surfaces using the density functional method. In contrast to other theoretical works [419, 453, 454, 476, 477], for description here of the strong heterogeneity of the electronic system in the interfacial region of interaction of atoms of the substrate and the alkali metals we have moved beyond the framework of the local density approximation and take into account the correction for the heterogeneity of the electron density for both kinetic and exchange–correlation energies. To calculate the the influence of the electron–ion interaction on the energy characteristics we used the Ashcroft pseudopotential. We have developed a multiparameter version of the functional of the electron density with test functions which take into account, along with the heterogeneity of the electron density distribution, the existence of the equilibrium distance between the adatoms and the substrate and the equilibrium effective thickness of a monolayer of the adsorbate, determined from the condition of the minimum interfacial energy of the interaction.

9.1.1. The methodology and results of calculation of the energy characteristics of adsorption

The model of a homogeneous background
To study the adsorption characteristics we use a model which is a semi-infinite metal with a monatomic film of the adsorbate. Consider a plane boundary of the metal, the outer normal to which is the z axis, so that the region of space occupied by the metal corresponds to $z < -D$. The average electron charge density of the metal substrate is denoted n_1. The adsorbate film with the charge density n_2 occupies an area of $D < z < D + h$, where h is the thickness of the film. Between the adsorbent and adsorbate in this model there is a vacuum gap of width $2D$ (Fig. 9.1). Thus, the electron density $n(\mathbf{r})$ and electrostatic potential $\phi(\mathbf{r})$ can be regarded as functions only of coordinate z and we shall use them in the future.

We represent the total energy functional in the form of a gradient expansion

$$E[n(z)] = \int_{-\infty}^{\infty} \left\{ w_0[n(z)] + w_2[n(z), |\nabla n(z)|^2] + \right.$$

$$\left. + w_4[n(z), |\nabla n|^2, n(|\nabla|^4] + O(|\nabla|^6)) \right\} dz, \quad (9.1)$$

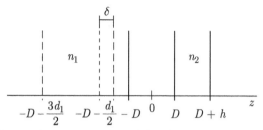

Fig. 9.1. Geometric representation of the distribution of surface layers.

where $w_0[n(z)]$ is the energy density of a homogeneous electron gas (5.2)–(5.5), $w_2[n(z),|\nabla n|^2]$ is the first term of the gradient expansion, taking into account the inhomogeneity of the electron gas for both the kinetic (5.6) as well as for the exchange and correlation energies (5.11) (in the Vashishta-Singwi approximation) and $w_4[n(z),|\nabla n|^2,|\nabla|^4]$ is the gradient correction of the fourth order of the kinetic energy density (6.13) and exchange–correlation energy density (6.14).

Consideration of the amendments in the case of contact of the transition metals with alkali metals was carried out according to the method described in section 6.2.

As test functions for the potential $\phi(z)$ and electron density $n(z)$ we have obtained solutions of the linearized Thomas–Fermi equation

$$\frac{d^2}{dz^2}\phi(z) = 4\pi\left[n_+(z) - n_1\left(1 - \frac{\phi(z) - \phi(-\infty)}{\alpha n_1^{2/3}}\right)^{3/2}\right] \approx$$

$$\approx 4\pi n_+(z) - 4\pi n_1 + \beta^2[\phi(z) - \phi(-\infty)], \qquad (9.2)$$

where

$$\alpha = 0.5(3\pi^2)^{2/3}, \beta = (6\pi\bar{n}^{1/3}/\alpha)^{1/2};$$

$$n(\beta,z) = \begin{cases} n_1\left(1 - 0.5e^{\beta(z+D)}\right) + \\ \qquad +0.5n_2e^{\beta(z-D)}\left(1 - e^{-\beta h}\right), & z < -D; \\ 0.5n_1e^{-\beta(z+D)} + \\ \qquad +0.5n_2e^{\beta(z-D)}\left(1 - e^{-\beta h}\right), & |z| < D; \\ n_2\left(1 - 0.5e^{\beta(z-D-h)}\right) - \\ \qquad -0.5\left(n_2 - n_1e^{-2\beta D}\right)e^{-\beta(z-D)}, & D < z < D+h; \\ 0.5e^{-\beta(z-D-h)}\left(n_1e^{-\beta(2D+h)} - \right. \\ \qquad \left. -n_2e^{-\beta h} + n_2\right), & z > D+h; \end{cases}$$

$$(9.3)$$

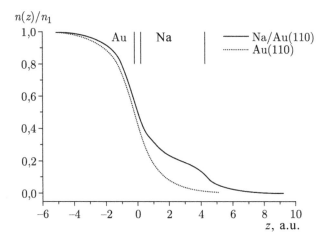

Fig. 9.2. The distribution of electron density $n(z)$ in the Na/Au (110) system at the coverage parameter $\Theta = 0.6$ (solid line) and the distribution of the electron density $n(z)$ on the pure gold surface (dashed line).

$$\phi(\beta, z) = -\frac{4\pi n(\beta, z)}{\beta^2}. \tag{9.4}$$

Here β is the variational parameter.

Figure 9.2 shows as an example a graph of the function of the distribution of electron density $n(\beta, z)$, calculated by the formula (9.3), for the Na/Au (110) system.

The value of interfacial energy in the model of the homogeneous background is the difference between the total energy of the system when the electrons are distributed in accordance with the function $n(z)$ (9.3), and the positive background:

$$n_+(z) = \begin{cases} n_1, & z < -D; \\ n_2, & D < z < D+h; \\ 0, & |z| < D, z > D+h. \end{cases} \tag{9.5}$$

Thus, the surface energy can be written as

$$\sigma_0(\beta) = \int_{-\infty}^{+\infty} \{w[n(\beta, z)] - w[n_+(z)]\} dz. \tag{9.6}$$

Next, the procedure for carrying out the minimization

$$\left(\frac{\partial \sigma_0(\beta)}{\partial \beta}\right)_{\beta=\beta_{min}} = 0, \tag{9.7}$$

is used to determine the value of β_{min}, and from (9.6) – the value of the interfacial energy $\sigma_0(\beta_{min})$.

Accounting for the discreteness of the crystal lattice. The Ashcroft model.
Amendments to the interfacial energy associated with the account of
discreteness in the distribution of ions over the lattice sites of the metals
were calculated using the Ashcroft pseudopotential model (5.17), averaged
over the crystal planes parallel to the metal surface. The σ_{ei} amendment
will then be calculated in following the following way:

$$\sigma_{ei} = \int_{-\infty}^{+\infty} \delta V(z)\left[n(\beta,z) - n_{+}(z)\right]dz, \tag{9.8}$$

where $\delta V(z)$ is the sum averaged over the crystal planes of the ionic
pseudopotentials less the potential of the homogeneous background of
the positive charge. For the quantity $\delta V(z)$ we obtained the following
expressions:

$$\delta V(z) = 2\pi n_1 \left\{ d_1(r_{c1} - |z + D + d_1/2|)\theta(r_{c1} - |z + D + d_1/2|) - \left[z + D + d_1\theta(-z - D - d_1/2)\right]^2 \right\} \tag{9.9}$$

for $z < -D$ (in the first near-surface ionic plane: $-D - d_1 < z < -D$)

$$\delta V(z) = 2\pi n_2 \left\{ h(r_{c2} - |z - D - h/2|)\theta(r_{c2} - |z - D - h/2|) - \left[z - D - h + h\theta(-z + D + h/2)\right]^2 \right\} \tag{9.10}$$

when $D < z < D + h$ (in the adsorbed monatomic film).

As a result, using the distribution (9.3) for the values of $n(\beta, z)$, and
in the summation over the ionic planes with $z = -D - (2k + 1)d_1/2$, $k =$
0, 1, 2, ..., the periodicity of the potential $\delta V(z - d_1) = \delta V(z)$, we obtain
from (9.8) the following expression for the electron–ion component of the
interfacial energy:

$$\sigma_{ei} = \frac{2\pi}{\beta^3}\left\{\left[n_1^2 - n_1 n_2 e^{-2\beta D}\left(1 - e^{-\beta h}\right)\right]\left(1 - \frac{\beta d_1 e^{-\beta d_1/2}}{1 - e^{-\beta d_1}}\,\mathrm{ch}(\beta r_1)\right) + \right.$$
$$\left. + \left(2n_2^2 - n_1 n_2 e^{-2\beta D}\right)\left(1 - e^{-\beta h}\right)\left(1 - \frac{\beta h e^{-\beta h/2}}{1 - e^{-\beta h}}\,\mathrm{ch}(\beta r_2)\right)\right\}. \tag{9.11}$$

Using the method presented in Chapter 6 the following expression was
obtained for the interfacial energy of the ion–ion interaction:

$$\sigma_{ii} = \sqrt{3}\,\frac{Z_1^2}{c_1^3}\exp\left(-\frac{4\pi d_1}{\sqrt{3}c_1}\right) +$$
$$2\sqrt{3}\,\frac{Z_2^2}{c_2^3}\exp\left(-\frac{4\pi h}{\sqrt{3}c_2}\right)\left[1 - \exp\left(-\frac{4\pi h}{\sqrt{3}c_2}\right)\right] -$$

$$-2\sqrt{3}\frac{Z_1 Z_2}{(c_1 c_2)^{3/2}}\exp\left(-\frac{2\pi}{\sqrt{3}}(\frac{d_1+D}{c_1}+\frac{h+D}{c_2})\right)\left[1-\exp\left(-\frac{4\pi h}{\sqrt{3}c_2}\right)\right], \quad (9.12)$$

where Z is the charge of the ions, c is the distance between the nearest neighbours in planes parallel to the surface (subscript 1 refers to the substrate, the index 2 – to the adsorbate), d_1 is the interplanar distance in the lattice of the metal substrate, h is thickness of the adsorbed film.

To understand the range of surface phenomena and processes of interaction in inhomogeneous electron systems requires information on the structural distortion of the surface layer of metals. To account for the effects of relaxation of the substrate surface in the area of its contact with the adsorbate, consider what changes will be caused by taking into account the displacement of one ionic surface plane of the substrate by the value δ ($\delta > 0$ corresponds to the expansion) with respect to the position of this plane in the undisturbed material, characterized by the coordinate $z = -D - d_1/2$. We assume that in this case the trial function for the electron density $n(\beta, z)$ in (9.3) continues to be characterized only by the parameter β and does not acquire an explicit dependence on the relaxation parameter δ. In this case, the component of the interfacial energy $\sigma_0(\beta)$ (9.6), reflecting the contribution from the electronic system within the model of the homogeneous background, also does not depend explicitly on δ. As for the contribution σ_{ii}, arising from the electrostatic interaction of the ions with each other, we obtained the following expression:

$$\sigma_{ii} = \sqrt{3}\frac{Z_1^2}{c_1^3}\exp\left(-\frac{4\pi(d_1-2\delta)}{\sqrt{3}c_1}\right)+2\sqrt{3}\frac{Z_2^2}{c_2^3}\exp\left(-\frac{4\pi h}{\sqrt{3}c_2}\right)\times$$

$$\times\left(1-\exp\left(-\frac{4\pi h}{\sqrt{3}c_2}\right)\right)-$$

$$-2\sqrt{3}\frac{Z_1 Z_2}{(c_1 c_2)^{3/2}}\exp\left(-\frac{2\pi}{\sqrt{3}}\left(\frac{d_1-2\delta+D}{c_1}+\frac{h+D}{c_2}\right)\right)\times$$

$$\times\left(1-\exp\left(-\frac{4\pi h}{\sqrt{3}c_2}\right)\right). \quad (9.13)$$

Component $\sigma_{ei}(\beta, \delta)$, which takes into account the interaction of electrons with the ions, can be represented as

$$\sigma_{ei}(\beta,\delta) = \sigma_{ei}(\beta,0)+\Delta\sigma_{ei}(\beta,\delta), \quad (9.14)$$

where

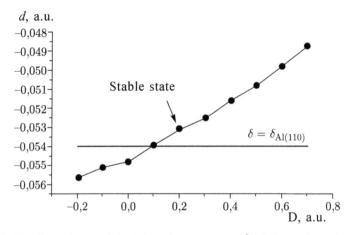

Fig. 9.3. The dependence of the relaxation parameter δ of the surface plane of the substrate on the half-width of the gap D for the Cs/Al(110) system at $\Theta = 0.25$.

$$\sigma_{ei} = \frac{2\pi}{\beta^3}\left\{\left[n_1^2 - n_1 n_2 e^{-2\beta D}\left(1-e^{-\beta h}\right)\right]\left[1-\frac{\beta d_1 e^{-\beta d_1/2}}{1-e^{-\beta d_1}}\mathrm{ch}(\beta r_{c1})\right]+\right.$$

$$\left.+\left(2n_2^2 - n_1 n_2 e^{-2\beta D}\right)\left(1-e^{-\beta h}\right)\left(1-\frac{\beta h e^{-\beta h/2}}{1-e^{-\beta h}}\mathrm{ch}(\beta r_{c2})\right)\right\},\qquad(9.15)$$

$$\Delta\sigma(\beta,\delta) = \frac{2\pi}{\beta^2}n_1^2 d_1\left(1-\frac{n_2}{n_1}e^{-2\beta D}\right)\left(1-e^{\beta\delta}\right)e^{-\beta d_1/2}\mathrm{ch}(\beta r_{c1})+$$

$$+2\pi n_1^2 d_1\delta^2,\qquad(9.16)$$

Here r_{c1}, r_{c2} are the Ashcroft pseudopotential parameters.

As a result, the total interfacial energy $\sigma(\beta, \delta)$ as a function of variational parameters is determined by the sum of the contributions given by the expressions (9.6), (9.13) and (9.14). The values of variational parameters β_{min} and δ_{min} are found from the condition of minimality.

As an example, Fig. 9.3 shows a plot of the dependence of the relaxation parameter δ on the half-width of the gap D for the Cs/Al(110) system at the coverage parameter $\Theta = 0.25$. The results of calculation of the parameter δ showed that its value for the metal + metal film bimetallic system in its steady state (for Cs/Al(110) equilibrium half-width of the gap is $D = 0.2$ a.u.) differs little from those for a clean metal surface ($\delta_{Al(110)} = 0.054$ a.u.). Consequently, the calculations give a slight difference in the values of the energy characteristics of the adsorption system with the relaxation of the substrate surface taken and not taken into account. As an example, Fig. 9.4 shows plots of the adsorption energy as a function of the half-width of the gap D for the above system with and without relaxation of the substrate surface. It should be noted that this pattern is also observed

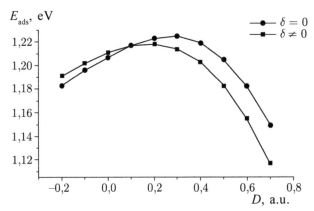

Fig. 9.4. Graphs of the adsorption energy as a function of the half-width of the gap D for the Cs/Al(110) system at $\Theta = 0.25$, calculated taking into account the relaxation surface of the substrate and without relaxation.

in the experimental data, according to which the actual relaxation of the substrate surface with activated adsorption is rather small [257, 274, 293, 415, 454]. A comparison of the adsorption energy as a function of the coverage parameter Θ calculated with and without relaxation of the substrate surface taken into account will be shown in Fig. 9.9a.

The electron density of the adsorbed film and recording it relaxation
The electron density of the adsorbed film n_2 is one of the most important parameters in this model. It should reflect not only the electronic properties of the adsorbate, but also its symmetry in the coating, as well as the symmetry of the surface edge substrate on which adsorption takes place.

In accordance with [379], the electron density of the film can be written as

$$n_2 = Z_2 n_{s2} / h, \tag{9.17}$$

where $Z_2 = 1$, since the alkali metals are monovalent, n_{s2} is the surface concentration of the adsorbate (the number of adsorbed atoms on the unit area).

The surface concentration n_{s2} is a function of the parameters c_1 and c_2, which characterize the symmetry of the surface of the substrate on which adsorption takes place, and the distribution of atoms in the adsorbed layer. Along with this the surface concentration n_{s2} is also a function of the parameter Θ characterizing the degree of filling of the substrate surface with the adatoms:

$$\Theta = n_{s2} / n_{s1}, \tag{9.18}$$

where n_{s1} characterizes the surface concentration of the substrate atoms.

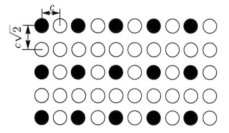

Fig. 9.5. Schematic representation of the structure of the coating of the (110) surface of the substrate with a face-centered cubic lattice by adatoms (black circles). The case with $\Theta = 0.25$ is shown.

The degree of filling Θ the substrate surface by the adatoms is given so that for $\Theta = 1$ the number of adatoms is equal to the number of atoms in the surface face of the substrate. In this model, it is assumed that the adatoms, lying on the substrate, repeat the symmetry of its surface face. In this case, the parameter Θ can be expressed in terms of the parameters c_1 and c_2, which characterize the distance between the nearest ions of contacting metals:

$$\Theta = c_1^2 / c_2^2. \qquad (9.19)$$

As an example, Fig. 9.5 shows the structure of the coating with $\Theta = 0.25$ on the (110) substrate with a face-centered cubic lattice.

We consider four variants of the model to describe the adsorption of alkali metal atoms on metal substrates.

1. The case of the absence of the gap between the substrate and the film ($D = 0$) and at a fixed film thickness ($h = d_2 = $ const), given by the bulk value of the interplanar distance d_2 between the most closely packed faces for the adsorbed metal. This approach is classical in the theory of adsorption. It is the basis for most of the early theoretical works on the adsorption of alkali metals, e.g. Lang [379]. However, in contrast to [379], where the consideration was carried out in the approximation of the model of the uniform background for the film and the substrate, we take into account the discreteness of the crystal structures of the substrate and the adsorbate.

2. The case of a fixed film thickness ($h = d_2 = $ const), but with the equilibrium gap $D = D_{min}$. The equilibrium gap width was found from the minimum total interfacial energy $\sigma(\beta_{min}, D_{min})$. This approach has significant advantages over the former since it allows to determine the equilibrium state of the system at the given filling of the surface with the adatoms.

In contrast to [379], where the thickness of the adsorbed film was considered fixed and equal to the interplanar distance d_2 between the most closely packed faces of the alkali metal crystal, and the gap between the substrate and the film was not introduced, in this chapter we take into account the relaxation of thickness h and also, according to (9.17), the relaxation of electron density n_2 of the adsorbed film. Thus, the thickness of the film can be written as

$$h = d_2 + \gamma, \qquad (9.20)$$

where γ is the variational parameter characterizing the relaxation of the film thickness. The parameter values γ_{min} can be determined from the minimum of the total interfacial energy. This allows us to identify the following cases.

3. The case of the absence of the gap between the substrate and the film ($D = 0$), but with the equilibrium film thickness $h = h_{min}$. The equilibrium film thickness is determined by the formula (9.20) for a known value of the variational parameter γ_{min}.

4. The case of the equilibrium gap $D = D_{min}$ and the equilibrium film thickness $h = h_{min}$. This approach, summing the above three cases, can be used to predict the formation of stable adlayers with a metallic bond or indicate their absence. Equilibrium h and D are found from the minimum total interfacial energy $\sigma(\beta_{min}, \gamma_{min}, D_{min})$.

Calculation of energy characteristics of adsorption systems

The energy characteristics of adsorption systems include the interfacial energy of the contact of the metal with the monatomic metal film, adhesion energy, the energy of adsorption and the work function of electrons from the substrate surface. The method of calculation of the electron work function will be described in section 9.1.2. Here, we consider methodology for calculating the energy of adsorption.

Knowing the interfacial energy of the metal contact with the film, as well as the surface energy of each component of a bimetallic system, it is easy to find the energy of adhesion of the system as a work that needs to be carried out to remove the substrate and the film from each other to infinity, i.e.

$$E_a(2D) = \sigma(\infty) - \sigma(2D). \qquad (9.21)$$

The energy of adhesion for this system is primarily an intermediate value used to calculate the energy of adsorption. The energy of adsorption can be measured experimentally [45, 48, 59, 97, 108, 117, 125, 131–133, 138, 139, 160, 161, 163, 175–177, 221, 227]. Its measure is the heat of evaporation of adsorbed atoms, which also corresponds to the work required to remove the adsorbed particle from the surface of the substrate under equilibrium conditions at 0 K. In this regard, the adsorption energy can be compared with the specific energy of adhesion, i.e. the energy of adhesion attributable to one adsorbed atom:

$$E_{ads}(\Theta) = E_a / n_{s2}, \qquad (9.22)$$

where the surface concentration of the adsorbate n_{s2} is a function of the

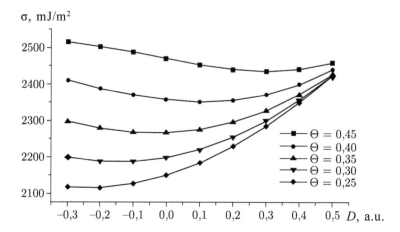

Fig. 9.6. Interfacial energy $\sigma(\beta_{min}, \gamma_{min}, D)$ as a function of the half-width of the gap D, calculated for the Na/W(111) system.

parameters c_1 and c_2, respectively, characterizing the symmetry of the substrate surface on which adsorption takes place, and the location of the atoms in the film. According to (9.18) the surface concentration n_{s2} is also directly a function of the coverage parameter Θ.

The distance of the adsorbed film from the substrate $2D$, corresponding to the equilibrium state of the system, was determined from the minimum of the total interfacial energy. Calculations show that with decreasing Θ the equilibrium distance $2D_{min}$ gradually decreases. As an example, Fig. 9.6 shows plots of the interfacial energy as a function of the half-width of the gap D, calculated in the case of model 4 ($\sigma(\beta_{min}, \gamma_{min}, D)$), for different values of Θ for the Na/W(111) system. The minimum in the interfacial energy appears when $\Theta = 0.45$ (equilibrium half-width of the gap $D_{min} = 0.3$) and with decreasing Θ is gradually shifted to negative values of D. The resulting dependence of the equilibrium distance as a function of Θ is in accordance with experimental studies of the adsorption of alkali metals [293]. Figure 9.6 also shows that for $\Theta < 0.35$ the equilibrium values of the half-width of the gap D_{min} become negative, which indicates the possibility of penetration of adatoms into the surface layer of the substrate in such coatings. A similar situation is observed for other systems.

Figure 9.7 shows as an example the graphs of interfacial energy σ as a function of the half-width of the gap D, calculated in the framework of the model case 4 ($\sigma(\beta_{min}, \gamma_{min}, D)$), for the Na/Ta(100) and K/Ta(100) systems with $\Theta = 0.45$. It can be seen that the interfacial energy of the Na/Ta(100) system has a minimum at $D = 0.1$, pointing thus to the steady-state of the adlayer of sodium atoms on the Ta(100) face at the given Θ. The interfacial energy of the K/Ta(100) system at $\Theta = 0.45$ has no minimum, therefore, the steady state of the films of potassium atoms on the Ta(100) face for this Θ is not realized.

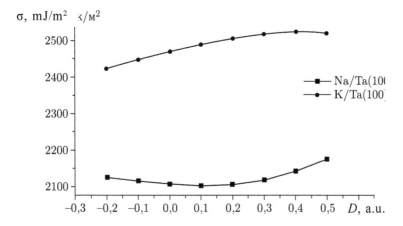

Fig. 9.7. Interfacial energy $\sigma(\beta_{min}, \gamma_{min}, D)$ as a function of the half gap D, calculated for the Na/Ta(100) and K/Ta(100) systems at $\Theta = 0.45$.

Figure 9.8 shows plots of the adsorption energy, calculated for each of the four cases, as a function of the degree of filling Θ by the sodium adatoms of the most friable (110) face of aluminum (*a*) and gold (*b*) substrates. From their analysis we can draw the following conclusions. The energy of adsorption monotonically decreases with increasing Θ. Although the energy of adsorption, taken in pairs in cases 4 and 2, 3 and 4, differs very little, in the cases 2 and 4 there are areas of changes of Θ in which a continuous monolayer coating of the alkali metal is unstable for positive values of the gap. Calculations have shown, for example, that the monolayer coating of Na atoms of the (110) face of the gold substrate in case 4 is stable only in the range $0.5 \le \Theta \le 0.6$, in the case 2 – in the range $0.5 \le \Theta \le 1.0$, and in the cases 1 and 3 is stable for all values of Θ.

For the considered wide range of systems (alkali metals Li, Na, K, Rb, Cs on the (111), (110), (100) face of metals Al, Cu, Au, Cr, Mo, W, Ta) studies of generalized conditions of the case 4 showed that monatomic films cannot form on the close-packed faces (there is no minimum interfacial energy). Monatomic films can form only on the loose faces of the metal substrates. At the same time, the monatomic films on substrates of transition metals (Cr, Mo, W, Ta) can form in a wider range of changes of Θ than for simple (Al) or noble (Cu, Au) metals. The calculations substantiate only the observed 'island' adsorption of alkali metals on the close-packed faces [133, 293, 454, 477].

Figure 9.9 shows as an example the graphs of the adsorption energy of alkali metals on the surfaces of Al(110) and W(111), calculated for the case 4, for the range of existence of the stable coating. The figure shows that with increasing radius of the alkali metal ion the range of existence of the stable coating is displaced to lower values of Θ. For example, for the Li/Al(110) system, it corresponds to $0.45 \le \Theta \le 0.55$, while for K/Al(110) to $0.23 \le \Theta \le 0.33$. Thus, the larger the radius of the alkali metal ion, the

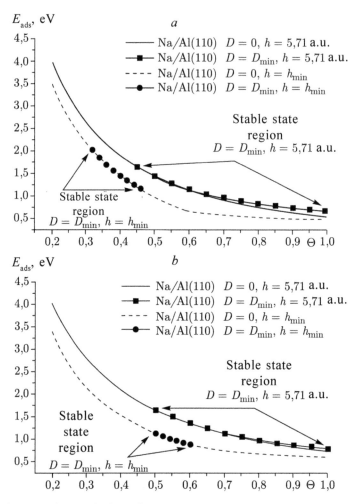

Fig. 9.8. Dependence of adsorption energy E_{ads} on the coverage parameters Θ of the surface calculated for four modelling cases for the systems: a – Na/Al(110), b – Na/Au(110).

smaller is the degree of coverage of the substrate Θ required to achieve the specified energy of adsorption. For example, the adsorption energy of 1.5 eV is achieved in the K/Al(110) system at $\Theta = 0.29$, and the Li/Al(110) system at $\Theta = 0.49$. The adsorption energy for a substrate made of aluminum was calculated without taking into account relaxation (solid line) and taking into account the relaxation of the surface ionic plane (dashed lines). Figure 9.9a shows that the if the surface relaxation of the substrate is taken into account the adsorption energy decreases only slightly.

Figure 9.10 shows plots of the adsorption energy calculated for the cases 1–3 as a function of the degree of coverage Θ by the sodium adatoms of the most close-packed (110) face of the tungsten substrate. To compare

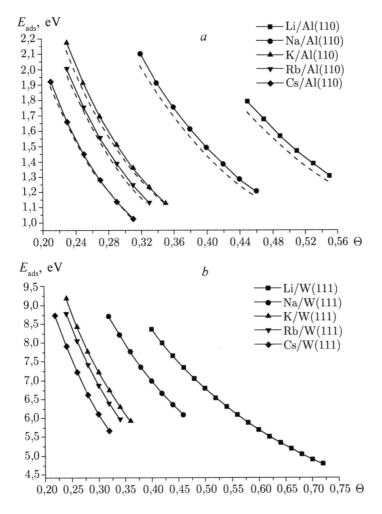

Fig. 9.9. Calculated (from the case 4) dependence of the adsorption energy of the alkali metal atoms on the degree of coverage Θ of the metal surfaces: a – Al(110), b – W(111) in the range of existence of a stable coating.

the calculated values of the adsorption energy with the experimentally measured values the figure also shows the heat of adsorption for this system (at Θ < 0.65), obtained by the authors of the experimental work [161]. The figure shows that the values of the adsorption energy, obtained in the case 3 (taking into account the relaxation of the film thickness), are in better agreement with experiment than in the cases 1 and 2 (without taking into account the relaxation of the film thickness). In this case the best fit calculated and experimental values obtained at high values of Θ, whereas at low Θ calculations in all cases give much higher values of adsorption energy. In our model, this discrepancy is explained by the fact that the adsorption energy cannot be determined by formula (9.22) at low

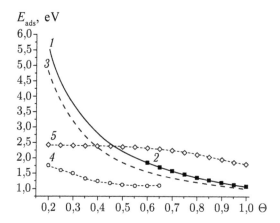

Fig. 9.10. The adsorption energy E_{ads} depending on the degree of coverage Θ by sodium atoms of the (100) close-packed surface face of the tungsten single crystal, calculated for three model cases: $1 - D = 0$, $h = 5.71$ a.u.; $2 - D = D_{min}$, $h = 5.71$ a.u., $3 - D = 0$, $h = h_{min}$ and obtained experimentally: $4 - $ [161], $5 - $ [177].

coverages $(\Theta \ll 1)$ and, consequently, low values of the surface concentration of adatoms n_{s2}.

9.1.2. The methodology and results of calculation of the electron work function of electrons from the metal surface modified by the adsorbate

Adsorption of alkali metals on metal surfaces is characterized by a very large change in the atomic geometry of the adsorbate and the electron work function from the surface depending on the degree of coverage Θ [415, 453, 476]. In earlier studies (see for example [17, 21, 327, 379]) on the adsorption of alkali metals, it was shown that the work function of electrons from the surface of the substrate decreases rapidly to a few electron volts with increasing Θ and, passing through a minimum $(0.1 \leq \Theta_{min} \leq 0.3)$, gradually approaches the value of the work function of the alkali metal. This phenomenon is explained by the effect of depolarization of the surface dipole barrier of the substrate by the alkali metal atoms [215]. However, since 1991 as a result of the development of theoretical methods for describing the adsorption and experimental studies, it became clear [293, 453, 476] that all interactions and reactions of alkali metal atoms on metal surfaces are more complex than previously supposed. Calculations of the surface characteristics should be carried out taking into account the full atomic structure of the surface relaxation effects atoms from their ideal positions, as well as processes of surface reconstruction. In particular, it was found [415, 453, 476] that the change in the work function as a function of coating by the alkali metal $\Delta\Phi(\Theta)$ may differ significantly from the shape of the curve $\Delta\Phi(\Theta)$ in the traditional model. For example, it can not demonstrate the minimum or at least have a minimum at a much higher degree of coverage. This section is devoted to explaining the latest data of experimental studies on the basis of the multi-parameter model of adsorption proposed by us [140, 151, 152].

Work function Φ in the 'jelly' model, as shown in § 5.2, is determined as follows:

$$\Phi = \Phi_0 - \mu, \qquad (9.23)$$

where $\mu = E_F$ is the chemical potential defined by the relation (5.48).

The expression for the dipole potential barrier in the model the homogeneous background Φ_0 (5.42) must be supplemented by the amendments to the electron–ion interaction, taking into account the discrete distribution of the charge of the ions in the crystal lattice. The effect of the electron–ion interaction on the work function is due to the difference in electrostatic interactions of ions with the electron density in the ground state and in a state with one electron removed. Following [381], an additional contribution to the potential barrier has the form

$$\Phi_{ei} = \frac{-\displaystyle\int_{-D-d_1}^{\infty} \delta V(z)\delta n_E(z)dz}{\displaystyle\int_{-D-d_1}^{\infty} \delta n_E(z)dz}. \qquad (9.24)$$

Since the expression (9.24) is uniform with respect n_E, then n_E can be regarded equal to the surface charge density of a system situated in a weak electric field with strength E_z, in a semi-infinite model of a homogeneous background. For $n_E(z)$, we obtained the following expression:

$$n_E(\beta,z) = \begin{cases}
\begin{aligned}
& n_1(1-0,5e^{\beta(z+D)}) + \\
& \quad +0,5n_2 e^{\beta(z-D)}\left[1-\left(1-\dfrac{\beta E_z}{4\pi n_2}\right)e^{-\beta h}\right], \quad |z| < D;
\end{aligned} \\[2em]
\begin{aligned}
& 0,5n_1 e^{-\beta(z+D)} + \\
& \quad +0,5n_2 e^{\beta(z-D)}\left[1-\left(1-\dfrac{\beta E_z}{4\pi n_2}\right)e^{-\beta h}\right], \quad |z| < D;
\end{aligned} \\[2em]
\begin{aligned}
& n_2 - \left[1-0,5\left(1+\dfrac{\beta E_z}{4\pi n_2}\right)e^{\beta(z-D-h)}\right] - \\
& \quad -0,5[n_2 - n_1 e^{-2\beta D}]e^{-\beta(z-D)}, \quad D < z < D+h;
\end{aligned} \\[2em]
\begin{aligned}
& 0,5e^{-\beta(z-D-h)}\left[n_1 e^{-\beta(2D+h)} - \right. \\
& \quad \left. -\left(n_2 e^{-\beta h} +1-\dfrac{\beta E_z}{4\pi n_2}\right)n_2\right], \quad z > D+h.
\end{aligned}
\end{cases} \qquad (9.25)$$

Finding $\delta n_E(z) = n(z) - n_E(z)$ and integrating (9.24) using the expressions (9.3) (9.9), (9.10) and (9.25), we obtain the electron–ion amendment to the potential barrier of the metal substrate

$$\Phi_{ei}^s = -\frac{4\pi n_1}{\beta^2}\frac{e^{-2\beta D}e^{-\beta d_1/2}\left[\beta d_1 \mathrm{ch}(\beta r_1) - 2\mathrm{sh}(\beta d_1/2)\right]}{2-e^{-\beta(2D+d_1)}} \qquad (9.26)$$

and for the adsorbate

$$\Phi_{ei}^a = -\frac{4\pi n_2}{\beta^2}\frac{e^{-\beta h/2}\left[\beta h \mathrm{ch}(\beta r_2) - 2\mathrm{sh}(\beta h/2)\right]}{2-e^{-\beta(2D+h+d_1)}}. \qquad (9.27)$$

Thus, the final expression for the electron work function from the substrate surface with the adsorbed monoatomic film of the alkali metal can be represented as

$$\Phi(\Theta) = \Phi_0(\Theta) - \mu + \Phi_{ei}^s(\Theta) + \Phi_{ei}^a(\Theta). \qquad (9.28)$$

Based on this methodology, we calculated the work function of the electrons from the surface of a wide range of simple, noble and transition metals adsorbed with the monatomic film of the alkali metals adsorbed on them. Figure 9.11 shows graphs of the electron work function $\Delta\Phi(\Theta) = \Phi(\Theta) - \Phi_m$ from the surfaces of Al(110) and W(111), due to the adsorption of the atoms of the alkali metal for each of the four cases in our adsorption model.

These graphs show that the cases 1 and 2 predict a very small effect of the monolayer of the alkali metal on the work function from the surface of the substrate. This is difficult to physically interpret and does not correspond to the experimental results. In the third case for the Na/Al(110) system, the results of calculations demonstrate a sharp increase in the work function from the surface of the substrate with $\Theta > 0.6$, which is caused by a sharp non-physical increase in the thickness of the film at the given values of Θ. This indicates the instability of the monolayer coating in this range of Θ, as also demonstrated by calculations in the framework of case 4. A similar behaviour $\Delta\Phi(\Theta)$ is predicted for the Cs/W(111) system, although in the case of 3 the increase in the work function with increasing Θ is smoother.

Figure 9.12 shows the graphs of the work function of electrons from the surfaces of Al(110) and W(111) due to the effects of absorption, calculated in the case 4, for the range of existence of the stable coating of the alkali metal. We see that in the region of existence of the stable monatomic film of the alkali metal the work function decreases with increasing values of Θ for all alkali metals. The adatoms of the potassium subgroup (K, Rb, Cs) reduce the work function of electrons from the substrate more appreciably than the lithium or sodium atoms. For example, when the degree of coating Θ for the K/Al(110) system is 0.33, the difference between the value of the

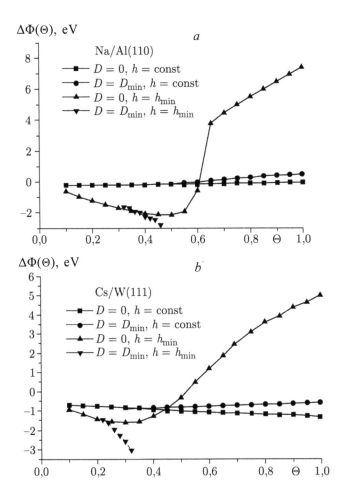

Fig. 9.11. The dependence of the electron work function $\Delta\Phi$ from the surface of metals: a – Al(110), b – W(111), on the degree of coverage Θ of these surfaces, calculated for the four modelling cases.

work function and the corresponding value for the clean surface Al(110) ($\Phi_m = 6.9$ eV [146]) is equal to $\Delta\Phi = -3.2$ eV, whereas for the Na/Al(110) sysem it is $\Delta\Phi = -1.7$ eV. In addition, the lowest values of the work function due to the adatoms of the potassium subgroup were achieved at lower values of Θ than for the lithium or sodium atoms. For example, the minimum value of the work function for the Cs/Al(110) system is reached at $\Theta = 0.31$ ($\Delta\Phi = -3.3$ eV), while the minimum value of the work function for the Li/Al(110) system is achieved at $\Theta = 0.55$ ($\Delta\Phi = -1.8$ eV).

The calculated values of the electron work function for the Cs/W(111) system were compared with experimental results published in [58, 164, 176, 379] for the minimum value $\Phi_{min} = 1.5$ eV, which is achieved when $\Theta \approx 0.25$. Given the fact that our calculated value of the electron work function from the clean W(111) surface is equal to 4.45 eV, and $\Delta\Phi_{min} = -3.03$ eV due to the adsorption of cesium atoms, then the value predicted by us, i.e. $\Phi_{min} = 1.42$ eV, agrees quite well with the experimental results. For the Na/W(111) system the experimental value $\Phi_{min} = 2.15$ eV when $\Theta \approx 0.5$ [107], and the value calculated by us $\Phi_{min} = 2.35$ eV is also in good agreement with the experimental results. For the Li/W(111) system, the experimental value $\Phi_{min} = 2.9$ eV at $\Theta > 0.65$ [163] and our value $\Phi_{min} = 2.55$ eV are also in good agreement with experimental results. Unfortunately, we failed to find experimental

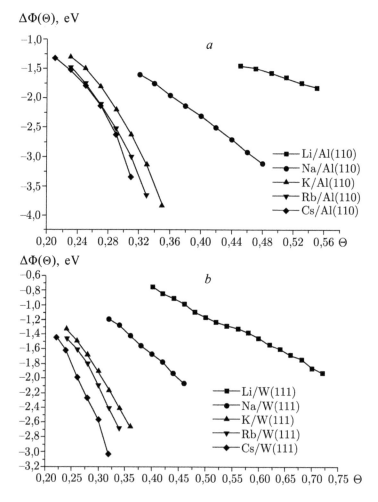

Fig. 9.12. Calculated (for the case 4) dependence of the work function of electrons $\Delta\Phi$ from the surface of metals: a – Al(110), b – W(111) – on the degree of coverage Θ of these surfaces by various alkali atoms metals in the range of existence of a stable coating.

values for the work function of the electron from the loose edges of other systems.

The results of calculation of the energy characteristics of adsorption can be explained as follows. At very low coverages ($\Theta < 0.1$), alkali metals are uniformly distributed on the surface of the substrate mainly in the form of positive ions [22, 453, 476]. With increasing degree of coverage Θ of the substrate by the alkali metal the adsorbate–adsorption interaction (interaction between the adatoms in the monolayer) is intensified and weakens the substrate–adsorption interactions (the interaction between atoms in the surface layer of the substrate and adsorbate atoms). This in turn leads to a slight increase in the distances between the substrate and the adsorbate (increased width of the gap $2D$, see, e.g., in Fig. 9.6). It should be noted that this is the main feature of the interaction in polyatomic systems: if the valence electrons are distributed among many bonds, then each bond becomes weaker and this increases the bond length, i.e. the formation of lateral bonds in the adsorbed layer will reduce the substrate–adsorption bonds and increase their length. This mutual competition between adsorbate–adsorption and substrate–adsorption bonds can lead to interesting surface formations from the alkali metal atoms and to structural phase transitions [22, 415, 453, 476].

In the close-packed surfaces of metals, as follows from the experimental and theoretical studies, alkali metals find it advantageous to break up into individual islands (with a metallic bond within the island), which grow in size with increasing coverage [293, 476]. Thus, at a certain value of Θ these surfaces are characterized by a structural phase transition from the uniform distribution of adatoms on the surface with a predominantly ionic bond to the close-packed islands with a metallic bond, which is confirmed by our calculations in the generalized conditions of the model case 4 [140, 153].

In the adsorption of alkali metals on the loose edges of metals the situation may be somewhat different. In this case, for example, at some value Θ we can obtain a monatomic adsorbed layer with a metal bond, as indicated by our calculations. It is also important to note the right boundary of the range of parameter Θ where the film ends and which corresponds to the smallest value of the work function of electrons from the substrate surface (see Fig. 9.12). Perhaps a phase transition takes place here and is associated with the change of symmetry of the adlayer or partition of the adlayer into close-packed islands where the work function starts to increase, approaching the value of the work function of electrons from the surface crystals of alkali metal. It is also possible that for a given value Θ a second adsorbed layer of the alkali metal already starts to form.

The calculations of the work function values as well as of the adsorption energy indicate a strong similarity in the adsorption properties of alkali metal of the potassium subgroup (K, Rb, Cs) and a significant difference between the properties of sodium and lithium, which is also confirmed by other theoretical and experimental studies of these systems [21, 175–178, 379, 476].

The main results of the study of adsorption of alkali metals on metal surfaces in the framework of the developed multi-parameter adsorption model can be formulated as follows.

1. The calculations have revealed that stable monatomic alkali metal films cannot form on the close-packed faces of the metal substrates without structural changes in the distribution of alkali metal atoms both along the surface and throughout the surface region. The existence of the stability regions of the parameter Θ for loose faces of metal substrates has been confirmed. The formation of monatomic films on the loose edges of the transition metal substrates can take place over a wider range of the parameter Θ than for simple or noble metals.

2. Self-consistent calculation of the displacement of the surface ion plane of the substrate showed that accounting for the effects of surface relaxation of the substrate leads only to a slight (less than 6%) decrease in the values of the adsorption energy for the alkali of metals.

3. Calculations of the emission of electrons from the substrate surface with the adsorbed metal film showed that in the range of existence of a stable monatomic film of the alkali metal on the loose edges of the metal substrate the work function decreases with increasing coverage Θ. It has been revealed that an increase of the radius of the alkali metal ion is accompanied by a reduction of the minimum value of the work function and its shift to lower values of Θ.

9.2. The model of activated adsorption of metal atoms on metallic surfaces

As a result of the development of theoretical methods of description of adsorption and experimental studies, it has become clear [293, 416, 452, 453] that all the interactions and reactions of metal atoms on metal surfaces are more complex than assumed so far. In particular, the study of the distribution of metal atoms on the surface layer of the substrate revealed the effects of expulsion of the substrate atoms by the metal adatoms to the surface with the implementation of substitutional adsorption (Fig. 9.13), as well the effects of mixing adatoms with substrate atoms with the formation of surface binary solutions (activated adsorption) [257, 293, 415, 454, 457, 474, 477].

This section presents the methodology and results of calculation of the energy of the activated adsorption of metal atoms on metal substrates within the framework of the density functional method. In contrast to other theoretical studies [363, 419, 427, 453, 454, 475–477, 506], to describe the emerging strong inhomogeneity of the electronic system in the interfacial region of interaction of the substrate atoms and adsorbate atoms we took

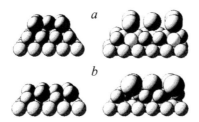

Fig. 9.13. Geometric illustration of the atomic structures of non-activated adsorption (*a*) and substitutional adsorption (*b*) on the surface of Al(111) for sodium adatoms (left) and potassium adatoms (right) [453].

into account the corrections for the heterogeneity of the electron density for both kinetic and for exchange–correlation energies.

We have developed a multiparameter version of the functional of electron density with test functions which take into account, along with the heterogeneity of the distribution of electron density, the existence of the equilibrium distance between the adatoms and the substrate determined from the condition of the minimum interfacial energy of interaction.

9.2.1. The basic equations. Methods of calculating the energy adsorption

Consider a semi-infinite metal with the mean charge density n_1, limited by the flat surface and occupying the region $z < -D$. The adsorbate film with thickness h occupies the region $D < z < D + h$. Between the substrate and the adsorbate in this model there is a vacuum gap with width $2D$. As a result of processes of mutual redistribution of the adsorbate and the substrate atoms, characteristic of activated adsorption, the electron density of the surface of the substrate will be different from the bulk electron density n_1, and the surface itself may be subject to a variety of reconstructions. At the same time considerable reconstruction usually takes place in the first surface layer of the substrate [293], which in this context is a separate region with an average charge density n_2 and thickness l (Fig. 9.14).

The positive charge of the background, therefore, is distributed in accordance with the formula

$$n_+(z) = \begin{cases} n_1, & z < -D-l; \\ n_2, & -D-l < z < -D; \\ n_3, & D < z < D+h; \\ 0, & |z| < D, z > D+h. \end{cases} \quad (9.29)$$

The solution of the linearized Thomas–Fermi equation with the use of the boundary conditions, reflecting the continuity of the electrostatic

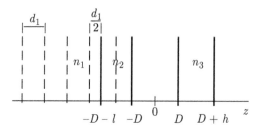

Fig. 9.14. Geometric representation of the distribution of surface layers.

potential $\phi(z)$ and its first derivative $d\phi/dz$ at $z = -D - l$, $z = \pm D$ and $z = D + h$, and the finiteness of the potential at $z \to \pm\infty$, gives for $\phi(z) = -4\pi n(z)/\beta^2$ the following expression for the density of the electron distribution $n(\beta, z)$ in a given system:

$$n(\beta, z) = \begin{cases} n_1\left(1 - 0,5e^{\beta(z+D+l)}\right) + 0,5n_2 e^{\beta(z+D)}\left(e^{\beta l} - 1\right) + \\ \quad + 0,5n_3 e^{\beta(z-D)}\left(1 - e^{-\beta h}\right), & z < -D - l; \\[2mm] 0,5n_1 e^{-\beta(z+D+l)} + n_2\left(1 - 0,5e^{\beta(z+D)} - \\ \quad - 0,5e^{-\beta(z+D+l)}\right) + 0,5n_3 e^{\beta(z-D)}\left(1 - e^{-\beta h}\right), & -D-l < \\ & < z < -D; \\[2mm] 0,5n_1 e^{-\beta(z+D+l)} + 0,5n_2 e^{-\beta(z+D)}\left(1 - e^{-\beta l}\right) + \\ \quad + 0,5n_3 e^{\beta(z-D)}\left(1 - e^{-\beta h}\right), & |z| < D; \\[2mm] 0,5n_1 e^{-\beta(z+D+l)} + 0,5n_2 e^{-\beta(z+D)}\left(1 - e^{-\beta l}\right) + \\ \quad + n_3\left(1 - 0,5e^{-\beta(z-D)} - 0,5e^{\beta(z-D-h)}\right), & D < z < \\ & < D+h; \\[2mm] 0,5n_1 e^{-\beta(z+D+l)} + 0,5n_2 e^{-\beta(z+D)}\left(1 - e^{-\beta l}\right) + \\ \quad + 0,5n_3 e^{-\beta(z-D)}\left(e^{\beta h} - 1\right), & z > D+h. \end{cases}$$

$$(9.30)$$

Here β is a variational parameter. From a physical point of view, the value of β^{-1} is a typical thickness of the surface layer in which the electron density dramatically changes.

The interfacial energy of the interaction of the substrate with the metal substrate, calculated within the jelly model σ_0, is defined as in the previous chapter by the relations (6.3)–(6.6).

Amendments to the interfacial energy associated with a given discrete distribution of the positive charge of the film and the substrate are calculated by the Ashcroft pseudopotential model:

$$\sigma_{ei} = \int_{-\infty}^{+\infty} \delta V(z)\left[n(z) - n_{+}(z)\right] dz. \tag{9.31}$$

Here $\delta V(z)$ has the meaning of the mean value over the planes of the sum of ionic pseudopotentials, less the potential of the homogeneous background of the positive charge. The following expressions applies to this value:

– at $D < z < D + h$ (in the region of the adsorbed film)

$$\delta V(z) = 2\pi n_3 \left\{ h(r_{c3} - |z - D - h/2|)\theta(r_{c3} - |z - D - h/2|) - \right.$$
$$\left. -\left[z - D - h + h\theta(-z + D + h/2)\right]^2 \right\}; \tag{9.32}$$

– at $-D - l < z < -D$ (in the first subsurface ion plane)

$$\delta V(z) = 2\pi n_2 \left\{ l(r_{c2} - |z + D + l/2|)\theta(r_{c2} - |z + D + l/2|) - \right.$$
$$\left. -\left[z + D + l\theta(-z - D - l/2)\right]^2 \right\}; \tag{9.33}$$

– for $z < -D - l$ ($k = 0, 1, 2, \ldots$ – number of ionic planes in the volume of the substrate)

$$\delta V(z) = 2\pi n_1 \left\{ d_1(r_{c1} - |z + D + l + (2k+1)d_1/2|) \times \right.$$
$$\times \theta(r_{c1} - |z + D + l + (2k+1)d_1/2|) -$$
$$\left. -\left(z + D + l + kd_1 + d_1\theta\left[-z - D - l - (2k+1)d_1/2\right]\right)^2 \right\}. \tag{9.34}$$

Using the distribution (9.30) for $n(\beta, z)$, as well as summing over the planes $z = -D - l - (2k + 1)d_1/2$ in (9.34) and using the periodicity of the potential $\delta V(z - d_1) = \delta V(z)$, we obtain

$$\sigma_{ei} = \frac{2\pi}{\beta^3}\left[\left(n_1^2 - n_1 n_2(1 - e^{-\beta l}) - n_1 n_3 e^{-\beta(2D+l)}(1 - e^{-\beta h})\right) \times \right.$$
$$\times\left(1 - \frac{\beta d_1 e^{-\beta d_1/2}}{1 - e^{-\beta d_1}}\,\mathrm{ch}(\beta r_{c1})\right) +$$
$$+\left[2n_2^2 - n_2 n_1 - n_2 n_3 e^{-2\beta D}(1 - e^{-\beta h})\right](1 - e^{-\beta l})$$
$$\times\left(1 - \frac{\beta l e^{-\beta l/2}}{1 - e^{-\beta l}}\,\mathrm{ch}(\beta r_{c2})\right) +$$
$$+\left(2n_3^2 - n_3 n_1 e^{-\beta(2D+l)} - n_3 n_2 e^{-2\beta D}(1 - e^{-\beta l})\right)(1 - e^{-\beta h}) \times$$
$$\left.\times\left(1 - \frac{\beta h e^{-\beta h/2}}{1 - e^{-\beta h}}\,\mathrm{ch}(\beta r_{c3})\right)\right]. \tag{9.35}$$

For amendments to the interfacial energy associated with the interaction of metal ions, we obtained the following expression:

$$\sigma_{ii} = \sqrt{3}\frac{Z_1^2}{c_1^3}\exp\left(-\frac{4\pi d_1}{\sqrt{3}c_1}\right) + 2\sqrt{3}\frac{Z_2^2}{c_2^3}\exp\left(-\frac{4\pi l}{\sqrt{3}c_2}\right)\times$$

$$\times\left[1-\exp\left(-\frac{4\pi l}{\sqrt{3}c_2}\right)\right] + 2\sqrt{3}\frac{Z_3^2}{c_3^3}\exp\left(-\frac{4\pi h}{\sqrt{3}c_3}\right)\left[1-\exp\left(-\frac{4\pi h}{\sqrt{3}c_3}\right)\right] -$$

$$-2\sqrt{3}\frac{Z_1 Z_2}{(c_1 c_2)^{3/2}}\exp\left[-\frac{2\pi}{\sqrt{3}}\left(\frac{d_1}{c_1}+\frac{l}{c_2}\right)\right]\left[1-\exp\left(-\frac{4\pi l}{\sqrt{3}c_2}\right)\right] -$$

$$-2\sqrt{3}\frac{Z_2 Z_3}{(c_2 c_3)^{3/2}}\exp\left[-\frac{2\pi}{\sqrt{3}}\left(\frac{l+D}{c_2}+\frac{h+D}{c_3}\right)\right]\times$$

$$\times\left[1-\exp\left(-\frac{4\pi l}{\sqrt{3}c_2}\right)\right]\left[1-\exp\left(-\frac{4\pi h}{\sqrt{3}c_3}\right)\right] -$$

$$-2\sqrt{3}\frac{Z_1 Z_3}{(c_1 c_3)^{3/2}}\exp\left[-\frac{2\pi}{\sqrt{3}}\left(\frac{d_1 + D + l/2}{c_1}+\frac{h+D+l/2}{c_3}\right)\right]\times$$

$$\times\left[1-\exp\left(-\frac{4\pi h}{\sqrt{3}c_3}\right)\right]. \tag{9.36}$$

Here Z_1, Z_2, Z_3 are the charges of the ions, c_1, c_2, c_3 are the distances between the nearest ions in planes parallel to the surface.

As a result, the total interfacial energy can be written as

$$\sigma(\beta, D) = \sigma_0(\beta, D) + \sigma_{ei}(\beta, D) + \sigma_{ii}(D). \tag{9.37}$$

The value of the variational parameter β is found from the requirements of its minimality, i.e.

$$\left(\frac{\partial\sigma(\beta, D)}{\partial\beta}\right)_{\beta=\beta_{min}} = 0. \tag{9.38}$$

The solution of equation (9.38) defines the parameter β_{min} as a function of the gap $2D$ and structural parameters of the substrate and the coating.

The result of solving this variational problem is the total interfacial energy of the system $\sigma[\beta_{min}(D), D]$. Knowing this energy, it is easy to find the energy of adhesion of the system as a work that is necessary to displace the substrate and the film from each other to infinity, i.e.

$$E_a(2D) = \sigma(\infty) - \sigma(2D). \tag{9.39}$$

The energy of adsorption can then be calculated by the formula

$$E_{ads} = E_a / n_s^a, \tag{9.40}$$

where the surface concentration of adatoms n_s^a is a function of the parameter Θ, characterizing the degree of filling of the surface of the substrate with the adatoms,

$$\Theta = n_s^a / n_{s1},$$
(9.41)

where n_{s1} characterizes the surface concentration of atoms of the non-reconstructed substrate.

9.2.2. Description of surface binary solutions

Substitutional adsorption is characterized by the fact that the adsorbed atoms can push the metal atoms near the surface of the substrate into the film and take their place. In such substitutional processes the subsurface region of the substrate is exposed to various reconstructions [293]. Each adsorbate has its own temperature below which substitutional adsorption is not possible [293, 453]. However, already at temperatures close to room temperature there is only substitutional adsorption since the adatoms of almost all metals at these temperatures have sufficient energy for substitution processes. Therefore, substitutional adsorption is also called activated adsorption. In this chapter, it is assumed that the adatoms cause only the reconstruction of the first surface layer of the substrate and the substituted atoms of this layer are distributed in the film. In this way, the first surface layer of the substrate and the film are characterized by the formation of a mixture of substrate and adsorbate atoms, which in its properties resembles a binary solution of two metals [453].

When considering the binary solid solutions we use the following terminology. We denote the adsorbate as A, the substrate as S. Then for the composition of the binary solution of the formula

$$A_p S_{1-p},$$
(9.42)

where the subscript p denotes the relative fraction of the atoms of the adsorbate in solution. Ion charge Z and the cutoff radius r_c for the binary solution are determined by the formulas

$$Z = pZ^a + (1-p)Z^s,$$
(9.43)

$$r_c = \left(\frac{pZ^a(r_c^a)^3 + (1-p)Z^s(r_c^s)^3}{pZ^a + (1-p)Z^s} \right)^{1/3},$$
(9.44)

where the superscript a characterizes the parameters related to the adsorbate, the subscript s – to the substrate.

The electron density of the film can be expressed as [379]

$$n_3(\Theta, p) = Z_3 n_{s3} / h,$$
(9.45)

where the ion charge of the film Z_3 is given by (9.43), n_{s3} is the surface concentration of atoms in the film, h is the thickness of the film,

$$h = pd^a + (1-p)d^s, \qquad (9.46)$$

d^a, d^s are the distances between the most closely packed planes in crystals of the adsorbate and the substrate, respectively.

Similar formulas are used for the binary solution in the surface layer of the substrate (the relative fraction of the atoms of the adsorbate in the solution indicated by the letter p': $A_{p'}S_{1-p'}$) with the electron density and the thickness of the surface layer, given by the relations

$$n_2(\Theta, p') = Z_2 n_{s2} / l, \qquad (9.47)$$

$$l = p'd^a + (1-p')d_1, \qquad (9.48)$$

where the ion charge of the layer Z_2 is given by (9.43), n_{s2} is the surface density of atoms in the layer. Assuming a uniform distribution of the adatoms in the surface layer of the substrate with repetition of its symmetry, the filling parameter Θ can be expressed through the parameters of binary solutions p and p':

$$\Theta = pq + p'q', \qquad (9.49)$$

where the parameters q and q' are given by

$$q = n_{s3} / n_{s1}, \quad q' = n_{s2} / n_{s1} \qquad (9.50)$$

and characterize the degree of filling by the atoms of the film and the surface layer, respectively.

Parameters Θ, q and q' are related to each other by the relation

$$\Theta = q + q' - 1. \qquad (9.51)$$

The degree of filling Θ of the substrate surface by the adatoms is defined in such a way that for $\Theta = 1$ the number of adatoms is equal to the number of atoms in the unreconstructed surface faces of the substrate. The parameters c_2 and c_3, defining the minimum distance between the atoms in the surface layer and in the film respectively are determined through the parameter c_1 by the relations

$$c_2 = c_1 \sqrt{q'}, c_3 = c_1 \sqrt{q}. \qquad (9.52)$$

The surface concentrations n_{s2} and n_{s3} are functions of these parameters and, moreover, reflect the symmetry of the face of the substrate on which adsorption takes place. For example, for the (111) face of the substrate with the fcc lattice, the surface concentrations in the layer and the film are determined as

$$n_{s2} = \frac{2}{\sqrt{3}c_2^2}, \quad n_{s3} = \frac{2}{\sqrt{3}c_3^2}. \tag{9.53}$$

The surface concentration of adatoms n_s^a can also be determined by the concentrations n_{s2} and n_{s3} using the ratio

$$n_s^a = p'n_{s2} + pn_{s3}. \tag{9.54}$$

9.2.3. Analysis of the results of calculating adsorption energy

Figure 9.15 shows as an example the graphs of the calculated adsorption energy as a function of the degree of coverage Θ by the adatoms of alkali metals Li, Na, K, Rb, Cs with the most close-packed (111) face of the aluminium substrate. For this case we chose the following values of the parameters of binary solutions of: $p = 0.5$, $q' = 1.0$ (the surface layer of the substrate has no vacancies). The values of the parameters q and p' are unambiguously determined by (9.49) and (9.51), according to which $p' = 0.5\Theta$, $q = \Theta$. The figure shows that adsorption energy decreases monotonically with increasing thickness of the coating and the adsorption energy for a given value of the parameter Θ takes the lowest values for the adsorbates of the potassium subgroup and the maximum – for lithium. For example, when $\Theta = 0.3$ the adsorption energy of the $Cs_{0.5}Al_{0.5}/Al(111)$ system is 1.65 eV, whereas for the $Li_{0.5}Al_{0.5}/Al(111)$ system it is equal to 2.61 eV, i.e. almost 1 eV higher. This pattern is easily explained by the difference in ionic radii of the alkali metal ions and the substrate. For the Cs–Al pair it is about 2.3 a.u., and for Li–Al only 0.4 a.u. Thus, lithium penetrates much easier into the surface layer of the substrate and forms the strongest bond with the aluminum atoms. In addition, the larger the radius of the alkali metal ion the lower the degree of coverage of the substrate Θ that is required to achieve a given value of adsorption energy. For example, the

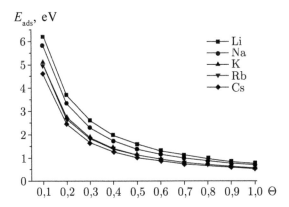

Fig. 9.15. The calculated dependence of the adsorption energy E_{ads} on the degree of coverage Θ for the Al(111) surface for $p = 0.5$, $q = \Theta(p' = 0.5\Theta, q' = 1)$.

adsorption energy of 1.4 eV is achieved in the $Rb_{0.5}Al_{0.5}/Al(111)$ system at $\Theta = 0.4$, and in the $Na_{0.5}Al_{0.5}/Al(111)$ system when $\Theta = 0.5$.

Figure 9.16 shows plots of the dependence of the adsorption energy E_{ads} of the atoms of sodium and potassium on the composition p of the solution in the (A_pS_{1-p}) film at $\Theta = 0.3$ for the close-packed (111) face of the aluminium substrate. From analysis we can draw the following conclusions. For values of q close to Θ (with $q \geq 4\Theta$), the adsorption energy increases monotonically with decreasing number of adatoms in the film ($p = 1$ corresponds to non-activated adsorption). This indicates

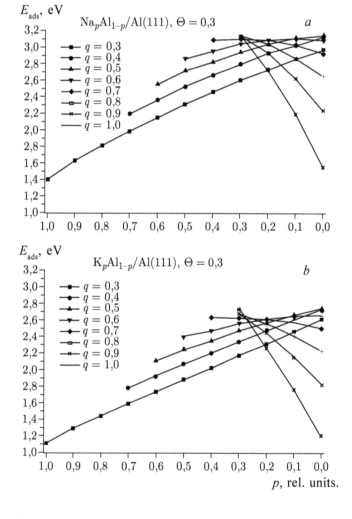

Fig. 9.16. The calculated dependence of the adsorption energy E_{ads} on the composition p of the film at the coverage parameter $\Theta = 0.3$ of the Al(111) surface for the atoms: a – sodium, b – K.

that for these values of the parameter q the processes of substitution of the aluminum atoms in the surface layer by the adatoms are energetically more favourable (reduced interfacial interaction energy per one adsorbed atom, taking into account the substitution processes, is lower than in the non-activated surface position of the adatoms). For example, when $q = 0.4$, $q' = 0.9$ for the Na_pAl_{1-p}/Al(111) system at $p = 0.6$ ($p' = 0.07$) $E_{ads} = 2.36$ eV and for $p = 0.1$ ($p' = 0.29$) $E_{ads} = 3.02$ eV. On the contrary, for values of parameter q close to unity, the adsorption energy decreases monotonically with a decrease in the number of adatoms in the film. This is because the surface layer of the substrate is already saturated with alkali metal atoms, and the substitutional processes are energetically unfavorable. For example, when $q = 0.9$, $q' = 0.4$ for the Na_pAl_{1-p}/Al(111) system $p = 0.2$ ($p' = 0.3$) $E_{ads} = 2.92$ eV, while for $p = 0$ ($p' = 0.75$) $E_{ads} = 2.23$ eV. In addition, the monolayer coating of alkali metal atoms ($q = 0.3$, $p = 1$ – non-activated adsorption), as shown by the calculations, has the lowest value of the adsorption energy. For the Na/Al(111) system it is 1.4 eV.

The calculations also show that the maximum energy of adsorption for the system replacement Na_pAl_{1-p}/Al(111) is equal to 3.14 eV for the parameter values $q = 0.8$, $p = 0.3$ ($q' = 0.5$, $p' = 0.12$) and $q = 0.5$, $p = 0$ ($q' = 0.8$, $p' = 0.38$), although there is a number of other states with similar values of the adsorption energy. The behaviour of the adsorption energy for the K_pAl_{1-p}/Al(111) system was similar but was characterized by lower values in the corresponding states ($E_{ads} = 1.10$ eV for non-activated adsorption and maximum adsorption energies for substitutional adsorption $E_{ads} = 2.72$ eV for the parameter values $q = 1.0$, $p = 0.3$ ($q' = 0.3$, $p' = 0$)). These results, indicating that the activated adsorption is energetically more advantageous than non-activated adsorption for all types of alkali metal atoms, were confirmed in experimental studies of adsorption of sodium on aluminum by the method of fine structure of X-ray radiation (SEXAFS) [454] and the adsorption of potassium on aluminum by low-energy electron diffraction (LEED) [274]. These studies clearly demonstrated the implementation of substitutional adsorption at room temperature.

Figure 9.17 shows plots of the adsorption energy E_{ads} of sodium atoms on the composition of the binary solution in the (A_pS_{1-p}) film with the coverage parameter $\Theta = 0.6$ and $\Theta = 0.9$ for the (111) aluminium substrate. At $\Theta = 0.6$ the maximum value of adsorption energy is 1.3 eV, and is obtained at the following parameters of the binary solutions:

– $q = 1$, $p = 0.6$ ($q' = 0.6$, $p' = 0$) – a condition in which part of the aluminum atoms are moved from the surface layer to the film with the formation of a binary solution;

– $q = 0.6$, $p = 0$ ($q' = 1$, $p' = 0.6$) – a condition in which all the sodium atoms are located in the surface layer of the substrate with the formation of a binary solution;

– $q = 0.6$, $p = 0.1$ ($q' = 1$, $p' = 0.54$) – the state with the formation of binary solutions in the surface layer and in the film.

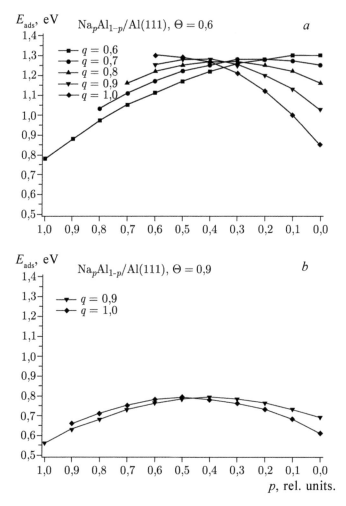

Fig. 9.17. The calculated dependence of the adsorption energy E_{ads} of sodium atoms on the composition p of the film for the Al(111) surface with the coverage parameters $\Theta = 0.6$ (*a*) and $\Theta = 0.9$ (*b*).

When the coverage parameter is $\Theta = 0.9$, the maximum adsorption energy is even lower and is equal to 0.79 eV, and it can also be implemented in several states: $q = 0.9$, $p = 0.4$ ($q' = 1$, $p' = 0.54$) or $q = 1$, $p = 0.5$ ($q' = 0.9$, $p' = 0.44$).

From the analysis of the graphs shown in Figs. 9.16 and 9.17 we can make another very important conclusion: as the coverage parameter Θ increases the difference ΔE_{ads} between the maximum value of adsorption energy and the minimum value, characteristic of the non-activated adsorption, greatly decreases. For example, for Na_pAl_{1-p}/Al(111) system at $\Theta = 0.3$, when $\Theta = 0.6$ and $\Theta = 0.9$ we have, respectively: $\Delta E_{ads} = 1.74$ eV,

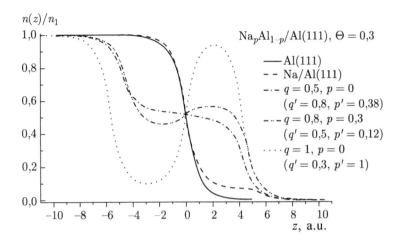

Fig. 9.18. Electron density distribution for the Na_pAl_{1-p}/Al(111) system at $\Theta = 0.3$ for different values of the parameters p, q (p', q'), specifying the composition of the film and the surface layer.

$\Delta E_{ads} = 0.52$ eV, $\Delta E_{ads} = 0.23$ eV. This indicates that the strongest bond of the adatoms with the atoms of the substrate is formed at small values of Θ. At the same time, at high values of the coverage, the adatoms, mixed with the substrate atoms, find it energetically more favourable to adsorb in several layers without constructing each of them to $q = 1$; this is also observed experimentally [273, 293].

Figure 9.18 shows plots of the electron density distribution (9.30) for the Na_pAl_{1-p}/Al(111) system at $\Theta = 0.3$ for different values of the parameters p, q (p', q'). The solid line corresponds to the distribution of electron density on the surface of pure Al(111). The dashed line denotes the electron density distribution for a system with a monolayer coating (non-activated adsorption). The distributions of electron density corresponding to the state of the system with the maximum adsorption energy, are shown in the figure by dash-dot lines with one or two points. The dotted line denotes the electron density distribution for the state of the system in which all the sodium atoms are located in the surface layer of the substrate, and all the aluminum atoms from this layer are shifted into the film. These graphs show that the adsorption processes with maximum energy are characterized by such transfer of the negative charge from the surface layer to the adsorbate layer that the electron density inside the substrate and the density of the uncompensated positive charge in the surface layer are close to 0.5.

In the adsorption of atoms of transition metals substitution processes of the substrate atoms are more favourable than in the alkali metals since in this case the difference of the radii of the atoms of the adsorbate and substrate atoms is much smaller. As a consequence, the adsorption of transition metals is characterized by large values of the coverage parameter

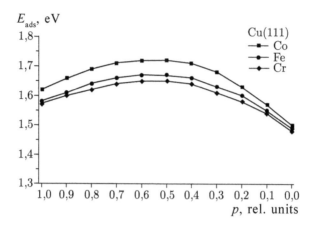

Fig. 9.19. The calculated dependence of the adsorption energy E_{ads} of transition metal atoms on the composition p of the film for the Cu(111) surface for the coverage parameter $\Theta = 1$.

($\Theta \geq 1$), in contrast to the alkali metals ($\Theta < 1$). Figure 9.19 shows plots of the adsorption energy of atoms of transition metals (Co, Fe, Cr) on the composition p of the solution in the film at $\Theta = 1$ for the close-packed surface (111) of the copper substrate. From the expressions (9.49)–(9.51) it follows that at $\Theta = 1$ with $q = q' = 1$, $p' = 1 - p$ the states of the system can be characterized by a single parameter p characterizing the composition of the binary solution in the adsorbed film. The figure shows that the adsorption energy transition-metal atoms increases with p decreasing *from* 1 to 0.5, and this is due to processes of substitution of the atoms of the surface layer of the substrate. These substitution processes for these values of parameter p are energetically favourable, while further penetration of the transition metal atoms into the surface layer with values of $0 \leq p \leq 0.5$ is energetically unfavourable, as evidenced by the decrease in the adsorption energy of the system. Calculations show that the state of the systems with the maximum mixing of the transition metal atoms and the atoms of the surface layer ($p \approx 0.5$ and $p' \approx 0.5$) is most advantageous. It corresponds to the maximum value of adsorption energy: the following values were obtained for Co/Cu(111), for Fe/Cu(111) and Cr/Cu(111), respectively: $E_{ads}^{max} = 1.72$ eV, $E_{ads}^{max} = 1.67$ eV, $E_{ads}^{max} = 1.65$ eV. Note that the maximum adsorption energy is slightly higher that the energy of non-activated adsorption ($p = 1$): for Co/Cu(111), for Fe/Cu(111) and Cr/Cu(111) we have respectively: $E_{ads} = 1.62$ eV, $E_{ads} = 1.58$ eV, $E_{ads} = 1.57$ eV – the difference is $\Delta E_{ads} \leq 0.1$. These calculations are quite comparable with the results of experimental studies of adsorption structures of the transition metals (Co, Fe) [432, 457] in the monolayer coatings with $p = 1$ (metastable state) dominate at low temperatures ($T < 250$ K) and at higher temperatures substitutional processes take place.

The main results of the study of adsorption of atoms of transition and alkali metals on metal surfaces using the proposed model of activated adsorption suggest the following conclusions.

1. The calculations of the energy characteristics of adsorption showed that the mixing processes of adatoms with the metal surface atoms of the substrate is energetically more favourable in comparison with the monolayer coating for all values of Θ. As a result, the activated adsorption of metal atoms is energetically more favourable than non-activated.

2. It was found that the non-activated adsorption of alkali metals has the lowest value of the adsorption energy and is a metastable state.

3. The adsorption energy decreases monotonically with increasing coverage parameter Θ. It is shown that for each value of Θ the adatoms of the potassium subgroup (K, Rb, Cs) are characterized by lower values of adsorption energy than lithium or sodium adatoms.

4. It has also been established that for the alkali metals at $\Theta = 0.33$ the most beneficial adsorption structure is $\sqrt{3} \times \sqrt{3}$, and at $\Theta = 0.5$ the 2×2 and 2×1 structures can be compete with respect to energy, while for the transition metals with $\Theta = 1$ the structure 2×2 is most suitable. These predictions are consistent with the experimental studies.

9.3. Effect of adsorption of metal atoms on the work function of the electron from metal surfaces

Recent experimental studies of the surface distribution of adatoms revealed effects such as ejection by the adsorbate atoms of the substrate atoms to the surface with occurrence of substitutional diffusion and the formation of 'islands' of the adsorbate and substrate atoms [273, 274, 293, 415, 416, 454]. The surface of the substrate undergoes a significant reconstruction, which significantly affects the work function of electrons [274, 415, 453, 454, 476]. This section is devoted to explaining these recent experimental results on the basis of our proposed model of activated adsorption [155–158].

In this model, the adsorption system is given in the form of a monatomic adsorbate film, separated by a vacuum gap from a semi-infinite metal substrate. The introduction of the vacuum gap whose size is determined from the minimum total interfacial energy of interaction of the adsorbate film and the substrate allows to consider effects of the relative displacement of ionic planes in the surface layer and the adsorbate layer. At the same time, the relative redistribution of the adsorbate atoms and the atoms of the first surface layer of the substrate with the form binary solid solutions, which is typical for substitutional adsorption, was taken into account [293].

This section presents the methodology and results of calculation of the change of the electron work function from the surface of metal substrates due to the adsorption of the atoms of alkali and transition metals within the density functional method.

9.3.1. The basic equations. Methods of calculating the work function

Consider a semi-infinite metal bounded by a flat surface and covering an area of $z < -D$. The film of the adsorbate occupies region $D < z < D + h$. Between the film and the surface there is a vacuum gap of width $2D$. As a result of processes of mutual redistribution of atoms of the adsorbate and the substrate atoms, the electron density of the surface of the substrate will be different from its bulk electron density n_1, and the surface itself may be subject to various reconstructions. At the same time significant reconstruction takes place usually in the first surface layer of the substrate [293] which in this connection is treated as a separate region with an average charge density n_2 and thickness l (Fig. 9.14).

The effect of adsorption on the electron work function from the surface is conveniently characterized by a change in the value of the work function $\Delta\Phi$, defined by the expression

$$\Delta\Phi(\Theta) = \Phi(\Theta) - \Phi_m. \tag{9.55}$$

Here Φ_m is the electron work function from the clean surface of the substrate, and the value $\Phi(\Theta)$ is determined by the expression

$$\Phi(\Theta) = \Phi_D(\Theta) - \mu, \tag{9.56}$$

where Φ_D is the dipole potential barrier, μ is the chemical potential, Θ is the degree of filling of the substrate surface by the adatoms. The height of the dipole potential barrier Φ_0 in the 'jelly' model can be written as follows:

$$\Phi_0 = \frac{4\pi n_1}{\beta^2}. \tag{9.57}$$

The chemical potential μ of the electron gas taking into account the exchange–correlation and pseudopotential corrections takes the form

$$\mu = 0.5(3\pi^2 n_1)^{2/3} - \left(\frac{3n_1}{\pi}\right)^{1/3} - \frac{0.056 n_1^{2/3} + 0.0059 n_1^{1/3}}{(0.079 + n_1^{1/3})^2} -$$
$$-0.4 Z_1^{2/3}\left(\frac{4\pi n_1}{3}\right)^{1/3} + 4\pi n_1 r_1^2. \tag{9.58}$$

The additional contribution to the potential barrier by taking into account the electron interaction and the charge distribution of the ions in the sites of the crystal is determined from the relation

$$\Phi_{ei} = \frac{-\displaystyle\int_{-D-l-d_1}^{\infty} \delta V(z)\delta n_E(z)dz}{\displaystyle\int_{-D-l-d_1}^{\infty} \delta n_E(z)dz}, \tag{9.59}$$

where d_1 is the interplanar distance in the crystal substrate, $\delta V(z)$ has the meaning of the mean in respect of the planes of the sum of the ion potentials (minus the potential of a homogeneous semi-infinite positive background, see equation (9.32), (9.33) and (9.34)), $\delta n_E(z) = n(z) - n_E(z)$ with $n_E(z)$, equal to the surface charge density of a system in a weak electric field strength E_z. For $\delta n_E(z)$, we obtained the following expression:

$$\delta n_E(z) = \begin{cases} \dfrac{\beta E_z}{8\pi}\exp[\beta(z-D-h)], & z \le D+h; \\[2mm] \dfrac{\beta E_z}{8\pi}\exp[-\beta(z-D-h)], & z \ge D+h. \end{cases} \tag{9.60}$$

Integrating (9.59) and using the expressions (9.32), (9.33), (9.34) and (9.60), we obtain the electron–ion correction to the potential barrier:

$$\Phi_{ei}^{m} = -\frac{4\pi n_1}{\beta^2}\frac{e^{-\beta(2D+l+h+d_1/2)}}{2-e^{-\beta(2D+l+h+d_1)}}\left[\beta d_1\mathrm{ch}(\beta r_{c1})-2\mathrm{sh}(\beta d_1/2)\right] \tag{9.61}$$

for the metal substrate;

$$\Phi_{ei}^{s} = -\frac{4\pi n_2}{\beta^2}\frac{e^{-\beta(2D+h+l/2)}}{2-e^{-\beta(2D+l+h+d_1)}}\left[\beta l\,\mathrm{ch}(\beta r_{c2})-2\mathrm{sh}(\beta l/2)\right] \tag{9.62}$$

for the surface layer;

$$\Phi_{ei}^{a} = -\frac{4\pi n_3}{\beta^2}\frac{e^{-\beta h/2}}{2-e^{-\beta(2D+l+h+d_1)}}\left[\beta h\,\mathrm{ch}(\beta r_{c3})-2\mathrm{sh}(\beta h/2)\right] \tag{9.63}$$

for the adsorbate film.

It should be noted that the work function from the surface of the noble and transition metals is strongly influenced by the effects of surface relaxation [35, 146]. Considering the effects of surface relaxation of the atoms of the substrate involves additional corrections to the dipole potential barrier Φ_δ [146]:

$$\Phi_{\delta}^{m} = \frac{4\pi n_1}{\beta^2}\frac{e^{-\beta(2D+l+h+d_1/2)}}{2-e^{-\beta(2D+l+h+d_1)}}\left[2(1-e^{\beta\delta})\mathrm{sh}(\beta d_1/2)-\beta^2\delta d_1\right]+$$

$$+\Phi_{ei}^{m}(e^{\beta\delta}-1)-4\pi n_1 d_1\delta, \tag{9.64}$$

where δ is the relaxation parameter. As a result, the dipole potential barrier is determined by the sum of the contributions

$$\Phi_{D} = \Phi_0 + \Phi_{ei}^{m} + \Phi_{\delta}^{m} + \Phi_{ei}^{s} + \Phi_{ei}^{a}. \tag{9.65}$$

In the implementation of substitutional adsorption in the first surface layer and the adsorbate film solid solutions form from the atoms of the substrate and the adsorbate. Changes of the parameters of the film and of the first surface layer of the substrate, depending on their composition, can be characterized by (9.43)–(9.50).

According to the results of recent experimental studies [293], in substitutional adsorption the substrate atoms, pushed to the surface, form both binary islet coatings and monotype coatings while maintaining the symmetry of the surface face of the substrate in the distribution. The parameters Θ, q and q' are associated with each other by the relationship

$$\Theta = q + q' - (1 - k), \tag{9.66}$$

where the parameter k takes into account the relative share of the substrate atoms pushed to the surface, which are not involved in the formation of the binary coatings. The degree of filling Θ of the substrate surface by the adatoms is defined in such a way that for $\Theta = 1$ the number of the adatoms is equal to the number of atoms in the surface faces of the unreconstructed substrate.

9.3.2. Analysis of the results of calculation of the work function

Table 9.1 shows the results of calculation of the electron work function changes from the most close-packed (111) face of the aluminium substrates due to the adsorption of sodium atoms with the coverage parameters $\Theta = 0.33$ and $\Theta = 0.5$. It is seen that in all types of coatings the electron work function decreases in comparison with the work function of the clean Al(111) surface ($\Phi_m = 4.19$ eV [146]). The maximum decrease in work function by 1.57 eV corresponds to state of the system with the monolayer surface (the case of non-activated adsorption) at $\Theta = 0.33$ and the minimum ($\Delta\Phi = -1.22$ eV) – with the surface binary solid solution with the symmetry (2×2) at $\Theta = 0.5$.

According to experimental data [293, 415, 454] with the parameter coverage $\Theta \le 0.33$ the Na/Al(111) system is characterized by the formation of island structures with the symmetry $\sqrt{3} \times \sqrt{3}$ (Fig. 9.20). The energetically favourable formation of these structures in comparison coated with the monolayer coating is supported by higher values of the adsorption energy $E_{ads} = 1.63$ eV, obtained by us on the basis of model calculations (see Table 9.1). In these structures there is one sodium atom for every two aluminium atoms. At the same time, every third substituted atom is not involved in the formation of a binary coating in the island. In our model, this structure is consistent with the values of the parameters $k = 0.33$, $p = 0.95$, $q = 0.33$ ($p' = 0.02$, $q' = 0.70$). The calculated value $\Delta\Phi$ for this structure, equal to -1.30 eV, agrees well with the experimental value $\Delta\Phi = -1.7$ eV [415].

When the coverage parameter is $\Theta = 0.5$, two competing structures with close values of the adsorption energy can form: the surface binary solid solution with the symmetry 2×2 ($E_{ads} = 1.04$ eV), when all the substituted aluminum atoms are involved in the formation of a binary coating film, and island structures with the symmetry of 2×1 ($E_{ads} = 1,07$ eV), when

Table 9.1. The results of calculation of changes of the values of electron work function from surface of Al(111) due to the adsorption of sodium, adsorption energy of sodium atoms on the surface of Al(111), as well as the experimental values $\Delta\Phi$ for a number of structures

Θ, rel. units, coating type	k rel. units.	p rel. units.	q rel. units.	p' rel. units.	q' rel. units.	E_{ads} eV	$\Delta\Phi_{theor}$ eV	$\Delta\Phi_{exp}$ eV
0.33; monolayer	0	1	0.33	0	1	1.40	−1.57	−
without relaxation	0	1	0.33	0	1	1.19	−1.02	−
DFT-LDA [197]	0	1	0.33	0	1	1.41	−1.7	−
0.33, $\sqrt{3}\times\sqrt{3}$	0.33	0.95	0.33	0.02	0.70	1.63	−1.30	−1.7
without relaxation	0.33	0.95	0.33	0.02	0.70	1.43	−0.65	−1.7
DFT-LDA [197]	0.33	0.95	0.33	0.02	0.70	1.58	−1.6	−1.7
0.50; monolayer	0	1	0.50	0	1	0.92	−1.46	−
without relaxation	0	1	0.50	0	1	0.81	−1.00	−
0.50; 2 × 2	0	0.90	0.50	0.05	1	1.04	−1.22	−1.2
without relaxation	0	0.90	0.50	0.05	1	0.93	−0.87	−1.2
DFT-LDA [197]	0	0.90	0.50	0.05	1	−	−1.16	−1.2
0.50, 2 × 1	0.50	0.90	0.50	0.10	0.50	1.07	−1.29	−1.3
without relaxation	0.50	0.90	0.50	0.10	0.50	0.95	−0.67	−1.3

substituted every second aluminum atom is not involved in the formation of a binary coating in the island. For a binary solution (2 × 2) we obtained the value $\Delta\Phi = -1.22$ eV, which is in good agreement with the experimental value $\Delta\Phi = -1.2$ eV [453]. For the islands (2 × 1) calculations give a value $\Delta\Phi = -1.29$ eV, which agrees well with the experimental value $\Delta\Phi = -1.3$ eV [415].

Figure 9.21 shows the graphs of the dependence of the change of the work function of the electron due to the adsorption of atoms of transition metals (cobalt, iron, chromium) on the composition p of the solution in the $A_p S_{1-p}$ film at $\Theta = 1$ for the most close-packed (111) face of the copper substrate. In the adsorption processes of the transition metal the process of substitution of the surface atoms of the substrate are more suitable than in

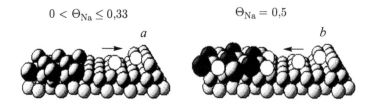

$$0 < \Theta_{Na} \leq 0{,}33 \qquad\qquad \Theta_{Na} = 0{,}5$$

a   b

Fig. 9.20. The proposed mechanism of redistribution of surface atoms with the formation of structures $\sqrt{3} \times \sqrt{3}$ (a) and the surface binary solution 2×2 (b) for the Na/Al (111) system [453].

the case of the alkali metals, since in this case difference between the radii of the atoms of the adsorbate and substrate atoms is much less. Therefore, the adsorption processes of transition metals are characterized by high values of the coverage parameters ($\Theta \geq 1$), in contrast to the adsorption of alkali metals ($\Theta < 1$). In addition, all substituted copper atoms are involved in the formation of binary coatings which corresponds to the value of $k = 0$. From the expressions (9.49) and (9.66) it also follows that for $\Theta = 1$ the parameters $q = q' = 1$, $p' = 1 - p$ and the state of the system can be defined by a single parameter p, characterizing the composition of the binary solution in the adsorbed film.

Figure 9.21 shows that at all values of p the electron work function increases in comparison with the work function of the clean Cu(111) surface ($\Phi_m = 4.88$ eV [146]). The maximum value of the work function corresponds to the system with the maximum possible mixing of the transition metal atoms and the substrate atoms ($p = p' = 0.5$), and the minimum – to the 'sandwich' structure ($p = 0$, $p' = 1$), when all adatoms penetrate to the surface layer, and all the copper atoms, coming out of this layer, form the outer monolayer coating. Calculations show that with increasing Cu concentration in the film the work function of the adsorption system is close to the value of the electron work function of the clean Cu(111) surface.

Table 9.2 shows the results of the calculation of changes in the values of the electron work function for the close-packed (100) face of the copper substrate due to the adsorption of the atoms of the transition metals (Co, Fe, Cr), with $\Theta = 1$ ($q = q' = 1$, $p' = 1 - p$). In this case, $\Delta\Phi$ also has positive values but is significantly lower than for adsorption on the Cu(111) face. It should be noted that the system with the 'sandwich' structure (Cu/Co/Cu(100), Cu/Fe/Cu(100), Cu/Cr/Cu(100)) and the clean Cu(100) surface have similar properties. For example, we found that the change of the work function of the system is $\Delta\Phi$(Cu/Co/Cu(100)) = –0.10 eV, $\Delta\Phi$(Cu/Fe/Cu(100)) = 0.10 eV, $\Delta\Phi$(Cu/Cr/Cu(100)) = 0.17 eV, i.e. close to zero, which is also confirmed by the results of [432].

Figure 9.22 shows the graphs of the dependence of the change of the work function of the electron due to the adsorption of transition metals (Co, Fe, Cr) on the composition p of the solution at $\Theta = 1$ for the most close-packed (110) tungsten substrate. In this case, as seen from the figure, the

Table 9.2. The results of calculation of changes of the values of electron work function from the surface of Cu(100) due to the adsorption of transition metal atoms and the adsorption energy of transition metal atoms on the Cu(100) surface

Θ cover type	p rel. units	q rel. units	q rel. units	p' rel. units	$\Delta\Phi$, eV			E_{ads}, eV		
					Co	Fe	Cr	Co	Fe	Cr
1, monolayer	1	1	0	1	0.33	0.43	0.47	1.70	1.67	1.65
without relaxation	1	1	0	1	−0.51	−0.26	−0.20	1.04	1.02	1.01
DFT-GGA [432]	1	1	0	1	0.53	−	−	1.75	−	−
1, 2 × 2	0.5	1	0.5	1	1.03	1.21	1.26	1.85	1.80	1.78
without relaxation	0.5	1	0.5	1	−0.34	−0.15	−0.03	1.09	1.06	1.05
DFT-GGA [432]	0.5	1	0.5	1	0.40	−	−	1.48	−	−
1 sandwich	0	1	1	1	−0.10	0.10	0.17	1.65	1.63	1.62
without relaxation	0	1	1	1	−1.36	−1.10	−1.02	1.00	0.99	0.98
DFT-GGA [432]	0	1	1	1	0.11	−	−	1.27	−	−
GGA – generalized gradient expansion										

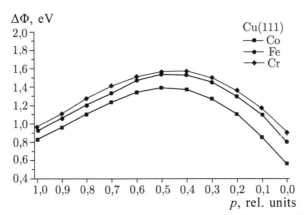

Fig. 9.21. The calculated dependence of the electron work function $\Delta\Phi$ due to the adsorption of atoms of transition metals, on the composition p of the solution in the film at $\Theta = 1$ Cu(111).

adsorption of atoms cobalt, iron and chromium lowers the value of function of the electron compared to the clean surface work function W(110) ($\Phi_M = 5,48$ eV [146]), which is confirmed by the results of experimental studies [18, 227].

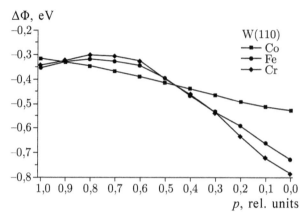

Fig. 9.22. The calculated dependence of the electron work function $\Delta\Phi$ determined by the adsorption of atoms of transition metals on the composition p of the solution in the film at $\Theta = 1$ of the W surface (110).

The studies lead to the general conclusion that the adsorption of transition metal atoms with a valence greater than the valency of the intrinsic atoms of the substrate, for all reconstruction processes in the surface region causes an increase of the work function of the electrons from the surface, whereas the inverse ratio of valences of the atoms of the adsorbate and the substrate leads to a decrease of the work function.

The results of studies of the effect of adsorption of the atoms of the transition and alkali metals on the electron work function of metal surfaces in the developed model of activated adsorption suggest the following conclusions.

1. For all types of coatings the adsorbates of the alkali metal atoms reduce the work function, while the adsorption of the atoms of transition metals can both increase the electron work function (valence of the atoms of the adsorbate atoms is higher than the valence of the substrate atoms), and lower it (the valence of the adsorbate atoms is less than the valences of the substrate atoms).

2. The dependence of the changes in the values of the electron work function on the composition of the film of the adsorbate and the surface layer was determined. The predicted changes in the work function values for the structures that formed during the adsorption of alkali metals on close-packed face of the aluminium substrate are in good agreement with the experimentally measured values of the work function.

10

FERROMAGNETISM OF ULTRATHIN FILMS OF TRANSITION METALS

Introduction

The properties of ultrathin magnetic films are the subject of intense research [498], which is largely determined by the potential applications in microelectronics and computer technology of ferromagnetic films as magnetic recording media and storing information in memory devices [38, 414, 443]. The magnetic films have a number of unique features that enhance the recording density of information storage and performance. Thin magnetic films have been used to develop various devices in the microwave range: filters, amplitude modulators, power limiters, and phase manipulators. Along with this, the study of the physical properties of ferromagnetic films is fundamental in nature, causing the development of both the physics of magnetic phenomena and the physics of surface phenomena.

To date, the magnetic ordering in ultrathin films of Fe, Co, Ni has been studied in many experimental works (see review [498]), which found that the long-range ferromagnetic order is established in them at some effective thickness of the films. However, the nature and laws of this phenomenon in ultrathin films are not entirely clear. The main difficulty in generalizing and adequate description of experimental results is related to the complex nature of the growth process of these films the morphology and properties of which depend heavily on a set of factors, in particular, the type of substrate (material, crystallinity, orientation of the surface, its cleanliness, temperature, etc.) and growth conditions. To ensure that the experiments are reproducible, it is necessary to carefully monitor multiple parameters in the course of these experiments. Note that the experimental studies of the formation of submonolayer metal films and the distribution of metal atoms on the surface of the substrate [419, 453, 475] revealed the effect of ejection of metal atoms by the substrate atoms on the surface with the implementation of substitutional adsorption and also surface phase transitions with the formation of 'islands' of adsorbed metal atoms. As regards the development of physical ideas concerning the mechanism of formation of stable ultrathin films and establishment in them of the magnetic order, the essential point is to consider the topology of these films, i.e.

their structure, taking into account the geometry and magnetic properties of 'islands' consisting of atoms of Fe, Co, Ni.

This chapter focuses on the theoretical description of the adsorption of magnetic ions of transition metals Fe, Co, Ni on metal surfaces of paramagnetic materials and on identifying the conditions of formation of ferromagnetic submonolayer films. The question of the influence of magnetic ordering on the energy characteristics of adsorption is studied and the conditions of implementation of the 'island' adsorption in a system of interacting magnetic atoms on a metal surface are investigated. The influence of the ferromagnetic ordering in submonolayer films of transition metals Fe, Co, Ni on the energy characteristics of both non-activated adsorption and substitutional adsorption processes, accompanied by mixing of atoms of the adsorbate and the surface layer of the substrate, is described.

The material of the chapter is based on the results obtained by the authors of this book in [141, 142, 193, 194].

10.1. Adsorption of transition metal ions on metal surfaces with formation of submonolayer ferromagnetic films

Consider a semi-infinite metal with mean charge density n_1, bounded by an infinite flat surface and occupying region $z < -D$. The adsorbate film with thickness h with charge density n_2 occupies the region $D < z < D + h$. Between the film and the substrate in this model there is a vacuum gap of width $2D$. The positive charge of the background is thus distributed according to the relation

$$n_0(z) = n_1\theta(-z-D) + n_2\theta(z-D)\theta(D+h-z), \qquad (10.1)$$

where $\theta(z)$ is a step function. The solution of the linearized Thomas-Fermi equation with boundary conditions reflecting the continuity of the potential $\varphi(z)$ and its first derivative $d\varphi/dz$ at $z = \pm D$ and $z = D + h$, and the conditions of finiteness of the potential for $z \to \pm\infty$, allows at $\varphi(z) = -4\pi n(z)/\beta^2$ to obtain the following expression for the density of the electron distribution $n(z)$ in the system:

$$n(z) = \begin{cases} \dfrac{n_1}{2}(2 - e^{\beta(z+D)}) + \dfrac{n_2}{2}e^{\beta(z-D)}(1 - e^{-\beta h}), & z < -D; \\[2ex] \dfrac{n_1}{2}e^{-\beta(z+D)} + \dfrac{n_2}{2}e^{\beta(z-D)}(1 - e^{-\beta h}), & |z| < D; \\[2ex] \dfrac{n_2}{2}(2 - e^{\beta(z-D-h)}) - & \\ \quad -\dfrac{1}{2}(n_2 - n_1 e^{-2\beta D})e^{-\beta(z-D)}, & D < z < D+h; \\[2ex] \dfrac{1}{2}e^{-\beta(z-D-h)}(n_1 e^{-\beta(2D+h)} - & \\ \quad -n_2 e^{-\beta h} + n_2), & z > D+h. \end{cases} \qquad (10.2)$$

Here β is a variational parameter. The quantity $1/\beta$ is the characteristic thickness of the surface layer in which the electron density changes dramatically.

In view of the need to describe the temperature variations of magnetization in the adsorbed ferromagnetic film we define the interfacial interaction energy per unit area of contact as an integral over z from the bulk density of the free energy of the electron gas:

$$\sigma_0 = \int_{-\infty}^{+\infty} \left\{ f[n(z,\beta)] - f[n_0(z)] \right\} dz. \tag{10.3}$$

In the 'jelly' model the volume density of the free energy of the inhomogeneous electron gas can be represented as a gradient expansion:

$$f[n(z)] = w_0[n(z)] + w_2[n(z), |\nabla n(z)|^2] +$$
$$+ w_4[n(z), |\nabla n(z)|^4] - T(s_{id} + s_{order}), \tag{10.4}$$

where

$$w_0[n(z)] = w_{kin} + w_{cul} + w_x + w_c \tag{10.5}$$

is the energy density of the homogeneous electron gas, including successively the kinetic, Coulomb, exchange and correlation energy, s_{id}, s_{order} are the entropic contributions to the free energy taking into account the temperature changes in the entropy for the electron gas and the effects of magnetic ordering in the electronic subsystem, respectively.

When considering the magnetically ordered state in metals in the model of collectivized electrons we take into account the redistribution of electrons over single-particle states due to the influence of the internal magnetic field of exchange nature [43]. In this case, the electron density of quasi-particles with spin 'up' n_+ is different from the electron density of quasi-particles with spin 'down' n_-. The distribution of quasi-particles in each of the subsystems of electronic states can be characterized by its Fermi level with energy $\varepsilon_{F_\pm}(n_\pm)$. The energies of the electronic system will then be as follows (in atomic units):

– kinetic

$$w_{kin} = 0.3(6\pi^2)^{2/3} \left(n_+^{5/3}(z) + n_-^{5/3}(z) \right) +$$
$$+ k_B^2 T \left(\frac{\pi^2}{4} \right) \left(\frac{n_+(z)}{\varepsilon_{F_+}} + \frac{n_-(z)}{\varepsilon_{F_-}} \right), \tag{10.6}$$

– Coulomb

$$w_{cul} = 0.5\varphi(z)n(z), \tag{10.7}$$

– exchange

$$w_x = -0.75(6/\pi)^{1/3}[n_+^{4/3}(z) + n_-^{4/3}(z)], \qquad (10.8)$$

– correlation

$$w_c = -0.056\frac{n_-^{1/3}(z)\cdot n_+(z)}{0.079 + n_+^{1/3}(z)} - 0.056\frac{n_+^{1/3}(z)\cdot n_-(z)}{0.079 + n_-^{1/3}(z)}. \qquad (10.9)$$

The densities n_+ and n_- can be expressed in terms of relative magnetization $m(T) = M(T)/M(T = 0)$ of electrons as follows:

$$n_\pm(z) = n(z)\frac{1 \pm m}{2},$$
$$n_\pm(z) = \int N(\varepsilon)f(\varepsilon \mp \lambda m)d\varepsilon, \qquad (10.10)$$

where $N(\varepsilon)$ is the density of electron states, $f(\varepsilon)$ is the function of the Fermi–Dirac distribution, λ is the exchange parameter for the collectivized electrons.

Entropic contributions to free energy are given by

$$S_{id} = k_B^2 T\left(\frac{\pi^2}{2}\right)\left(\frac{n_+(z)}{\varepsilon_{F_+}} + \frac{n_-(z)}{\varepsilon_{F_-}}\right),$$

$$S_{order} = k_B\frac{n(z)}{2}[\ln 4 - (1+m)\ln(1+m) - (1-m)\ln(1-m)]. \qquad (10.11)$$

Gradient corrections to the density of the kinetic and exchange-correlation energies, taking into account the heterogeneity of the electron gas in the surface region are determined by the following relations:

$$w_2 = \sum_{n=n_+,n_-}\frac{1}{72}\frac{|\nabla n|^2}{n} + w_{2,xc}\left(n, |\nabla n|^2\right),$$

$$w_{2,xc}[n_\pm] = \frac{A(n_\pm)B^2(n_\pm)|\nabla n_\pm|^2}{3^{4/3}\pi^{5/3}n_\pm^{4/3}},$$

$$A(n_\pm) = 0.4666 + 0.3735k_{F_\pm}^{-2/3}(n_\pm),$$

$$B(n_\pm) = -0.0085 + 0.3318k_{F_\pm}^{1/5}(n_\pm), \qquad (10.12)$$

$$k_{F_\pm} = (3\pi^2 n_\pm)^{1/3},$$

where the first term in w_2 is the correction to the heterogeneity of the kinetic energy, a $w_{2,xc}$ denotes the correction for heterogeneity for the exchange and correlation energies in the Vashishta–Singvi approximation (VS), k_F is the Fermi wave vector. The VS approximation is used most widely for the majority of metals. As shown in chapters 5 and 6, to improve the quantitative agreement of the values of the surface and interfacial energies for the transition and noble metals, in calculations of the surface characteristics for these metals it is also necessary to take into account fourth-order terms in powers of $|\nabla n(z)|$ to the densities of kinetic and exchange-correlation energies:

$$w_4 = w_{4,\text{kin}}(z) + w_{4,\text{xc}}(z),$$

$$w_{4,\text{kin}}(z) = \sum_{n=n_+,n_-} \frac{1.336}{540(3\pi^2 n)^{3/2}} \times$$

$$\times \left[\left(\frac{\nabla^2 n}{n} \right) - \frac{9}{8} \left(\frac{\nabla^2 n}{n} \right) \frac{|\nabla n|^2}{n} + \frac{1}{3} \frac{|\nabla n|^4}{n} \right], \qquad (10.13)$$

$$w_{4,\text{xc}}(z) = \sum_{n=n_+,n_-} 1.98 \cdot 10^{-5} \exp\left(-0.2986 n^{-0.26}\right) \left(\frac{\nabla^2 n}{n} \right)^2.$$

Accounting for the discreteness in the distribution of ions leads to corrections in the electrostatic interaction energy due to both ion–ion and electron–ion interactions. As a result, the interfacial interaction energy can be written as

$$\sigma = \sigma_0 + \sigma_{ei} + \sigma_{ii}, \qquad (10.14)$$

where σ_0 is the contribution of the electronic system in the framework of the 'jelly' model, σ_{ii} is the correction to the energy of the electrostatic interaction of ions, σ_{ei} is the correction to the energy associated with the difference in electrostatic interactions of electrons with discrete ions and with uniform 'jelly' background. To calculate σ_{ii} we can use the method described in Chapter 5. As a result we obtain the following expression for the interfacial energy of the ion–ion interaction:

$$\sigma_{ii} = \sqrt{3} \frac{Z_1^2}{c_1^3} \exp\left(-\frac{4\pi d_1}{\sqrt{3}c_1} \right) +$$

$$+ 2\sqrt{3} \frac{Z_2^2}{c_2^3} \exp\left(-\frac{4\pi h}{\sqrt{3}c_2} \right) \left[1 - \exp\left(-\frac{4\pi h}{\sqrt{3}c_2} \right) \right] -$$

$$- 2\sqrt{3} \frac{Z_1 Z_2}{(c_1 c_2)^{3/2}} \exp\left[-\frac{2\pi}{\sqrt{3}} \left(\frac{d_1 + D}{c_1} + \frac{h + D}{c_2} \right) \right] \times$$

$$\times \left[1 - \exp\left(-\frac{4\pi h}{\sqrt{3}c_2} \right) \right], \qquad (10.15)$$

where Z_1, Z_2 are the charges of the ions, c_1 is the distance between the nearest ions of the substrate in a plane parallel to the surface, c_2 is the distance between neighbouring ions in the adsorbate layer.

The electron–ion component of the interfacial energy σ_{ei} is given accordance with chapter 5 by the expression

$$\sigma_{\mathrm{ei}} = \frac{2\pi}{\beta^3} \left\{ \left[n_1^2 - n_1 n_2 e^{-2\beta D} \left(1 - e^{-\beta h}\right) \right] \left(1 - \frac{\beta d_1 e^{-\beta d_1/2}}{1 - e^{-\beta d_1}} \mathrm{ch}(\beta r_1) \right) + \right.$$

$$\left. + \left(2n_2^2 - n_1 n_2 e^{-2\beta D}\right) \left(1 - e^{-\beta h}\right) \left(1 - \frac{\beta h e^{-\beta h/2}}{1 - e^{-\beta h}} \mathrm{ch}(\beta r_2) \right) \right\}, \quad (10.16)$$

where r_1 and r_2 are the Ashcroft pseudopotential cutoff radii for the ions of the substrate and the film respectively, d_1 is the distance between the ionic planes of the substrate material.

The equilibrium distribution of magnetization in the system can be determined from the thermodynamic requirement of the minimum interfacial energy of interaction

$$\left(\frac{\partial \sigma}{\partial m} \right) = 0, \quad (10.17)$$

however, in [141, 142] to obtain more adequate quantitative values of the surface characteristics, taking into account the effects of magnetic ordering in the adsorbate film, it was proposed to apply the known temperature dependence of relative magnetization m in the molecular field approximation, obtained from solutions of the well-known equation [43]

$$m = B_S \left\{ 3mST_c / \left[(S+1)T\right] \right\}, \quad (10.18)$$

determined by the Brillouin function

$$B_S(x) = \frac{2S+1}{2S} \mathrm{cth} \left(\frac{2S+1}{2S} x \right) - \frac{1}{2S} \mathrm{cth} \left(\frac{x}{2S} \right), \quad (10.19)$$

where S is the spin of magnetic ions, T_c is the Curie temperature.

Although the relation (10.19) was introduced to describe the magnetic properties of the localized spins, it can, in contrast to the ratios obtained in the Stoner theory for the ferromagnetism of collectivized electrons, well describe the observed temperature dependence of relative magnetization of ferromagnetic materials (such as transition metals Fe, Co, Ni with spins $S_{\mathrm{Fe}} \approx 1.11$, $S_{\mathrm{Co}} \approx 0.86$, $S_{\mathrm{Ni}} \approx 0.30$, corresponding to the fractional effective magnetic moments of ions in these metals [43]) with the exception of the critical fluctuation region and the low temperature range in which $m(T)$ is described by the spin-wave approximation. The functional relation (10.19) is used due to the fact that in this specification it is not attempted to substantiate the ferromagnetism submonolayer films and determine the main characteristics of the magnetic ordering, and instead it is intended to investigate the effect of ferromagnetism on the adsorption properties of these films which justifies the use of the phenomenological approach with the more realistic behaviour of the magnetization in the range of finite temperatures which is not covered by the Stoner theory [100].

The critical temperature $T_c^{(s)}$ of magnetic ordering of the monatomic film depends on the parameter Θ, i.e. $T_c^{(s)} = T_c^{(s)}(\Theta)$, and differs from the critical temperature of the bulk magnetic $T_c^{(b)}$. The value of $T_c^{(s)}(\Theta)$ in the molecular field approximation can be estimated as follows:

$$T_c^{(s)}(\Theta) \cong \Theta T_c^{(b)} \frac{z_{surf}}{z_{bulk}}, \qquad (10.20)$$

where z_{surf} is the number of nearest neighbours in the ferromagnetic film, and z_{bulk} – in the bulk ferromagnet (the well-known relation $T_c = 2zJS(S + 1)/3k_B$ [43] is used, which at $\Theta = 1$ immediately gives $T_c^{(s)}(\Theta)/T_c^{(b)} = z_{surf}/z_{bulk}$, and using the relations (9.18) and (9.19) we obtain (10.20) for submonolayer films). At the same time as the critical temperature of magnetic ordering of the bulk ferromagnet $T_c^{(b)}$, it is proposed to use its experimental value.

In accordance with the method of the density functional the value of the variational parameter β in the trial function (10.2) is obtained from the requirement of the minimum total interfacial energy of the system, i.e.,

$$\left(\frac{\partial \sigma(\beta, D)}{\partial \beta} \right)_{\beta = \beta_{min}} = 0. \qquad (10.21)$$

The solution of equation (10.21) defines the parameter β_{min} as a function of the size of the gap and the structural parameters for the substrate and the coating. The result of solving this variational problem is the total interfacial energy of the system $\sigma(\beta_{min}(D), D)$. Knowing this value, it is easy to find the adhesion energy of the system as the work that must be carried out to remove the substrate and film from each other to infinity, i.e.

$$E_a(2D) = \sigma(\infty) - \sigma(2D). \qquad (10.22)$$

The adhesion energy is used to calculate the adsorption energy as the work necessary to remove the adsorbed particles from the surface of the substrate. For the adsorption energy we use the specific adhesion energy per one adsorbed atom:

$$E_{ads}(\Theta) = E_a / n_{s2}, \qquad (10.23)$$

where the surface concentration of the adsorbate n_{s2} is a function of the coverage parameter Θ and the parameters c_1 and c_2, which characterize the symmetry of the surface and the arrangement of atoms in the film. With the use of relations (9.17)–(9.19) and the approximation to which they correspond, in calculating the total interfacial energy we can separate the energy characteristics of adsorption as a function of the coverage parameter Θ, the film thickness h and the size of the gap D between the substrate and the adsorbed film.

In studying the effect of the ferromagnetic ordering on the adsorption energy, as in the case of the study of non-activated adsorption of atoms of alkali metals (chapter 9), four model options were studied.

1. The case of the absence of the gap between the substrate and the film ($D = 0$) and a fixed film thickness ($h = d_2$ = const), given by the bulk value for the interplanar distance d_2 of the adsorbed metal. This approach is classical in the theory of adsorption.

2. The case of a fixed film thickness ($h = d_2$ = const), but with the equilibrium gap D_{min}. The value D_{min} is determined from the minimum total interfacial energy. Introduction of the equilibrium vacuum gap simulates the effective influence of the atomic roughness of the substrate using the one-dimensional version of the density functional $n(z)$. When $D = D_{min}$ the strength of adhesion interaction between the substrate and the film per unit area of the interface section is $F_a(D_{min}) = 0$. Therefore, the equilibrium state of the system at a given filling of the surface with the adatoms can be determined.

The effects of relaxation of the film thickness h can also be taken into account

$$h = d_2 + \gamma, \tag{10.24}$$

where γ is the variational parameter characterizing the relaxation of the film thickness and is determined from the minimum total interfacial energy. The introduction of these effects simulates the influence of the potential formed by the ions of the substrate on the distribution of the adsorbate ions in the set of the energetically non-equivalent positions of the ions on the surface (e.g. interstitial positions and the positions of the ions of the adsorbate ions above the substrate ions) in the one-dimensional version of the density functional $n(z)$. The following cases can therefore be defined.

3. The case of the absence of the gap between the substrate and film ($D = 0$), but with the equilibrium film thickness $h = h_{min}$, defined by (10.24) with the known value of the variational parameter γ_{min}.

4. The case of the equilibrium gap $D = D_{min}$ and the equilibrium film thickness $h = h_{min}$. Their values are determined from the minimum total interfacial energy $\sigma(\beta_{min}, \gamma_{min}, \Theta, D_{min})$. This approach allows us to predict the formation of stable coatings or indicate their absence.

The results of calculating the characteristics of adsorption and analysis

In [141, 142, 194], the authors carried out calculations of the adsorption energy of the ions of Fe, Co and Ni on the copper substrate with different orientations of its surface facets. However, the investigations carried out in generalized conditions of the model case 4 of the dependence of the interfacial energy on the coverage parameter Θ, varying in the range $0.2 \le \Theta \le 1$, showed that stable monatomic films of Fe, Co and Ni cannot form on the close-packed surface (111) and (100) copper substrate (no minimum interfacial energy). It was shown that the minimum interfacial energy, and, consequently, the formation of submonolayer films of Fe, Co and Ni can be take place only on the loose edges of the copper substrate (110) and with higher Miller indices. This justifies the observed only 'island' adsorption of metal ions on the faces of close-packed metal surfaces [324, 419, 453, 475]. The results of the calculated adsorption energy of the Fe, Co and Ni ions on the loose (110) face of the copper substrates for all four simulation cases are presented below.

Figure 10.1 shows the calculated adsorption energy of iron ions (Fig. 10.1a), cobalt (Fig. 10.1b) and nickel (Fig. 10.1c) on the (110) surface face of copper in the framework of the case 1. In accordance with (10.20), the temperature of the ferromagnetic phase transition in an iron film at $\Theta = 1$ is estimated as $T_c^{(s)}(\Theta = 1) = 521$ K, in the cobalt film $T_c^{(s)}(\Theta = 1) = 464$ K and in the nickel film $T_c^{(s)}(\Theta = 1) = 209$ K. The graphs clearly show that taking into account the effects of ferromagnetic ordering makes a significant contribution to the change of adsorption energy. The adsorption energy in the paramagnetic phase ($m = 0$) is much lower than the energy of the completely ordered ferromagnetic phase ($m = 1$). Thus, the differences in energies for the paramagnetic and ferromagnetic ($T = 0$ K) states for the iron–copper, cobalt–copper and nickel-copper systems (with $\Theta = 1$) are about $2 \div 3$ eV. It suggests that the formation of a ferromagnetic film on the substrate surface significantly increases the adsorption energy in comparison with the paramagnetic film.

Calculations show that the adsorption energy for the Ni ions on a copper substrate at all deposition parameters T and Θ is characterized by higher values than for the Fe and Co ions, which reflects the best adhesion of films of nickel to copper in relation to the films of iron and cobalt on the same copper substrate.

Figure 10.1 shows that with decreasing temperature in relation to temperature T_c of the ferromagnetic phase transition the adsorption energy decreases rapidly with increasing Θ from its value in the paramagnetic phase to the values corresponding to the adsorption energy for the completely ordered state of the film with $m = 1$. Examination of the first model case shows that in the whole range of Θ the condition for the existence of a continuous monatomic adsorbate film on the (110) face is fulfilled.

The effective spins of magnetic ions of Fe, Co and Ni in the respective metals ($S_{Fe} \approx 1.11$, $S_{Co} \approx 0.86$, $S_{Ni} \approx 0.30$) were used as calculation parameters, but in the range of small $\Theta = 0.2 \div 0.4$ and at low temperatures it is preferred to use in calculations the values of the spins of the isolated ions with $S_{Fe} \approx 2.95$, $S_{Co} \approx 2.4$, $S_{Ni} \approx 1.6$.

Figure 10.1 shows the results of calculations of the adsorption energy of Fe and Co at $T = 100$ K and at $T = 200$ K, and Ni ions at $T = 100$ K and at $T = 150$ K at the spins of isolated ions, as well as when using for the effective spins in Fe and Co the linear extrapolation in respect of Θ from their values in metals $\Theta = 1$ to the values in isolated ions with $\Theta = 0.2$. It

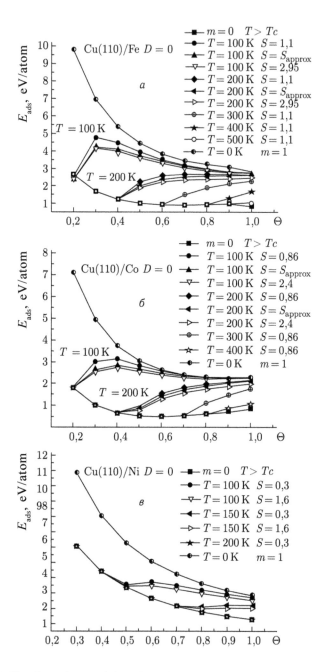

Fig. 10.1. The dependence of the adsorption energy on the coverage parameter Θ for systems: a – Cu (110)/Fe; b – Cu (110)/Co; c – Cu (110)/Ni, when considering in the model case 1.

is evident that the use of values of spins for the isolated ions affects only the Fe and Co ions at $T = 100$ K, leading to a decrease in the adsorption energy by the value of $\Delta E_{ads} \leq 0.5$ eV at $\Theta = 0.3 \div 0.4$.

Figure 10.2 shows the calculated adsorption energy of the ions of the same metals in the model case 2. It can be seen (Fig. 10.2*a*) that in contrast to the case 1, the stable submonolayer film for iron at temperatures $T = 200$ K and $T = 300$ K forms only at $\Theta < 0.46$ and $\Theta < 0.72$, respectively, and at Θ higher than the given values, the 'island' adsorption of magnetic iron ions is energetically more favourable. At $T = 400$ K and $T = 500$ K we obtain a stable coating over the entire range of Θ. For temperatures $200 \div 500$ K the adsorption energy first decreases with increasing parameter Θ in the range where the specific magnetization of the film is equal to zero (in this range the graphs merge with the graph corresponding to $m = 0$, $T > T_c$), and at specific magnetization different from zero, the adsorption energy begins to sharply increase with increasing Θ.

Similar effects are predicted for a cobalt film (Fig. 10.2*b*): for temperatures $T = 150$ K and $T = 200$ K, a stable magnetically ordered film is formed only when $\Theta < 0.42$ and $\Theta < 0.55$, respectively, and at Θ higher than the given values 'island' adsorption is energetically more favourable; for $T = 300$ K and $T = 400$ K a stable cobalt coating forms throughout the entire range of values Θ.

For a nickel film (Fig. 10.2*c*) at $T = 125$ K a magnetically stable film is formed when $\Theta < 0.675$ (already at $T \leq 120$ K the ferromagnetic component in E_{ads} disappears), and at Θ, higher than the given values, 'island' adsorption is energetically more advantageous; at $T = 150$ K and at $T = 200$ K, a stable coating is formed over the entire range of values Θ. At small Θ, where the specific magnetization of the film is zero, the adsorption energy decreases with increasing Θ, and then, beginning at certain values of Θ ($\Theta = 0.7$ for $T = 150$ K and $\Theta = 0.9$ for $T = 200$ K) the adsorption energy due to the effects of the ferromagnetic ordering in the film changes the nature of the changes (increases with increasing Θ for $T = 150$ K).

Figure 10.2 presents also the results of using in calculating the adsorption energy of the spin values for isolated spins of the ions and the procedures for the linear extrapolation with respect to Θ for the effective spin of magnetic ions in the systems: Cu(110)/Fe at $T = 200$ K and at $T = 300$ K (Fig. 10.2*a*), Cu (110)/Co at $T = 150$ K and at $T = 200$ K (Fig. 10.2*b*), Cu (110)/Ni at $T = 125$ K (Fig. 10.2*c*).

For th iron films taking these effects into account leads to changes of the adsorption energy $\Delta E_{ads} \cong 0.4$ for $T = 300$ K for $\Theta = 0.72$ and $\Delta E_{ads} \cong 0.3$ eV for $T = 200$ K for $\Theta = 0.46$, for films of cobalt leads to changes $\Delta E_{ads} \cong 0.6$ eV for $T = 200$ K for $\Theta = 0.55$ and $\Delta E_{ads} \cong 0.3$ eV for $T = 150$ K, when $\Theta = 0.44$, for films of nickel-in $\Delta E_{ads} \cong 0.3$ eV for $T = 125$ K for $\Theta = 0.67$. However, for the region of small values of Θ ($\Theta = 0.2 \div 0.4$), in which it is important to consider these effects, changes ΔE_{ads} did not exceed values of 0.3 eV for films of iron and cobalt.

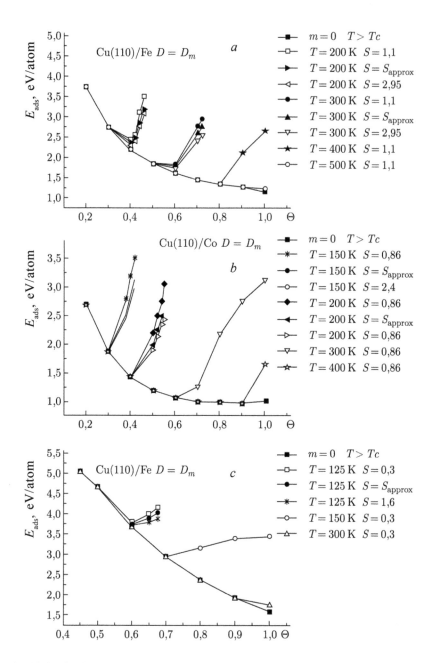

Fig. 10.2. The dependence of the adsorption energy on the coverage parameter Θ for systems: *a* – Cu (110)/Fe; *b* – Cu (110)/Co; *c* – Cu (110)/Ni, in the case of model 2.

In other cases, changes in the adsorption energy are too small or lie in the range of sufficiently large Θ, where the effective spins of the magnetic ions reflect the collective nature of the interaction and take values as in the respective metals.

Figures 10.3 and 10.4 present the results of calculations of E_{ads}, respectively, for the model variants 3 and 4 of description of adsorption. It can be seen that in these cases the stable submonolayer coating formed only in a very narrow range of variation of Θ: from $\Theta = 0.90 \div 0.92$ to 1 for an iron film, in the range $\Theta = 0.90 \div 0.96$ to 1 for a cobalt film, in the range $\Theta = 0.86 \div 0.96$ to 1 for a nickel film. In addition, the temperature at which a stable monolayer coating can form is high enough and close to the critical temperature.

For example, the minimum temperature at which the formation of a stable monatomic magnetic film of iron is observed, for the model case 3 is 460 K, and for the case 4 it is 480 K, while for the cobalt film it is 400 K (case 3) and 420 K (case 4), for the nickel film 160 K for the cases 3 and 4. In the rest of the range of variation of parameter Θ and at temperatures lower than noted above, the energetically more favourable 'island' adsorption takes place.

Analysis of the results of studies leads to the following conclusions.

1. The formation of stable submonolayer films of ferromagnetic metals can take place only on the loose edges of the metal substrates, which justifies the occurrence of only the 'island' adsorption of metal ions on the close-packed faces of the metal surfaces;

2. Considering the effects of ferromagnetic ordering in the adsorbed monatomic film of transition metals has a significant influence on the adsorption energy, leading to a marked increase of this energy;

3. Adsorption of transition metal ions on a metal substrate at the equilibrium film thickness and the equilibrium interfacial gap leads to the formation of a stable monatomic ferromagnetic film only in a narrow temperature range close to the magnetic phase transition temperature, and the values of Θ close to unity. In the rest of the range of values Θ and at temperatures that are substantially lower than the temperature of the phase transition the energetically more favorable 'island' adsorption takes place.

The further development of this technique has made it possible to describe more complex substitutional adsorption processes, presented in the next section.

Substitutional adsorption is characterized by the fact that the adsorbed metal atoms can push the surface substrate atoms into the film and take their place. Every adsorbate has its own temperature below which the substitutional adsorption is not possible [293, 453]. However, even at temperatures close to room temperature there is only substitutional adsorption taking place, as almost the adatoms of all metals have sufficient energy at this temperature for substitutional processes. Therefore, substitutional adsorption is also called *activated* adsorption. It has been

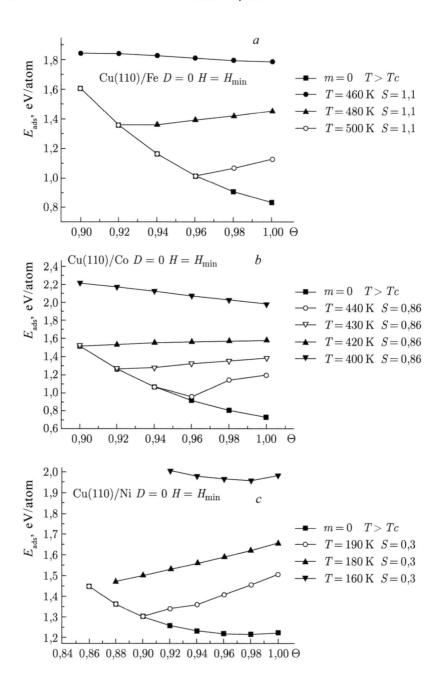

Fig. 10.3. The dependence of the adsorption energy on the coverage parameter Θ for systems: a – Cu (110)/Fe; b – Cu (110)/Co; c – Cu (110)/Ni when considering in the model case 3.

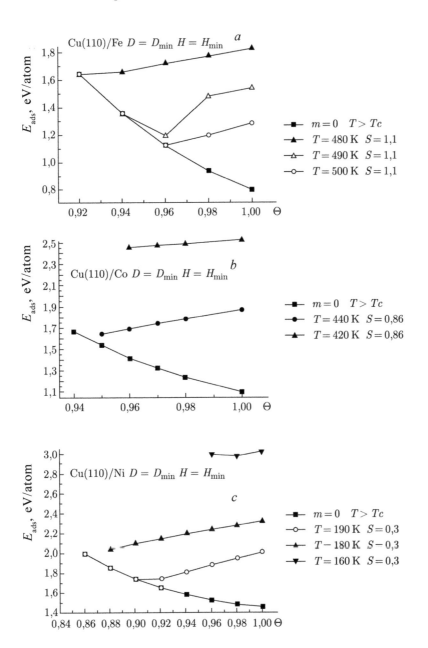

Fig. 10.4. The dependence of the adsorption energy on the coverage parameter Θ for systems: a – Cu (110)/Fe; b – Cu (110)/Co; c – Cu (110)/ Ni, when considering the model case 4.

established [25, 140, 449] that the processes of substitutional adsorption are energetically more favourable compared with the processes of activated adsorption.

10.2. Activated adsorption of magnetic ions. The influence of the effects of ferromagnetic ordering

In this section we consider the question of the influence of magnetic ordering on the energy characteristics of adsorption and investigate the conditions for the realization of activated (substitutional) adsorption characterized by processes of mutual mixing of atoms of the adsorbed material and the substrate.

Consider a semi-infinite metal with mean charge density n_1, bounded by the infinite flat surface and occupying the region $z < -D - l$. The film of the adsorbate with the charge density n_3 and thickness h occupies the region $D < z < D + h$. As a result of processes of mutual mixing of atoms of the adsorbate and the substrate, characteristic for activated adsorption, the surface of the substrate is exposed to various reconstructions.

The most significant reconstruction takes place in the first near-surface layer of the substrate. Therefore, this layer is separated to an area with mean charge density n_2 and thickness l. Between the film and the substrate in this model there is a vacuum gap with width $2D$. The positive charge of the background, therefore, is distributed in accordance with the formula

$$n_0(z) = n_1\theta(-z-D-l) + n_2\theta(z+D+l)\theta(-z-D) +$$
$$+n_3\theta(z-D)\theta(D+h-z). \qquad (10.25)$$

The distribution of the surface layers is clearly shown in Fig. 9.14.

The solution of the linearized Thomas–Fermi equation with the use of the boundary conditions, reflecting the continuity of the electrostatic potential $\phi(z)$ and its first derivative $d\phi/dz$ at $z = -D - l$, at $z = \pm D$ and $z = D + h$, and the finiteness of the potential for $z \to \pm\infty$, allows for the relationship $\phi(z) = -4\pi n(z)/\beta^2$ gives the expression for the density of the electron distribution $n(\beta, z)$ in the system given by (9.30).

Since it is necessary to describe the thermal processes of mixing of atoms in the surface layer, as well as temperature changes in magnetization of the ferromagnetic film we define the interfacial interaction energy falling on unit area of contact as an integral over z from the bulk density of the *free energy* of the electron gas:

$$\sigma_0 = \int_{-\infty}^{+\infty} \{f[n(z,\beta)] - f[n_0(z)]\}\, dz . \qquad (10.26)$$

In the 'jelly' model, the bulk density of the free energy of the inhomogeneous electron gas can be represented as a gradient expansion and is characterized, as in the previous section, by expressions (10.4)–(10.13).

Amendments to the interfacial energy of the electron–ion σ_{ei} and ion–ion interaction σ_{ii}, associated with the discrete distribution taking into account the positive charge of the film and the substrate are determined by the relations (9.35)–(9.36).

As a result, the total interfacial energy can be written as

$$\sigma(\beta, D) = \sigma_0(\beta, D) + \sigma_{ei}(\beta, D) + \sigma_{ii}(D) \tag{10.27}$$

The value of the variational parameter β is found from the minimality requirement

$$\left(\frac{\partial \sigma(\beta, D)}{\partial \beta} \right)_{\beta=\beta_{min}} = 0, \tag{10.28}$$

and the total interfacial energy of the system $\sigma(\beta_{min}(D), D)$ is then determined. Knowing it, we can find the adhesion energy

$$E_a(2D) = \sigma(\infty) - \sigma(2D). \tag{10.29}$$

The adsorption energy is represented by the specific adhesion energy per one adsorbed atom,

$$E_{ads} = E_a / n_s^a, \tag{10.30}$$

where the surface concentration of adatoms n_s^a is a function of the parameters c_1, and c_2 which characterize the symmetry of the substrate surface and distribution of atoms in the adsorbate layer.

The surface concentration n_s^a is also a function of the parameter Θ determining the degree of filling of the surface of the substrate by the adatoms:

$$\Theta = n_s^a / n_{s1}, \tag{10.31}$$

where n_{s1} characterizes the surface concentration of atoms of the non-reconstructed substrate.

We believe that the adatoms cause the reconstruction of only the first surface layer of the substrate and the substituted atoms of this layer are located in the film. Thus, the first surface layer of the substrate and the film are characterized by the formation of a mixture of substrate and adsorbate atoms which has properties similar to a binary solution of two of metals. To find the parameters of these binary solutions we use the relations (9.43)–(9.54).

To determine the temperature dependence of magnetization of the adsorbed film, we use, as in the previous section, the molecular field approximation and the relations (10.18) and (10.19). The critical temperature $T_c^{(s)}$ of magnetic ordering of the monatomic film depends on the coverage parameter Θ, i.e. $T_c^{(s)} = T_c^{(s)}(\Theta)$ and differs from the bulk critical temperature of the magnet $T_c^{(b)}$. The dependence $T_c^{(s)}(\Theta)$ and its relationship with $T_c^{(b)}$ is defined by (10.20).

The theoretical models constructed in [193] implemented two cases of the adsorption of magnetic ions of transition metals Fe, Co, and Ni on a paramagnetic copper substrate with the orientation of the (110) surface.

Figure 10.5 shows plots of the calculated adsorption energy of the Fe submonolayer film on the Cu (110) face, depending on the parameter of binary solution p, defining the relative share of adsorbate atoms in the film, for three values $\Theta = 1.0, 0.7, 0.5$.

There is a clearly noticeable shift of the maximum adsorption energy as a function of the parameter p from the values $p = 0.5 \div 0.6$, characterizing the temperature dependence of the energy advantage of almost half mixing of the adsorbate ions with the substrate ions for $\Theta = 1.0$, to a maximum at $p = 1$ for $\Theta = 0.5$, characterizing the manifestation of the energetically unfavorable substitutional adsorption with decreasing coverage Θ.

Similar conclusions follow from studying the behaviour of the graphs of adsorption energy for the Co submonolayer films (Fig. 10.6). However, a different picture is observed for the adsorption of the Ni ions (Fig. 10.7): position of the maximum adsorption energy practically for all $\Theta < 1$ is found at $p = 0$, which indicates the energy advantage for Ni/Cu (110) of the adsorption structure of the 'sandwich' type in which the adsorbate atoms are almost completely displaced from the first surface layer. When $\Theta = 1$ the Ni adsorption energy is characterized by the dependence on the composition of adsorbed film which is so weak that thermal fluctuations can completely mix the ions of the adsorbate and substrate. Such differences of Ni adsorption from the adsorption of Fe and Co are due, on the one hand, to the smallest ionic radius compared with the ions of Fe, Co and Cu, which contributes to easier penetration of Ni ions into the surface layer of Cu and, on the other hand, due to the fact thyat the fcc crystalline structures of metals Ni and Cu are closest in comparison to other investigated metals and their lattice constants are: $a_{Ni} = 3.52$ Å, $a_{Cu} = 3.61$Å.

Figure 10.8 shows a plot of values of the parameter p of the composition of the film at the maximum adsorption energy of ions of Fe, Co and Ni on a copper substrate with the variation of Θ. These values were used in calculating the adsorption energy for the energetically more favourable adsorption structure of the ions of Fe, Co and Ni on the Cu (110) face, depending on the parameter Θ of the film (Fig. 10.9).

Figure 10.9 shows by the dashed lines the graphs for the case of activated adsorption (no mixing and substitution of the substrate ions by the adsorbate ions), the solid lines show the case of activated adsorption in the system. The presented plots clearly shows that the energy of activated adsorption is much higher than that for non-activated adsorption ($\Delta E_{ads} \geq 2$ eV), especially at small values of Θ, which indicates the energy advantages of substitutional adsorption as compared to non-activated adsorption; in addition, the effects of ferromagnetic ordering have a significant impact

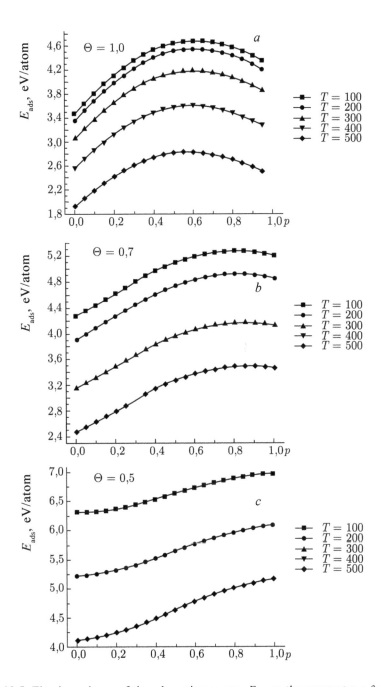

Fig. 10.5. The dependence of the adsorption energy E_{ads} on the parameter p for the Fe/Cu (110) system at $\Theta = 1.0$ (a), $\Theta = 0.7$ (b) and $\Theta = 0.5$ (c).

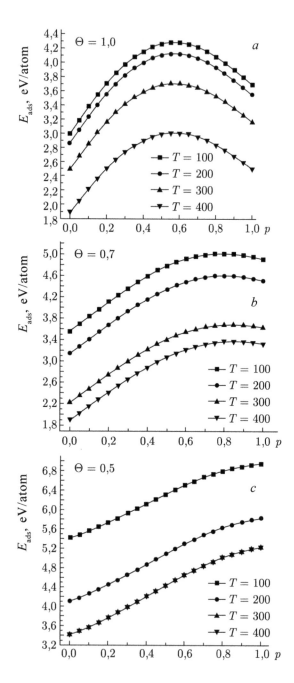

Fig. 10.6. The dependence of the adsorption energy E_{ads} on the parameter p for the Co/Cu (110) system at $\Theta = 1.0$ (*a*), $\Theta = 0.7$ (*b*) and $\Theta = 0.5$ (*c*).

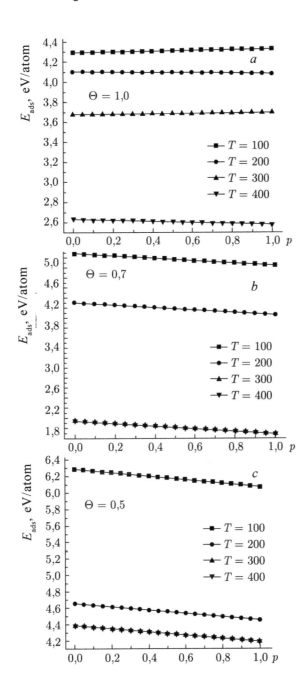

Fig. 10.7. The dependence of the adsorption energy E_{ads} on the parameter p for the Ni/Cu (110) system at $\Theta = 1.0$ (a), $\Theta = 0.7$ (b) and $\Theta = 0.5$ (c).

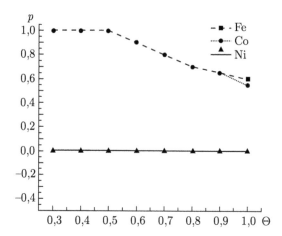

Fig. 10.8. The dependence of the maximum adsorption energy on the coverage parameter Θ.

on the adsorption energy, leading to its increase. Thus, in the paramagnetic phase $(m = 0, T > T_{c}^{(s)})$ the adsorption energy is less than in the cases with appreciable ferromagnetic ordering, for example, $T = 100$ K.

Note that the observed temperature changes in the adsorption energy are largely due to the temperature dependence of magnetization of the ferromagnetic film, as the effects of the magnetization without the influence of temperature on the energy of adsorption slightly.

Thus, in [193] it was shown that taking into account the effects of the spontaneous magnetization in submonolayer films of ferromagnetic materials has a significant impact on the value of the adsorption energy, and also stimulates the substitutional processes with increase of the coverage parameter Θ. The influence of the effects of ferromagnetic ordering on the implementation of various surface adsorption structures with the variation of Θ was determined.

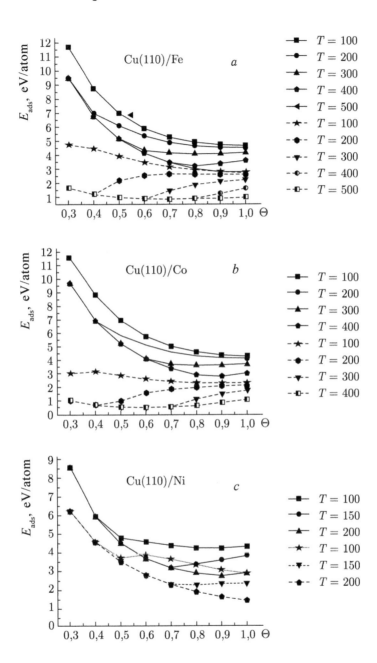

Fig. 10.9. The dependence of the adsorption energy E_{ads} of the parameter coverage Θ for for the systems Cu (110)/Fe, Cu (110)/Co and Cu (110)/Ni.

Appendix

DETEMINATION OF THE MINIMUM OF
THE FUNCTION OF n VARIABLES

Preliminary discussion. The surface energy has a complicated dependence on the variational parameters, so determination of its derivatives would cause some difficulties. Therefore, to determine the minimum surface energy as function of several variables, it is necessary to use methods in which only function values are used. Such methods are called *direct search methods*. In this appendix, we present one of the them, which is very effective. We consider the function of two variables.

Its constant level lines are represented are shown in Fig. A.1, and the minimum lies at the point (x_1^*, x_2^*). The simplest method is to find the method of coordinate descent. From point A we find the minimum along the direction of the x_1 axis, and thus find a point B, in which the tangent to the constant level line is parallel to the x_1 axis. Then, searching from point B in the direction of x_2, we get the point C, producing search parallel to the axis x_2, we get the point D, and so on. As a result, we arrive at the optimal point. Here to search along the axis we can use any of the one-dimensional methods.

This idea can also be applied to functions of n variables. Theoretically, this method is effective in case of a single minimum of the function. But in practice it is too slow. Therefore, more sophisticated methods that use more information on the basis of the already obtained values of the function have been developed.

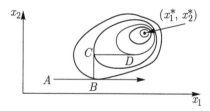

Fig. A.1. Lines of constant-level functions of two variables.

Hooke–Jeeves method. This method was developed in 1961 but is still very effective and original. It is a sequence of *steps* of the exploring search around the *basic point*, beyond which, if successful, *search of the sample is carried out.*

Here is a description of this procedure (Fig. A.2).

1. Select the initial basic point $\mathbf{b}_1$ and step length h_j for each variable x_j, $j = 1, 2, \ldots, n$.

2. Compute $f(x)$ at the basic point $\mathbf{b}_1$ to obtain information on the local behaviour of the function $f(x)$. This information will be used to find a suitable search direction on the sample, by which we can hope to achieve a greater decrease of the values of the function. The values of the function $f(x)$ at the basic point $\mathbf{b}_1$ and its neighbourhood are found by the following procedure (Fig. A.3).

(a) Calculate the value of the function $f(\mathbf{b}_1)$ at the basic point $\mathbf{b}_1$.

(b) Each variable in turn is changed by adding the length of step. Namely, we calculate the value of the function $f(\mathbf{b}_1 + h_1\mathbf{e}_1)$, where $\mathbf{e}_1$ – unit vector along the axis x_1. If this leads to a decrease in the value of

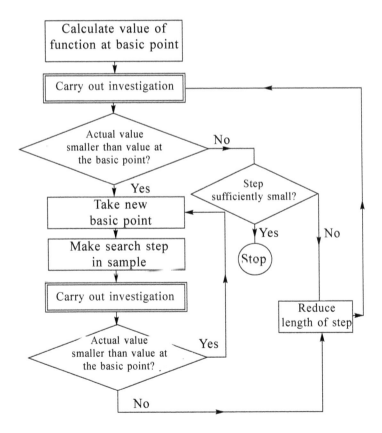

Fig. A.2. The overall block diagram of the Hooke–Jeeves method.

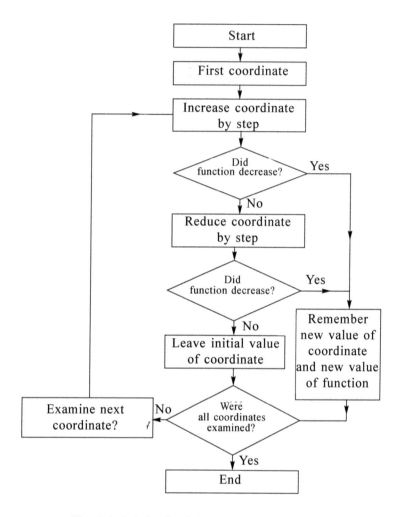

Fig. A.3. Subcircuit of the 'Run the study' block.

the function, then $\mathbf{b}_1$ is replaced by $\mathbf{b}_1 + h_1\mathbf{e}_1$. Otherwise, the value of the function $f(\mathbf{b}_1 - h_1\mathbf{e}_1)$ is calculated and if its value decreased, then $\mathbf{b}_1$ is replaced by $\mathbf{b}_1 - h_1\mathbf{e}_1$. If none of these changes reduce the value of the function, the point $\mathbf{b}_1$ remains unchanged. Next, we consider the changes in the direction of the x_2 axis, i.e., find the value of the function $f(\mathbf{b}_1 + h_2\mathbf{e}_2)$, etc. When we have examined all the variables, we have, generally speaking, the new basic point $\mathbf{b}_2$.

(c) If $\mathbf{b}_2 = \mathbf{b}_1$, i.e. the decrease in the function was not achieved, then the study is repeated around the same basic point $\mathbf{b}_1$, but with a reduced step length. In practice, the reduction of the step (steps) by to ten times of the initial length is satisfactory.

(d) If $\mathbf{b}_2 \neq \mathbf{b}_1$, then the search is performed on the sample.

3. When we search in the sample, we use the information obtained in the research process, and minimization of the function ends in the search in the direction specified by the sample. This procedure is performed as follows.

(a) It is reasonable to move from the basic point $\mathbf{b}_2$ in the direction $\mathbf{b}_2 - \mathbf{b}_1$, since the search in this direction has already led to a decrease in the value of the function. Therefore, we compute the function at the point of the sample $\mathbf{P}_1 = \mathbf{b}_2 + \mathbf{b}_2 - \mathbf{b}_1 = \mathbf{b}_1 + 2(\mathbf{b}_2 - \mathbf{b}_1)$. In the general case $\mathbf{P}_i = \mathbf{b}_i + 2(\mathbf{b}_{i+1} - \mathbf{b}_i)$.

(b) The study should then continue around the point $\mathbf{P}_1$ ($\mathbf{P}_i$).

(c) If the value in step 3(b) is less than the value at the basic point $\mathbf{b}_2$ (in general, $\overrightarrow{b_{i+1}}$), we get a new basic point $\mathbf{b}_3(\mathbf{b}_{i+2})$, followed by a repeat step 3(a). Otherwise, we do not search in the sample from the point $\mathbf{b}_2(\mathbf{b}_{i+1})$, and to continue study at the point $\mathbf{b}_2(\mathbf{b}_{i+1})$.

4. The process is completed when the step length (lengths of steps) is be reduced to a predetermined small value.

References

1. Ageev V.H., *Poverkhnost'*, 1982, No. 4, 1–19.
2. Agranovich V.M., *Zh. Eksp. Teor. Fiz.*, 1959, vol. 37, No. 2, 430–441.
3. Agranovich V.M., Ginzburg V.L., Crystal optics with spatial dispersion taken into account and excitons, Moscow, Nauka, 1979.
4. Ando T., et al., Electronic properties of two-dimensional systems, Springer-Verlag, 1985.
5. Artamonov O.M., et al., *Pis'ma Zh. Teor. Fiz.*, 1984, vol. 10, No. 10, 633–637.
6. Aseev A.L., et al., in: Problems of electronic materials, Novosibirsk, 1986, 109–127.
7. Afanas'yev A.M., et al., *Elektron. Promyshl.*, 1980, No. 11–12, 47–55.
8. Akhmanov S.A., et al., *Usp. Fiz. Nauk*, 1985, vol. 147, No. 4, 675–745.
9. Bazarov I.P., Thermodynamics, Moscow, Vysshaya shkola, 1976.
10. Bak T., Spectrometry of scattering of slow ions, in: Methods of analysis of surfaces, Moscow, Nauka, 1979, 102–136.
11. Barash Yu S., Ginzburg V.L., *Usp. Fiz. Nauk*, 1975, vol. 116, No. 1, 5–40.
12. Basin V.E., Adhesion strength, Moscow, Nauka, 1981.
13. Bennett A., A new study in the solid surface, Academic Press, 1977.
14. Berezinskiy V.L., *Zh. Eksp. Teor. Fiz.*, 1970, vol. 59, No. 3, 907–920.
15. Berlin A.A., Basin V.E., Fundamentals of adhesion of polymers, Moscow, Khimiya, 1974.
16. Beztesnnyi S.I., Mamonova M.V., Vestnik Omsk. Univ., 2008, No. 3, 15–18.
17. Bol'shov L.A., et al., *Usp. Fiz. Nauk*, 1977, vol. 122, No. 1, 125–160.
18. Bondarenko B.V., et al., *Fiz. Tverd. Tela*, 1969, vol. 11, No. 12, 3574.
19. Borzyak P.G., Kulyupin Yu.A., *Ukr. Fiz. Zh.*, 1979, vol. 24, No. 2. 204–214.
20. Bowden F.P., Tabor D., Friction and lubrication of solids, Moscow, Mashinostronie, 1968.
21. Braun O.M., *Fiz. Tverd. Tela*, 1980, V. 22, No. 6, 1649.
22. Braun O.M., Medvedev V.K., *Usp. Fiz. Nauk*, 1989, vol. 157, No. 4, 631–666.
23. Breger A.Kh., Zhukhovitskii A.A., Zh. Khim. Fiz., 1946, vol. 20, No. 4, 355–362.
24. Briksin V.V., et al., *Usp. Fiz. Nauk*, 1974, vol. 113, no. 1, 29–67.
25. Vakilov A.N., et al., Theoretical models and methods in surface physics, Publishing House of the Omsk State University, Omsk, 2005.
26. Vakilov A.N., et al., Vestnik Omsk. Univ., 1996, No. 1, 37–40.
27. Vakilov A.N., et al., *Fiz. Tverd. Tela*, 1997, vol. 39, No. 6, 964–967.
28. Vakilov A.N., et al., Poverkhnost', 1998, No. 10, 55–61.
29. Vakilov A.N., et al., The theoretical description of the adhesion and tribological properties of diamond-like films, in: Proceedings of the Fifth Workshop of the Russian Academy of Sciences – Ural Branch of Russian Academy of Sciences 'Thermodynamics and Materials', Novosibirsk, 2005, 166.
30. Vakilov A.N., et al., The development of computational methods of optimal choice of materials and coatings, in: Proc. of III International Workshop 'Contact interaction and dry friction', Moscow, 2005, Moscow State Technical University, 2005, 39–47.

31. Vakilov A.N., et al., Calculation methods for the selection of materials for non-lubricated friction sections, in: Proceedings of the III International Congress of Technology, 'Military equipment, weapons and dual-use technology', Part 1, Omsk, 2005, 249–251.

32. Vakilov A.N., et al., *Fiz. Met. Metalloved.,* 1995, vol. 79, No. 4, 13–22.

33. Vakilov A.N., Prudnikov V.V., *Fiz. Met. Metalloved.,* 1991, No. 8, 11–20.

34. Vakilov A.N., Prudnikov V.V., *Poverkhnost',* 1991, No. 12, 72–75.

35. Vakilov A.N., et al., *Fiz. Met. Metalloved.,* 1993, vol. 76, No. 6, 38–48.

36. Varshavskaya I.G., Tolmachev Yu.N., Materialovedenie, 1995, No. 10, 51–56.

37. Vasilevski M.I., *Poverkhnost',* 1985, No. 1, 32–39.

38. Vasilevski Yu.A., Magnetic recording media, Moscow, Nauka, 1989.

39. Vasil'ev B.V., et al., Usp. Fiz. Nauk, 1994, vol. 164, 375–378.

40. Vedula Yu.S., Naumovets A.G., in: Surface diffusion and spreading, Moscow, Nauka, 1969, 149–160.

41. Wilson K., Kohut G., The renormalization group and the ε-decomposition, Springer-Verlag, 1975.

42. Volotskii S.S., Autohesion and adhesion of high molecular polymers, Moscow, Rostekhizdat, 1960.

43. Vonsovskii S.V., Magnetism, Moscow, Nauka, 1971.

44. Gavrilyuk V.M., et al., *Fiz. Tverd. Tela,* 1967, vol. 9, No. 4, 1126.

45. Gavrilyuk V.M., Medvedev V.K., *Fiz. Tverd. Tela,* 1966, vol. 8, No. 6, 1811.

46. Garkunov D.N., Tribotechnology, Moscow, Mashinostroenie, 1985.

47. Garkunov D.N., Tribotechnology (wear and wear resistance), Moscow, MSKhA 2001.

48. Garnov A.V., et al., *Usp. Fiz. Nauk,* 1971, vol. 35, No. 2, 341–344.

49. Geguzin Ya.E., Kaganovskii Yu.S., Diffusion processes on the surface of the crystal, Moscow, Energoatomizdat, 1984.

50. Geguzin Ya.E., Ovcharenko N.N., *Usp. Fiz. Nauk,* 1962, vol. 76, 283.

51. Gibbs J.W., Thermodynamics. Statistical mechanics, Moscow, Nauka, 1982.

52. Guinier A., X-ray diffraction of crystals, Moscow, Fizmatgiz, 1961.

53. Glauberman L., *Zh. Fiz. Khim.,* 1949, vol. 23, No. 2, 115–123.

54. Goysa S.N., et al., *Poverkhnost',* 1986, No. 7, S. 28-36.

55. Golubev O.L., et al., *Fiz. Tverd. Tela,* 1967, 1971, vol. 13, No. 3, 761–768.

56. Gombash P., Statistical theory of the atom, McGraw-Hill, 1951, 398.

57. Gomoyunova M.B., *Usp. Fiz. Nauk,* 1982, vol. 136, No. 1, 105–148.

58. Gorbatyi, N.A., et al., *Fiz. Tverd. Tela,* 1968, vol. 10, No. 4, 1185.

59. Gorbatyi, N.A., Khashimova C., *Fiz. Tverd. Tela,* 1966, vol. 8, No. 5, 1441.

60. Gorodetsky A.E., In: Spectroscopy and electron diffraction in the study of solid surfaces, Moscow, Nauka, 1985, 222–285.

61. Gruverman S.L., Sukhman A.A., *Poverkhnost',* 1990, No. 12, 123–129.

62. Gryaznov B.T., et al., *Effekt Bezyzn. Tribotekhnologii,* 1998, No. 1, 12–20.

63. Gryaznov B.T., et al., *Effekt Bezyzn. Tribotekhnologii,* 1998, No. 1, 24–31.

64. Gryaznov B.T., et al., Technological methods to improve the durability of machines for microcryogenic technology, Novosibirsk, Nauka, 1999..

65. Gryaznov B.T., et al., *Effekt Bezyzn. Tribotekhnologii,* 2000, No. 1, 37–41.

66. Gryaznov B.T., et al., *Trenie i Iznos,* 1998, vol. 19, No. 4, 466–474.

67. Gryaznov B.T., et al., in: Cryogenic equipment and cryogenic technologies, Proc., No. 1, Part 2, Siberian Regional Branch of the International Academy of Refrigeration, Omsk, 1997, 25–35.

68. Gryaznov B.T., et al., *Trenie i Iznos,* 1998 Vol. 19, No. 4, 440-447.

69. Gryaznov B.T., et al., in: Cryogenic equipment and cryogenic technologies, Proc., No. 1, Part 2, Siberian Regional Branch of the International Academy of Refrigeration, Omsk, 1997, 3–18.

70. Gubanov, A.I., *Fiz. Tverd. Tela,* 1975, vol. 17, No. 4, 1089–1094.

71. Gubanov A.I., *Fiz. Met. Metalloved.,* 1976, vol. 41, No. 3, S. 457–463.

72. Gubanov, A.I., Dunaevskii, S.M., *Fiz Tverd. Tela,* 1979, vol. 19, No. 5, 1369–1373.

73. Gul' V.E., et al., Vysokomol. Soed., 1976, vol. A18, No. 1, 122–126.

74. Gupalo M.S., et al., *Fiz. Tverd. Tela,* 1980, vol. 22, No. 8, 2311.

75. Gupalo M.S., et al., *Fiz. Tverd. Tela,* 1979, vol. 21, No. 4, 973.

76. Gupalo M.S., et al., *Fiz. Tverd. Tela,* 1981, vol. 23, No. 7, 2076.

77. Davydov A.S., , Solid state theory, Moscow, Nauka, 1976.

78. Deryagin B.V., *Zh. Fiz. Khim.,* 1934, vol. 5, No. 9, 1165–1176.

79. Deryagin B.V., What is friction?, Moscow, Publishing House of the USSR Academy of Sciences, 1963.

80. Deryagin B.V., et al., Adhesion of solids, Moscow, Nauka, 1976.

81. Deryagin B.V., Fedoseyev D.V., Growth of diamond and graphite from the gas phase, Moscow, Nauka, 1977..

82. Deryagin B.V., et al., Surface forces, Moscow, Nauka, 1977.

83. Dzyaloshinskii, I.E., et al., *Usp. Fiz. Nauk,* 1961, vol. 73, No. 3, 381.

84. Dobretsov L.N., Gomoyunova M.V., Emission Electronics, Moscow, Nauka, 1966.

85. Davison, S., Levin G., Surface (Tamm) states, Moscow, Mir, 1973.

86. Zhizhin G.A., et al., Zh. Eksp. Teor. Fiz., 1980, vol . 79, No. 2, 561–574.

87. Zhurakovskii E.A., Nemchenko V.F., Kinetic properties and electronic structure of interstitial phases, Kiev, Naukova Dumka, 1989.

88. Zadumkin S.N., *Fiz. Met. Metalloved.,* 1961, vol. 11, No. 3, 331–346.

89. Zadumkin S.N., *Fiz. Met. Metalloved.,* 1962, vol. 13, No. 1, 24–32.

90. Zadumkin S.N., in: Surface phenomena in melts and solid phases arising from them , Nal'chik, Kab.-Balk. Publishing House, 1965, 12–29.

91. Zandberg, E., *Zh. Teor. Fiz.,* 1974, vol. 44, No. 9, 1809–1828.

92. Zandberg, E., Ionov N., in: Problems of modern physics, Leningrad, Nauka, 1980, 487–504.

93. Zandberg, E., et al., 1971, vol. 41, No. 11, 2420–2427.

94. Zandberg, E., et al., *Zh. Teor. Fiz.,* 1972, vol. 42, No. 1, 171–175.

95. Zenguil E. Surface physics, Academic Press, 1990.

96. Zimon A.D., Adhesion of the films and coatings, Moscow, Nauka, 1977.

97. Zubenko Yu.V., *Fiz. Tverd. Tela,* 1964, vol. 6, No. 1, 123.

98. Ziryanov G.K., Hoang A.T., Bulletin of the Leningrad University, Phys. Chem., 1969, No. 10, 69.

99. Izyumov Yu.A., Syromyatnikov V.N., Phase transitions and crystal symmetry, Moscow, Nauka, 1984.

100. Irkhin V.Yu, Irkhin Yu.P., Electronic structure, physical properties and correlation effects in *d*- and *f*-metals and their compounds, Cambridge International Science Publishing, 2008

101. Katznel'son A.A., et al., Electronic theory of condensed media, Moscow, Moscow State University Press, 1990.

102. Kascheev V.N., Processes in the area of friction contact of the metals, Moscow, Mashinostroenie, 1978.

103. Kirzhnits D.A., *Zh. Eksp. Teor. Fiz.,* 1957, vol. 32, 115–123.

104. Kirzhnits D.A., et al., *Usp. Fiz. Nauk,* 1975, vol. 117, No. 1, 3-47.

105. Kiselev V.F., Krylov O.V., Adsorption processes on the surface of semiconductors and

dielectrics, Moscow, Nauka, 1978.

106. Kittel' S., Quantum theory of solids, Moscow, Nauka, 1967.
107. Klimenko E.V., Medvedev V.K., *Fiz. Tverd. Tela,* 1968, vol. 10, No. 7, 1986.
108. Klimenko E.V., Naumovets E.G., *Fiz. Tverd. Tela,* 1973, vol. 15, No. 11, 3273.
109. Knyazev S.A., et al., *Usp. Fiz. Nauk,* 1985, vol. 146, No. 1, 73–104.
110. Kobeleva A.V., et al., *Dokl. Akad. Nauk SSSR,* 1978, vol. 243, No. 3, 692–695.
111. Kobeleva R.M., et al., *Fiz. Met. Metalloved.,* 1978, vol. 45, No. 1, 25–32.
112. Kobeleva R.M., et al., *Fiz. Met. Metalloved.,* 1979, vol. 48, No. 2, 251–259.
113. Kobeleva R.M., et al., *Fiz. Met. Metalloved.,* 1976, vol. 41, No. 3, 493–498.
114. Kobeleva R.M., et al., *Fiz. Met. Metalloved.,* 1974, vol. 38, No. 3, 640–643.
115. Kobeleva R.M., et al., *Koll. Zh.,* 1977, vol. 39, No. 2, 295–301.
116. Kompaneets A.S.. Pavlovski E.S., *Zh. Eksp. Teor. Fiz.,* 1956, vol. 31, No. 3, 427–438.
117. Konovalov N.D., Makukha V.I., *Fiz. Tverd. Tela,* 1967, vol. 9, No. 9, 2686.
118. Kostecki B.I., Friction, lubrication and wear in machines, Kiev, Tekhnika, 1970.
119. Kragel'skii I.V., Friction and wear, Moscow, Mashinostroenie, 1968.
120. Kragel'skii I.V., Vinogradova I.E., Friction coefficients, Moscow, Mashgiz, 1962.
121. Krasnov Yu.N., et al., Methods for testing and evaluation of the materials for sliding bearings, Moscow, Nauka, 1972, 137–139.
122. KremkoV M.V., Corpuscular low-energy diagnostics of solid surface, Tashkent, Fan, 1986.
123. Krivoglaz M.A., Smirnov A.A., The theory of ordered alloys, Fizmatgiz, 1958.
124. Kuznetsov V.A., *Fiz. Tverd. Tela,* 1971, vol. 13, No. 6, 1715.
125. Kuznetsov V.A., Tsarev, B.M., *Fiz. Tverd. Tela,* 1967, vol. 9, No. 9, 2524.
126. Lavrent'ev M.A., Shabat B.V., Methods of the theory of functions of complex variable, Moscow, Nauka, 1973, 89.
127. Lagalli M., Martin J., *Prib. Nauchn. Issled.,* 1983, vol. 54, No. 10, 3–21.
128. Landau L.D., Lifshitz E.M., Theoretical physics, in 10 volumes, vol. V. Statistical physics. Part 1, 5th ed., Fizmatlit, Moscow, 2010.
129. Landau L.D., Lifshitz E.M., Theoretical physics, in 10 volumes, vol. VII. Elasticity theory, Fizmatlit, Moscow, 2007.
130. Litovchenko V.G., Basic physics of semiconductor layered systems, Kiev, Naukova dumka, 1980.
131. Loginov M.V., et al., *Fiz. Tverd. Tela,* 1980, vol. 22, No. 5, 1411.
132. Loginov M.V., et al., *Fiz. Tverd. Tela,* 1980, vol. 22, No. 11, 3299.
133. Lozovyi Ya.B., *Fiz. Tverd. Tela,* 1986, vol. 28, No. 12, 3693.
134. Lyuksyutov I.F., et al., Two-dimensional crystals, Kiev, Naukova Dumka, 1988.
135. Ma Sh., Modern theory of critical phenomena, Academic Press, 1980.
136. Madelung O., Theory of solids, Moscow, Nauka, 1980.
137. Madelung O., Solid state physics. Localized states, Moscow, Nauka, 1985.
138. Makukha, V.I., *Fiz. Tverd. Tela,* 1967, vol. 9, No. 1, 150.
139. Makukha, V.I., Tsarev, B.M., *Fiz. Tverd. Tela,* 1966, vol. 8, No. 5, 1417.
140. Mamonova M.V., et al., *Fiz. Met. Metalloved.,* 2002, vol. 94, No. 5, 16–25.
141. Mamonova M.V., et al., *Fiz. Met. Metalloved.,* 2009, vol. 107, No. 5, 451–458.
142. Mamonova M.V., et al., *Fiz. Tverd. Tela,* 2009, vol. 51, No. 10, 2004–2010.
143. Mamonova M.V., et al., *Vestnik Omsk. Univ.,* 1996, No. 1, 41–43.
144. Mamonova M.V., et al., *Vestnik Omsk. Univ.,* 1996. No. 2. S. 44–46.
145. Mamonova M.V., et al., *Fiz. Met. Metalloved.,* 1998, vol. 86, No. 1, 5–14.
146. Mamonova M.V., et al., *Fiz. Met. Metalloved.,* 1998, vol. 86, No. 2, 33–39.
147. Mamonova M.V., Prudnikov V.V., *Vestnik Omsk. Univ.,* 1998, No. 1 (7), 22–26.
148. Mamonova M.V., Prudnikov V.V., *Izv. VUZ, Fizika,* 1998, vol. 41, No. 12, 7–12.

149. March N., Parrinello M., Collective effects in solids and liquids, Academic Press, 1986, 30.
150. Matveev A.V., et al., *Vestn. Omsk. Univ.*, 2000, No. 2, 30–32.
151. Matveev A.V., et al., *Vestn. Omsk. Univ.*, 2002, No. 1, 26–28.
152. Matveev A.V., et al., *Vestn. Omsk. Univ.*, 2002, No. 2, 23–25.
153. Matveev A.V,, et al., *Kondens. Sredy Mezhfaz. Granitsy,* 2002, Vol. 4, No. 3, 263–272.
154. Matveev A.V., et al., *Vestn. Omsk. Univ.*, 2003, No. 4, 34–36.
155. Matveev A.V,, et al., *Kondens. Sredy Mezhfaz. Granitsy,* 2003, Vol. 5, No. 4, 401–409.
156. Matveev A.V,, et al., *Vestn. Omsk. Univ.*, 2004, No. 1, 31–33.
157. Matveev A.V., et al., *Vestn. Omsk. Univ.*, 2004, No. 2, 26–28.
158. Matveev A.V., et al., *Fiz. Met. Metalloved.*, 2004, Vol. 97, No. 6, 26–34.
159. Matveev A.V., et al., *Poverkhnost',* 2005, No. 1, 28–34.
160. Medvedev V.K., *Fiz. Tverd. Tela,* 1968, Vol. 10, No. 11, 3469.
161. Medvedev V.K., et al., *Fiz. Tverd. Tela,* 1970, Vol. 12, No. 2, 375–385.
162. Medvedev V.K., Smereka T.P.,*Fiz. Tverd. Tela,* 1973, Vol. 15, No. 5, 1641.
163. Medvedev V.K., Smereka T.P., *Fiz. Tverd. Tela,* 1974, Vol. 16, No. 6, 1599.
164. Medvedev V.K., Yakivchuk A.I.,*Fiz. Tverd. Tela,* 1975, Vol. 17, No. 1, 14.
165. Flad E.M., (editor), Gas–solid interface, Moscow, Mir, 1970.
166. Myshkin N.K., Petrakovets M.I., Friction, lubrication, wear and tear, Moscow, Fizmatlit., 2007, 368 p.
167. Muller E., Tsong T., Field ion microscopy, Moscow, Metallurgiya, 1972.
168. Muller E., Tsong T., Field ion microscopy. Field ionization and field evaporation, Moscow, Nauka, 1980.
169. Naumovets A.G., *Ukr. Fiz. Zh.*, 1978, Vol. 23, No. 10, 1585–1607.
170. Naumovets A.G., in: Spectroscopy and electron diffraction in the study of solid surfaces, Moscow, 1985, 162–221.
171. Nevolin K., et al., *Poverkhnost',* 1983, No. 1, 79–83.
172. Nevolin K., et al., *Poverkhnost',* 1986, No. 7, 76–82.
173. Nesterenko B.A., Snitko O., Physical properties of atomically clean semiconductor surfaces, Kiev, Naukova Dumka, 1983.
174. Nefedov V.I., Cherepin V.T., Physical methods of studying the surface of solids, Moscow, Nauka, 1983.
175. Ovchinnikov A.P., *Fiz. Tverd. Tela,* 1967, Vol. 9, No. 2, 628.
176. Ovchinnikov A.P., Tsarev B.M., *Fiz. Tverd. Tela,* 1966, Vol. 8, No. 5, 1493.
177. Ovchinnikov A.P., Tsarev B.M., *Fiz. Tverd. Tela,* 1967, Vol. 9, No. 7, 1927-1934.
178. Ovchinnikov A.P., Tsarev B.M., *Fiz. Tverd. Tela,* 1967, Vol. 9, No. 12, 3559.
179. Rzhanov A.V., Basics ellipsometry, Moscow, Nauka, 1979.
180. Oura K., et al., Introduction to the physics of the surface, Moscow, Nauka, 2006, 496.
181. Paasch, G., Hitshold M., in: Achievements of the electron theory of metals, Volume 2, Ed. Tsishe P., Lehmann G., Academic Press, 1987, 466–540.
182. Pines D., Nozier F.. The theory of quantum liquids, Academic Press, 1967.
183. Partensky M.B., *Usp, Fiz. Nauk,* Vol. 128, No. 1, 69–106, Vol. 129, No. 3, 590.
184. Partensky M.B., Kuzema V.E., *Fiz. Tverd. Tela,* 1979, Vol. 21, No. 9, 2342–2844.
185. Partensky M.B., et al., Fiz. Met. Metalloved., 1976, Vol. 41, No. 2, 279-283.
186. Partensky M.B., Smorodinskii Ya.G., *Fiz. Tverd. Tela,* 1974, Vol. 16, No. 3, 644-647.
187. Patashinskii A.Z., Pokrovsky V.L., Fluctuation theory of phase transitions, Moscow, Nauka, 1982.
188. Petrov H.H., Abroyan I.A., Diagnostics of the surface using ion beams, Leningrad, Publishing House of Leningrad University, 1977.

189. Platzman F., Wolf P.A., Waves and interactions in solid-state plasma, Academic Press, 1975, 440.
190. Surface polaritons. Electromagnetic waves on surfaces and phase boundaries. Ed. V.M. Agranovich, D.L. Mills, Moscow, Nauka, 1985.
191. Pratton M., Introduction to the physics of the surface, Izhevsk, NITs RHD, 2000.
192. Prudnikov V.V., et al., Phase transitions and methods of computer simulation, Moscow, Fizmatlit, 2009.
193. Prudnikov V.V., et al., V*est. Omsk. Univ.*, 2009, No. 2, 77–83.
194. Prudnikov V.V., et al., V*est. Omsk. Univ.*, 2009, No. 2, 63-70.
195. Prudnikov V.V., Poterin R.V., *Fiz. Met. Metalloved.*, 1997, Vol. 84, No. 5, 48-56.
196. Prudnikov V.V., Poterin R.V., *Izv. VUZ, Fizika,* 1998, Vol. 41, No. 11, 10-15.
197. Pchelyakov O.P., et al., *Poverkhnost'*, 1982, No. 1, 147-149.
198. Pshenechnyuk E.A., Yumaguzin Yu.M., *Zh. Teor. Fiz.*, 2001, Vol. 71, No. 10. 99-103.
199. Development of methods for calculating the adhesion component of friction: NIR Report (final) Om~GU, V.V. Prudnikov, No. GR 01.930009970; No. 02940001313. Omsk, 1993.
200. Leaflet of Patinor Coatings Limited» (www.patinor.ru), Nanomechanics laboratory of the Moscow Power Engineering Institute (MEI) (nanomech-lab.chat.ru) of Belgorod plant Rhythm (www.fox.btrc.ru), the company Albatek (www.aljbatech.ru).
201. Rzhanov A.V., Electronic processes on semiconductor surfaces, Moscow, Nauka, 1971.
202. Riggs V., Parker M., Surface analysis by X-ray photoelectron spectroscopy, in: Methods of surface analysis, ed. A. Zanderna, Moscow, Nauka, 1979, 137-199.
203. Roldugin V.I., Physical chemistry of surfaces, Dolgoprudnyi, Intellekt, 2008.
204. Samoylovich A.G., *Zh. Eksp. Teor. Fiz.*, 1946, Vol. 16, No. 2, 135-150.
205. Properties, production and application of refractory compounds, a handbook, ed. T. Ya. Kosolapova, Moscow, Metallurgiya, 1986.
206. Semenchenko V.K., Surface phenomena in metals and alloys. Moscow, GNTTL, 1957..
207. Sergienko V.P., Methods for diamond and diamond-like films, Dep. in VINITI 1999, No. 021595.
208. Sokolov, A.A., et al., Quantum Mechanics, Moscow, Prosveshchenie, 1965, 440-446.
209. Spicer B., in: Electron and ion spectroscopy of solids Moscow, Nauka, 1981, 61-97.
210. Spitkovskii I.M., *Zh. Fiz. Khim.*, 1950. Vol. 24, No. 9. 1090-1093.
211. Tamm I.E., *Zs. Physik*, 1932, Vol. 76, 849; *Phys. Z. Sowjet.,* 1932, Vol. 1. 733; Collection of scientific papers, vol. 1, Moscow, Nauka, 1975, 216.
212. Teodorovich E.V., *Izv. VUZ Fiz.*, 1977, No. 10, 67-79.
213. The theory of an inhomogeneous electron gas, ed. S. Lundqvist, N. Murch, Springer-Verlag, 1987.
214. The theory of chemisorption, ed. J. Smith, Academic Press, 1983.
215. Tishin EA., Tsarev B.M., *Fiz. Tverd. Tela,* 1966, Vol. 8, No. 11, 3181.
216. Tompkins H. in: Methods of surface analysis, Academic Press, 1979, 542-570.
217. Uglov A.A., et al., Adhesion ability of films, Moscow, Radio i svyaz', 1987, 104.
218. Ukhov V.F., Kobeleva R.M., Electronic component of the surface energy of metals in an external electric field, Sverdlovsk, 1978, Dep. in VINITI, No. 495-78.
219. Ukhov V.F., Kobeleva R.M., in: Questions of physics of shaping and phase transformations, Kalinin: KSU, 1979, 34-40.
220. Ukhov V.F., et al., Electron-statistical theory of metals and ionic crystals, Moscow, Nauka, 1982.
221. Fedorus A.G., et al., *Fiz. Tverd. Tela*, 1969, Vol. 11, No. 1, 207.

222. Fedoseyev D.V., et al., Crystallization of diamond, Moscow, Nauka, 1984, 135.
223. Feigin V.A., et al., The surface energy and distribution of electrons in an external electric field, in: Physical chemistry of interfaces contacting phases, Kiev, Naukova Dumka, 1976, 38-42.
224. Feynman R.P., Statistical mechanics, Academic Press, 1975.
225. Physical quantities: a handbook, ed. I.S. Grigor'ev, E.Z. Melikhov, Moscow, Energoatomizdat, 1991.
226. Fomenko V.S., Emission properties of materials, Kiev, Naukova Dumka, 1970.
227. Fomenko V.S., Podchernyaeva I.A., Emission and adsorption properties of substances and materials, Handbook, Oxford, Clarendon Press, 1975.
228. Frenkel Ya.I., Introduction to the theory of metals, Moscow, Leningrad, GITTL, 1948.
229. Heine P., et al., Pseudopotential theory, Academic Press, 1973.
230. Khokonov Kh.B., Surface phenomena in melts and emerging solids, Khisinau, Shtiintsa, 1974.
231. Khrushchov M.M., Main types and patterns of wear, Moscow, MATI, 1968.
232. Tsarev B.M., Contact potential difference, Moscow, Gostekhteoretizdat,1955.
233. Cherepin V.T., Vasiliev M.A., Methods and apparatus for analyzing the surface of materials, Kiev, Naukova Dumka, 1982.
234. Chernov A.A., Modern crystallography, Vol. 3, Moscow, Nauka, 1980.
235. Shrednik V.N., Snezhko E.V., *Fiz. Tverd. Tela,* 1964, Vol. 6, No. 6, 1501.
236. Shrednik V.N., Odishariya G.A., *Fiz. Tverd. Tela,* 1969, Vol. 11, No. 7 1844-1853.
237. Evarestov R.A., Quantum chemical methods in the theory of solids, Leningrad, Publishing House of Leningrad University, 1982.
238. Ehrlich G.,, News on the study of solid surfaces, Vol. 1, Moscow, Mir, 1978.
239. Yakovlev V.A., Zhizhin G.N., *Pis'ma Zh. Eksp. Teor. Fiz.*, 1974, Vol. 19, 333.
240. Yastrebov L.I., Katznel'son A., Pseudopotential theory of crystal structures, Moscow, Moscow State University Press, 1981.
241. Ageev V.N., Ionov N.I., *Progr. Surface Sci.,* 1975, Vol. 5, Pt. 1, 1–118.
242. Aisenberg S., Chabot R., J. Appl. Phys., 1971, Vol. 42, 2953.
243. Allan G., Lannoo M., Dobrzynski L., *Phil. Mag.*, 1974, Vol. 30, No. 1, 33–45.
244. Alvarado S.F., Campagna M., Hopster H., *Phys. Rev. Lett.*, 1982, Vol. 48, No. 1, 51–54.
245. Andreussi F., Gurtin M., *J. Appl. Phys.,* 1977, Vol. 48, 3798.
246. Aono M., et al., *Phys. Rev. Lett.,* 1983. Vol. 50, 1293.
247. Artmann K., *Zs. Physik*, 1952, Vol. 131, 244.
248. Bak P., *Solid State Commun.,* 1979, Vol. 32, 581.
249. Bardeen J., *Phys. Rev.,* 1947, Vol. 71, 717.
250. Barnes M.R., Willis R.F., *Phys. Rev. Lett.,* 1978, Vol. 41, 1729.
251. Bartelt N.C., et al., *Phys. Rev.* B, 1985, Vol. 32, No. 5, 2993–3002.
252. Bassett D. W. Observing surface diffusion at the atomic level, Surface mobilities on solid materials, New York, Plenum press, 1983, 63–108.
253. Bauer E., Poppa H., *Thin Solid Films*, 1972, Vol. 12, No. 1, 167–185.
254. Bauer E., et al., *J. Appl. Phys.,* 1977, Vol. 48, No. 9, 3773–3787.
255. Becker R., et al., *Phys. Rev. Lett.,* 1985, Vol. 55, No. 19, 2032–2034.
256. Becker R., et al., *Phys. Rev. Lett.,* 1985, Vol. 55, No. 19, 2028–2031.
257. Berndt W., et al., *Surf. Sci.,* 1995, Vol. 330, 182–192.
258. Bianconi A., Bachrach R., *Phys. Rev. Lett.,* 1979, Vol. 42, No. 2, 104–108.
259. Binnig G., Fuchs H., Stall E., *Surface Sci.,* 1986, Vol. 169, No. 2/3, L296–L300.
260. Binnig G., Rohrer H. *Physica* B, 1984, Vol. 127, No. 1, 37–45.
261. Binnig G.K., et al., *Surface Sci.,* 1984, Vol. 144, No. 2/3, 321–335.

262. Binnig G.K., et al., *Phys. Rev. Lett.,* 1982. Vol. 49, 57.
263. Binnig G.K., *Phys. Rev. Lett.*, 1983, Vol. 50, No. 3, 120–123.
264. Binnig G.K., et al., *Surface Sci.,* 1983, Vol. 131, L379.
265. Bishop D.J., Reppy J.D., *Phys. Rev. Lett.*, 1978, Vol. 40, No. 22, 1727–1730.
266. Bishop D.J., et al., *Phys. Rev. B*, 1982, Vol. 49, No. 25, 1861–1864.
267. Bonzel H.P., Pirug G., The Chemical physics of aolid Surface and heterogeneous ca-
    talysis. Vol. 6: Coadsorption, Promotors and Poisons. Ed. by King D.A., Woodruff D.,
    Amsterdam, Elsevier, 1993.
268. Bretz M., et al., *Phys. Rev. B*, 1973. Vol. 8, No. 3, 1589–1615.
269. Brongersma H.H., Buck T.M., *Nucl. Instr. Meth.*, 1978, Vol. 149, 569.
270. Brown R.C., March N.H., *Phys. Reports C*, 1976, Vol. 24, No. 2, 77.
271. Buchholz J.C., Lagally M., *Phys. Rev. Lett.*, 1975, Vol. 35, No. 7, 442–445.
272. Budd H.F., Vannimenus J., *Phys. Rev. Lett.*, 1973, Vol. 31, No. 19, 1218–1221.
273. Burchhardt J., et al., *Phys. Rev. Lett.,* 1995, Vol. 74, 1617–1620.
274. Burchhardt J., et al., *Phys. Rev. Lett.,* 1992, Vol. 69, No. 10. 1532–1535.
275. Burton W.K., Cabrera N., *Disc. Faraday Soc.,* 1949, Vol. 5, 33–39.
276. Burton W.K., et al., *Phil. Trans. Roy. Soc. London A*, 1951, Vol. 243, No. 866, 299–
    358.
277. Butz R., Wagner H., *Surface Sci.,* 1977, Vol. 63, No. 2, 448–459.
278. Campagna M., *J. Vac. Sci. Technol.*, 1985, Vol. A3, No. 3, 1491–1495.
279. Campbell J.H., Bretz M., *Phys. Rev. B,* 1985, Vol. 32, No. 5, 2993–3002.
280. Campuzano J.C., et al., *Phys. Rev. Lett.,* 1980, Vol. 45, 1649.
281. Cao Y., Conrad E. H., *Phys. Rev. Lett.,* 1990, Vol. 64, 447.
282. Carneiro K., *J. Phys.* (France), 1977. Vol. 38. Colloq. C-4. Suppl., No. 10. C4-1–C4-9.
283. Cates M., Miller D., *Phys. Rev. B,* 1983, Vol. 28, No. 6, 3615–3617.
284. Coleman R.V., et al., *Phys. Rev. Lett.,* 1985, Vol. 55, No. 4, 394–397.
285. Cowan P., et al., *Phys. Rev. Lett.,* 1980, Vol. 44, No. 25, 1680–1683.
286. Dash J., *Phys. Low Temp.*. 1975. Vol. 1, No. 7, 839–877.
287. Davies J.A., et al., *Nucl. Inst. and Methods,* 1976, Vol. 132, No. 1, 609–613.
288. Davis H.L., Noonan N.R., *J. Vac. Sci. Tech.*, 1982. Vol. 20, 842.
289. Davison S.G., Koutecky J., *Proc. Phys. Soc.*, 1966, Vol. 89, 237.
290. De Jongh L.J., Miedema A.R., *Adv. Phys.*, 1974, Vol. 23, No. 1, 1–260.
291. Deloy R., Prigogine I., Tension superficielle at adsorption, Liège, 1951.
292. Dev B.N., et al., *Surface Sci.,* 1985, Vol. 163, No. 2–3, 457–477.
293. Diehl R., Grath R., *Surf. Sci. Rep.,* 1996, V. 23, 43–171.
294. Doak R.B., et al., *Phys. Rev. Lett.,* 1983, V. 51, No. 7, 578–581.
295. Dobson P., et al., *Vacuum,* 1983, V. 33, No. 10, 593–596.
296. Doll K., *The European Physical Journal*, 2001, V. 21, No. 3, 389.
297. Dose V., *Progr. Surface Sci.,* 1983, Vol. 13, 225–283.
298. Drechsler M., *Japan J. Appl. Phys.*, 1974, Suppl. 2, pt. 2, 35.
299. Dreizler R.M., Eberhard K.U., Gross density functional theory: an approach to the
    quantum many-body problem, Berlin, Springer, 1990.
300. Duval X., Thorny A., *C. r. Acad. sci. B.* 1964, Vol. 259, No. 22, 4007–4009.
301. Ehrlich G., Stolt K., *Ann. Rev. Phys. Chem.*, 1980, Vol. 31, 603–637.
302. Eisenberger P., Marra W.C., *Phys. Rev. Lett.,* 1981, Vol. 46, No. 16, 1081–1084.
303. Feder R., *J. Phys. C*, 1981, Vol. 14, No. 15, 2049–2091.
304. Feder R., et al., *Surface Sci.,* 1983, Vol. 127, No. 1, 83–107.
305. Fedorus A.G., et al., *Phys. status solidi* (a), 1972. Vol. 13, No. 2, 445–456.
306. Feinstein L.R., Schoemaker D.P., *Surf. Sci.*, 1965, Vol. 3, 294.
307. Feldman L.C., Surface science: Recent progress and perspectives, ed. R. Vanselow,

Cleveland, CRC, 1980.

308. Ferrante J., Smith J.R., *Surface Sci.,* 1973, V. 38, No. 1, 77–92.
309. Ferrante J., Smith J.R., *Solid State Communs.,* 1977, Vol. 23, No. 8, 527–529.
310. Ferrante J., Smith J.R., *Phys. Rev.,* 1979, Vol. 1319, No. 8, 3911–3920.
311. Fink H.-W., Faulian Bauer E., *Phys. Rev. Lett.,* 1980, V. 44, No. 15, 1008–1011.
312. Frenkel I., *Phil. Mag.,* 1917, Vol. 33, No. 196, 297–322.
313. Frenken J., Veen van der J., *Phys. Rev. Lett.,* 1985, Vol. 54, No. 2, 134–137.
314. Fu C.L., et al., *Phys. Rev. Lett.,* 1985, Vol. 54, 2261.
315. Fujimoto F., Komaki K., *J. Phys. Soc. Japan,* 1968, Vol. 25, 1679.
316. Gallet F., et al., *Europhys. Lett.,* 1986, Vol. 2, 701.
317. Gaskell T., *Proc. Phys. Soc.,* 1961, Vol. 77, No. 6. 1182–1192; The collective treatment of many-body systems, *Proc. Phys. Soc.,* 1962, Vol. 80, No. 5. 1091–1100.
318. Gell-Mann M., Brueckner R., *Phys. Rev.,* 1957, Vol. 106, 364–368.
319. Gomer R., *Appl. Phys.,* 1986. Vol. A39, No. 1, 1–8.
320. Goodwin E.T., *Proc. Cambridge. Phil. Soc.,* 1939, Vol. 35, 205; 221; 232.
321. Gradmann U., *J. Magnetism and Magn. Materials,* 1977, Vol. 6, No. 1, 173–182.
322. Grazis D.C., et al., *Phys. Rev.,* 1960, Vol. 119, 5336.
323. Gronwald K.D., Henzler M., *Surface Sci.,* 1982, Vol. 117, No. 1/3, 180–187.
324. Gu E., et al., *Phys. Rev. B,* 1999, Vol. 60, 4092.
325. Gunnarsson O., et al., *Surf. Sci.,* 1977, Vol. 63, 348.
326. Gunnarsson O., et al., *Phys. Scripta,* 1980, Vol. 22, 165.
327. Gurney R.W., *Phys. Rev.,* 1935, Vol. 181, 479.
328. Haase J., *Appl. Phys. A,* 1985, Vol. 38, No. 3, 181–190.
329. Halperin B.I., Nelson D.R., *Phys. Rev. B,* 1979, Vol. 19, No. 5, 2457–2484.
330. Hartree D.R., *Proc. Camb. Philos. Soc.,* 1928, Vol. 24, 89.
331. Heiney A., et al., *Phys. Rev. B,* 1983, Vol. 28, No. 11, 6416–6434.
332. Heinrichs J., Kumar N., *Phys. Rev.,* 1975, Vol. B12, No. 14. 802–810; *Solid Stat. Communs.,* 1976, Vol. 18, No. 8, 961–963.
333. Heinz K., et al., *Phys. Rev. B,* 1985, Vol. 32, 6214.
334. Henzler M., *Appl. Surf. Sci.,* 1982, Vol. 11/12, 450.
335. Herman F., et al., *Phys. Rev. Lett.,* 1969, Vol. 22, No. 16, 807–811.
336. Hietschold M., et al., *Phys. Status Solidi (b),* 1980, Vol. 101, No. 2, 239–252.
337. Hietschold M., Paasch G., *Phys. Status Solidi (b),* 1975, Vol. 70, No. 2, 653–662.
338. Higgingbottom P., et al., Appl. *Surface Sci.,* 1985, Vol. 22–23, Pt. 1, 100–110.
339. Hirabayashi K., *Phys. Rev.,* 1971, Vol. B3, No. 12, 4023–4025.
340. Hirose Y., Proceedings and Abstracts of 1st Conf. on the New Diamond Science and Technology, Tokyo, 1988.
341. Hirose Y., Knodo N., Extended Abstracts of 35-th Japan Applied Physics Spring Meeting, Tokyo, 1988.
342. Hirose Y., Mitsuizumi M., *New Diamond,* 1988, Vol. 4, 34.
343. Hjelmberg H., *Surf. Sci.,* 1979, Vol. 81, 539.
344. Hoffmann T A., *Acta Phys. Acad. Sci. Hung.,* 1952, Vol. 2, 195.
345. Hohenberg P., Kohn W., *Phys. Rev.* 1964, Vol. 136, No. 3B, B864–B871.
346. Holland B.W., Woodruff D.P., *Surf. Sci.,* 1973, Vol. 36, 488.
347. Honio H., et al., *Phys. Rev. Lett.,* 1985, Vol. 55, No. 8, 841–844.
348. Hopfield J.J., *Phys. Rev.,* 1958, Vol. 112, No. 5, 1555–1567.
349. Hove M. van, Tong S., Surface crystallography by LEED, Berlin: Springer-Verlag, 1979.
350. Hsu T., Cowley J.M., *Ultramicroscopy,* 1983. Vol. 11, 239–250.
351. Hubbard J., *Proc. Roy. Soc. London,* 1957, Vol. A243, No. 1234, 336–352.

352. Hubbard J., *Proc. Roy. Soc.*, 1958, Vol. A243, No. 1234, 336–352.
353. Huff H.R., et al., *Surf. Sci.*, 1968, Vol. 10, 232.
354. Huntington H.B., et al., *Surf. Sci.*, 1975, Vol. 48, 187.
355. Ibach H.J., *J. Vac. Sci.*, 1972, Vol. 9, 713.
356. Ibach H., Mills D.L., Electron energy loss spectroscopy and surface vibrations, London: Academic Press, 1982.
357. Inaoko T., *J. Phys.: Cond. Matt.*, 1996, Vol. 8, 10241.
358. Inglesfield J.E., *Prog. Surf. Sci.*, 1985, Vol. 92, 365.
359. Jaeger R., et al., *Phys. Rev. Lett.*, 1980, Vol. 45, No. 23, 1870–1873.
360. Jona F., *J. Phys. C: Solid State Phys.*, 1978, Vol. 11, No. 21, 4271–4306.
361. Jones R.O., Gunnarsson O., *Reviews of Modern Physics*, 1989, Vol. 61, No. 3. 689–746.
362. Kahn L.M., Ying S.C., *Surf. Sci.*, 1976, Vol. 59, 333.
363. Kim H.W., et al., *Surf. Sci. Lett.*, 1999, Vol. 430, L515–L520.
364. Kiskinova M.P., Pirug G., Poisoning and promotion in catalysis based on surface science methods, Vol. 70, Studies in Surface Science and Catalysis, Amsterdam, Elsevier, 1992.
365. Kjems J., et al., *Phys. Rev. Lett.*, 1974, Vol. 32, No. 13, 724–727.
366. Kobelev A.V., et al., *Phys. status solidi (b)*, 1979, Vol. 96, No. 1, 169–1711.
367. Kohn W., Sham L.J., *Phys. Rev.*, 1965, Vol. 140, No. 4A, A1133–A1138.
368. Kolaczkiewicz J., Bauer E., *Phys. Rev. Lett.*, 1984, Vol. 53, No 5, 485–488.
369. Kosterlitz L.M., Thouless D.J., *J. Phys. C*, 1973, Vol. 6, No. 7, 1181–1203.
370. Kosterlitz J.M., *J. Phys. C*, 1974, Vol. 7, No. 6, 1046–1060.
371. Koutecky J., *Phys. Rev.* 1957, Vol. 108, 13.
372. Koutecky J., Tomasec M., *Phys. Rev.* 1960, Vol. 120, 1212.
373. Krane K.J., Raether H., *Phys. Rev. Lett.*, 1976, Vol. 37, 1355.
374. Kreibig U., Vollmer V. Optical properties of metal clusters, Berlin, Springer, 1995.
375. Kuk Y., Feldman L.C., *Phys. Rev. B*, 1984, Vol. 30, 5811.
376. Lagally M.G., Chemistry and physics of solid surfaces, Vol. 4, eds. Vanselow, Hoowe-Berlin, Springer-Verlag, 1982, 281–313.
377. Lander J.J., Morrison J., *J. Appl. Phys.*, 1963, Vol. 34, 1403.
378. Lang N.D., *Solid State Communs.*, 1969, Vol. 7, No. 15, 1047–1053.
379. Lang N.D., *Phys. Rev. B*, 1971, Vol. 4, No. 12, 4234–4245.
380. Lang N.D., *Solid State Phys.*, 1973, Vol. 28, No. 4, 225–300.
381. Lang N.D., Kohn W., *Phys. Rev.*, 1970, Vol. B1, No. 12, 4555–4508.
382. Lang N.D., Williams A.R., *Phys. Rev. Lett.*, 1975, Vol. 34, 531.
383. Lang N.D., Williams A.R., *Phys. Rev. B*, 1978, Vol 18, 616.
384. Langmuir l., *J. Am. Chem. Soc.*, 1932, Vol. 54, 2798.
385. Langreth D.C., Mehl M.J., *Phys. Rev. Lett.*, 1981, Vol. 47, No. 6, 446–450.
386. Levine J.D., *Phys. Rev.*, 1968, Vol. 171, 701.
387. Levine J.D., Davison S.G., *Phys. Rev.*, 1968, Vol. 174, 911.
388. Levy M., *Phys. Rev. A*, 1982, Vol. 26, 1200.
389. Linford R.G., Solid State Surface Science, Vol. 2, Ed. by M. Green, New-York, Marcel Dekker, 1973.
390. Lippmann B.A., *Ann. Phys.*, 1957, Vol. 2, 16.
391. Louie S.G., *Phys. Rev. Lett.*, 1979, Vol. 42, 476.
392. Love H.M., Wiederick H.D., *Can. J. Phys.*, 1969, Vol. 47, No. 6, 657–663.
393. Lundqvist B.I.,et al., *Surf. Sci.*, 1979, Vol. 89, 196.
394. Lundqvist S., Electrons at metal surfaces, Surface Science, Vienna, Intern. Atom. Energy Agency, 1975, 331–392.

395. Ma C.Q., Sahni V., *Phys. Rev.*, 1977, Vol. B16, No. 10, 4249–4255.
396. Madey T.E., ET AL., *Surface Sci.,* 1976, Vol. 57, No. 2, 580–590.
397. Madey T.E., et al., The determination of molecular structure at surfaces using angle resolved electron and photon-stimulated desorption, in: Desorption Induced by Electronic Transitions, Berlin, 1983, 120–138.
398. Maksimenko V.V., et al., *Phys. Stat. Sol.* (b), 1977, Vol. 82, 685.
399. Marks L.D., et al., *Phys. Rev. Lett.*, 1984, Vol. 52, No. 8, 656–658.
400. Marra W.C., et al., *Phys. Rev. Lett.*, 1982, Vol. 49, No. 16, 1169–1172.
401. Marschall N., et al., *Phys. Rev. Lett.,* 1971, Vol. 27, 95.
402. Matcha R.L., King S.C., *J. Am. Chem. Soc.*, 1976, Vol. 98, 3415–3420.
403. Maue A.W., *Zs. Physik*, 1935, Vol. 94, 717.
404. McRae E.G., *Surf. Sci.*, 1967, Vol. 8, 14.
405. McRae E.G., *Surf. Sci.*, 1984, Vol. 147, 663.
406. Mehrotre R., et al., *Solid State Communs.*, 1976, Vol. 10, No. 2, 199–201.
407. Meyerhof W.E., *Phys. Rev.,* 1947, Vol. 71, 727.
408. Moncton D.E., et al., *Phys. Rev. Lett.,* 1982, Vol. 49, No. 25, 1865–1868.
409. Monnier R., Perdew J., *Phys. Rev.,* 1978, Vol. B17, No. 6, 2595–2611.
410. Morgan D.V., *Contemp. Phys.*, 1975, Vol. 16, No. 3, 221–241.
411. Muller E.W., *Z. Physik*, 1951, Vol. 131, 136.
412. Murkland I., Anderson S., *Surf. Sci.*, 1966, Vol. 5. 197.
413. Muscat J.P., Newns D.M., *Phys. Rev.*, B, 1979, Vol. 19, 1270.
414. Nabiyouni G., Schwarzacher W., *J. Crystal Growth*, 2005, Vol. 275, No. 1–2, 1259.
415. Nagao T., et al., *Surf. Sci.*, 1995, Vol. 329, 269–275.
416. Naumovets A.G.. Phase transitions and adsorbate restructuring at metal surfaces, The Chemical Physics of Solid Surfaces. Vol. 7, Ed. by D. A. King and D. Woodruff, Amsterdam, Elsevier, 1994.
417. Naumovets A.G., Vedula Yu.S., *Surface Sci. Rep.*, 1984, Vol. 4, No. 7/8, 365–434.
418. Nelson D.R., Kosterlitz J.M., *Phys. Rev. Lett.,* 1977, Vol. 39, No. 9, 1201–1205.
419. Neugebauer J., Scheffler M., *Phys. Rev.* B, 1992, Vol. 46, 16067–16080.
420. Nicholas J. F. An atlas of models of crystal structures, New York: Gordon and Breach, 1965.
421. Nielsen M., Mc Tague J., *Phys. Rev.* B, 1979, Vol. 19, No. 6, 3096–3106.
422. Norskov J.K., et al., *Solid State Commun.*, 1978, Vol. 28, 899.
423. Nozieres P., Pines D., *Phys. Rev.*, 1958, Vol. 111, No. 2, 442–454.
424. Osakabe N., et al., *Surface Sci.,* 1980, Vol. 97, No. 2, 393–408.
425. Paasch G., et al., *Phys. Status Solidi* (b), 1972, Vol. 51, No. 1, 283–293.
426. Paasch G., Hietschld M., *Phys. Status Solidi* (b), 1977, Vol. 83, No. 1, 209–222.
427. Padilla-Campos L., et al., *Surf. Sci.*, 1997, Vol. 385, 24–36.
428. Park R.L., Lagally M.G., Solid State Physics: Surfaces, Orlando, Academic, 1985.
429. Pashley M.D., et al., *Surface Sci.,* 1985, No. 152/153, 27–32.
430. Pendry J.B., Low energy electron diffraction, London, Academic Press, 1974.
431. Penn D.R., *Phys. Rev.* B, 1976, Vol. 13, 5248.
432. Pentcheva R., Scheffler M., *Phys. Rev.* B, 2000, Vol. 61, No. 3, 2211–2220.
433. Perdew J.P., et al., *Phys. Rev. Lett.,* 1977, Vol. 38, No. 18, 1030–1033.
434. Perdew J.P., Monnier R., *Phys. Rev. Lett.,* 1976, Vol. 37, No. 19, 1286–1289.
435. Perdew J.P., Monnier R., *J. Phys. F: Metal Phys.,* 1980, Vol. 10, No. 11, L287–L201.
436. Peuckert V., *J. Phys.,* 1976, Vol. C9, 809.
437. Physics and Chemistry of Alkali Metal Adsorption, Ed. by Bonzel H.P., Bradshaw A.M., Ertl G., Amsterdam, Elsevier, 1989.
438. Pierce D.T., et al., *Phys. Rev.* B, 1982, Vol. 26, 2566.

439. Plummer E., et al. *Surface Sci.,* 1985, Vol. 158, No. 1-3, 58–85.
440. Plummer E., Gustafson T., *Science*, 1977, Vol. 198, No. 4313, 165–170.
441. Posternak M., et al., *Phys. Rev.* B, 1980, Vol. 21, 5601.
442. Rahman T.S., et al., *Phys. Rev.* B, 1984, Vol. 30, No. 2, 589–603.
443. Ranjbar M., et al., *Materials Science and Engineering*, 2006, Vol. B127, No. 1, 17.
444. Rhodin T.N., Gadzuk J.W., Treatise on Solid State Chemistry, Ed. N. Hannay, Vol. 6A, 343–484, New York, Plenum, 1976.
445. Rieder K.H., *Contemp. Phys.,* 1985, Vol. 26, No. 6, 559–578.
446. Robinson I.K., et al., *Phys. Rev. Lett.,* 1991, Vol. 67, 1890.
447. Robinson I.K., et al., *Phys. Rev.* B, 1986, Vol. 33, 7013.
448. Roelofs L.D., et al., *Phys. Rev. Lett.,* 1981, Vol. 46, No. 22, 1465–1468.
449. Rossitsa P., Scheffler M., *Phys. Rev.* B, 2003, Vol. 65, 155418.
450. Rudnick I. *Phys. Rev. Lett.*, 1978, Vol. 40, 1454–1455.
451. Salanon B., et al., *Phys. Rev.* B, 1988, Vol. 38, 7385.
452. Scharoch P., et al., *Phys. Rev.* B, 2003, Vol. 68, 4031–4035.
453. Scheffler M., Stampfl C., Theory of Adsorption on Metal Substrates, Handbook of Surface Science, Vol.2: Electronic Structure, Ed. K. Horn, M. Scheffler, Amsterdam, Elsevier, 2000, 286–356.
454. Schmalz A., et al., *Phys. Rev. Lett.,* 1991, Vol. 67, No. 16, 2163–2166.
455. Schrieffer J.R., Collective Properties of Physical Systems, Ed. B. Lundqvist, S. Lundqvist, Stockholm, Nobel Foundation, 1973.
456. Schulte F.K., *Z. Phys.*, 1977, Vol. B27, 303.
457. Shen J., et al., *Surf. Sci.*, 1995, Vol. 328, 32–46.
458. Shinn N., Madey T., *J. Vac. Sci. Technol.*, 1985, Vol. A3, No. 3, 1673–1677.
459. Shockley W., *Phys. Rev.*, 1939, Vol. 56, 317.
460. Shockley W., Pearson G., *Phys. Rev.,* 1948, Vol. 74, 232.
461. Singwi K.S., et al., *Phys. Rev.,* 1970, Vol. B1, No. 3, 1044–1053.
462. Smith J.R., *Phys. Rev.,* 1969, Vol. 181, No. 2, 522–529.
463. Smith J.R., *Phys. Rev. Lett.,* 1970, Vol. 25, No. 15, 1023–1025.
464. Smith J.R., et al., *Solid State Commun.*, 1977, Vol. 24, 279.
465. Smith J.H., Ferrante J., *Solid State Commun.*, 1977, Vol. 21, No. 12, 1059–1060.
466. Smith J.R., et al., *Phys. Rev. Lett.,* 1973. Vol. 30. 610.
467. Smith N.V., *Vacuum*, 1983, Vol. 33, No. 10–12, 803–811.
468. Smoluchowski R., *Phys. Rev.*, 1941, Vol. 60, 661–674.
469. Sokolov J., et al., *Solid State Commun.*, 1984, Vol. 49, 307.
470. Somorjai G.A., Chemistry in Two Dimensions: Surfaces, Ithaca, Cornell, 1981.
471. Sparnaay M.J., *Surface Sci. Rep.*, 1984, Vol. 4, No. 3/4, 101–269.
472. Specht E.D., et al., *Phys. Rev. B*, 1984, Vol. 30, No. 3, 1589–1592.
473. Spencer E.G., et al., *Appl. Phys. Lett.*, 1976.,Vol. 29, 118.
474. Stampfl C., et al., *Surf. Sci.*, 1993, Vol. 287–288, 418–422.
475. Stampfl C., et al., *Surf. Sci.*, 1994, Vol. 307/309, 8–15.
476. Stampfl C., Scheffler M., *Surf. Rev. and Lett.*, 1995, Vol. 2, 317–340.
477. Stampfl C., et al., *Phys. Rev. Lett.,* 1992, Vol. 69, No. 10, 1532–1535.
478. Statz H., *Zs. Naturforsch.*, 1950, Vol. 5a, 534.
479. Stefanou N., et al., *Phys. Rev.*, 1987, Vol. B36, No. 12, 6372.
480. Steiner M., et al., *Adv. Phys.*, 1976, Vol. 25, No. 2, 87–209.
481. Stensgaard I., et al., *Surf. Sci.*, 1978, Vol. 77, 513.
482. Stohr J., *J. Vac. Sci. Technol.*, 1979, Vol. 16, No. 1, 37–41.
483. Suzanne J., *Surf. Sci.,* 1973, Vol. 40, No. 2, 414–418.
484. Suzanne J., et al., *Phys. Rev. Lett.*, 1984, Vol. 52, No. 8, 637–639.

485. Swingler J.N., Inkson J. C., *Solid State Communs.*, 1977, Vol. 24, No. 4, 305–307.
486. Takayanagi K., et al., *Surf. Sci.*, 1985, Vol. 164, 367.
487. Tanishiro K., et al., *Ultramicroscopy*, 1983, Vol. 11, 95–102.
488. Tejedor C., Flores F., *Solid. State Communs.*, 1975, Vol. 17, No. 8, 995–998.
489. Telieps W., Bauer E., *Surface Sci.,* 1985, Vol. 162, No. 1–3, 163–168.
490. Theophilou A.K., Modinos A., *Phys. Rev.*, 1972, Vol. B6, No. 3, 801–812.
491. Thomas L.H., *Proc. Camb. Phil. Soc.*, 1927, Vol. 23, 542.
492. Fermi E., *Z. Phys.*. 1928, Vol. 48, 73.
493. Tilly J.C., et al., *Phys. Rev. B*, 1985, Vol. 31, 1184.
494. Tsong T.T., Progress in Surface Science.,Vol. 10, N.Y., Pergamon Press, 1980.
495. Tsong T.T., Sweeney J., *Solid State Commun.*, 1979, Vol. 30, 767.
496. Vannimenus J., Budd H., *Solid State Commun.*, 1975, Vol. 17, No. 10, 1291–1297.
497. Vashishta P., Singwi K.S., *Phys. Rev.*, 1972, Vol. B6, No. 3, 875–887.
498. Vaz C.A.F., et al., *Rep. Prog. Phys.*, 2008, Vol. 71, 056501 (78 pp).
499. Vibrational Spectroscopy of Adsorbates, Ed. by R.F. Willis, Berlin, Springer-Verlag, 1980.
500. Wagner C.D., et al., Handbook of X-ray Photoelectron Spectroscopy, Eden Prairic: Perkin-Elmer, 1978.
501. Wallis R., *J. Vac. Sci. Technol.*, 1985, Vol. A3, No. 3, Pt. 2, 1422–1427.
502. Warner C., Thermionic Conversion Specialists Conference, New York, Institute of Electrical and Electronics Engineers, 1972.
503. Weeks J.D., Chui S.T., *Phys. Rev. Lett.*, 1978, Vol. 40, No. 12, 733–736.
504. Weizsacker O.F., *Ztschr. Phys.* 1935. Vol. 96, No. 7, 431–458.
505. Wendelken J.F., Wang G.C., *Phys. Rev. B*, 1985, Vol. 32, 7542.
506. Wenzien B., et al., *Surf. Sci.*, 1993, Vol. 287–288, 559–563.
507. Wicborg E., Inglesfield J.E., *Solid State Comm.*, 1975, Vol. 16, 335.
508. Wigner E., *Phys. Rev.,* 1934, Vol. 46, No. 11, 1002–1011.
509. Wigner E., Seitz F. On the Constitution of metallic sodium, *Phys. Rev.* 1933. Vol. 43, No. 10. 804–810.
510. Wojciechowski K.F., *Surf. Sci.,* 1976, Vol. 55, 246.
511. Wojciechowski K.F., *Surf. Sci.*, 1979, Vol. 80, 253.
512. Wolfram T., Dewames R.E., *Progr. Surf. Sci.*, 1972, Vol. 2, 233.
513. Yagi K., et al., *Thin Solid Films*, 1985, Vol. 126, No. 1/2, 95–105.
514. Yang H.-N., et al., *Phys. Rev. Lett.,* 1989, Vol. 63, 1621.
515. Ying S.C., et al., *Phys. Rev. B*, 1975, Vol. 11, 1483.
516. Young A., *Phys. Rev. B*, 1979, Vol. 19, No. 4, 1855–1866.
517. Zeller R., et al., *Solid State Commun.*, 1982, Vol. 44, No. 5, 993.
518. Zhino Z., *Vacuum*, 1990, Vol. 40, No. 6, 505.

Index